Additive Manufacturing – Process Optimisation

Additive Manufacturing – Process Optimisation

Editor

Muhannad Ahmed Obeidi

MDPI

Basel • Beijing • Wuhan • Barcelona • Belgrade • Novi Sad • Cluj • Manchester

Editor
Muhannad Ahmed Obeidi
School of Mechanical and
Manufacturing Engineering,
Dublin City University
Dublin
Ireland

Editorial Office
MDPI
St. Alban-Anlage 66
4052 Basel, Switzerland

This is a reprint of articles from the Special Issue published online in the open access journal *Designs* (ISSN 2411-9660) (available at: https://www.mdpi.com/journal/designs/special_issues/additive_manufacturing_process_optimisation).

For citation purposes, cite each article independently as indicated on the article page online and as indicated below:

Lastname, A.A.; Lastname, B.B. Article Title. *Journal Name* **Year**, *Volume Number*, Page Range.

ISBN 978-3-7258-1031-4 (Hbk)
ISBN 978-3-7258-1032-1 (PDF)
doi.org/10.3390/books978-3-7258-1032-1

Cover image courtesy of Muhannad Ahmed Obeidi

Contents

About the Editor

Muhannad Ahmed Obeidi

Professor Muhannad Ahmed Obeidi is a lecturer in Advanced Manufacturing Processes in the School of Mechanical and Manufacturing Engineering at Dublin City University DCU, Ireland. He obtained his B.Sc. degree in mechanical engineering from Baghdad University in 1988. Dr Obeidi completed his master's and PhD in Mechanical and Manufacturing Engineering at DCU in 2014 and 2018, respectively. He worked as researcher at the Advanced Processing Technology Centre (APT) in DCU from January 2015 till March 2018. The main area of research was focused on the laser surface processing of metals including laser surface polishing, hardening, cladding, glazing, and texturing. Dr Obeidi was also involved in several industrial problem-solving projects. He also contributed in the establishing of the new Additive Manufacturing (AM) facility lab in DCU. Dr Obeidi has over 35 years of experience in Mechanical Engineering and materials processing. His current research focuses on the investigation and optimization of advanced manufacturing technologies including additive manufacturing (AM). His research is being carried out on several metal powders, including 316L stainless steel, cobalt chromium, aluminium, Inconel, and nitinol (NiTi) utilising different application in aerospace, automotive, tooling, and biomedical. It is important for manufacturers and AM users to understand the relationship between the laser input processing parameters and the output properties, which will help in the prediction of the part's performance and life cycle. In order to achieve this aim, an experimental model must be modified in which a statistical description of this relationship can be established.

designs

Editorial
Additive Manufacturing—Process Optimisation

Muhannad Ahmed Obeidi

School of Mechanical and Manufacturing Engineering, Dublin City University, DCU, D09 V209 Dublin, Ireland; muhannad.ahmedobeidi@dcu.ie

The realm of Additive Manufacturing (AM), often referred to as 3D printing, encompasses a broad spectrum of applications and methodologies, each contributing distinctively to the progress of this dynamic field. This book, a curated compilation of twelve articles, offers a comprehensive exploration of AM, delineating its multifaceted applications across varied sectors. The collection presents an academic discourse, analysing the intricate nuances and technical advancements within AM. It delves into the core aspects of design, material innovation, and process optimization, illustrating the field's adaptability and impact. The articles included in this Special Issue, selected for their contribution to the body of knowledge, reflect the depth and breadth of AM, from industrial applications in aerospace and medical fields to research-oriented pursuits in material development and prototyping. This Special Issue not only showcases the current state of AM but also provides insights into its evolving trends, underscoring its significance in contemporary and future technological landscapes.

"Design for Additive Manufacturing: Recent Innovations and Future Directions" presented by Egan et al. [1] delves into the intersection of design and AM technologies, highlighting advancements in diverse engineering applications, including aerospace, automotive, construction, and medicine. It underscores the importance of integrating material processes and constraints in the design phase to optimize AM's capabilities for specific applications.

In "A Review of the Current State of the Art of PEEK Composite Based 3D-Printed Biomedical Scaffolds", Surendran et al. [2] the focus shifts to biomedical applications, particularly the use of Polyether Ether Ketone (PEEK) composites in scaffold development for tissue engineering. This review illuminates the potential of PEEK-based composites as a next-generation bioactive material, emphasizing their biocompatibility and mechanical properties.

"An Assembly-Oriented Design Framework for Additive Manufacturing", Sossou et al. [3] offers a novel design perspective, concentrating on the assembly level. It discusses how AM can be utilized to design complex, performant products, exploring the concept of AM-based architecture minimization, including part consolidation and assembly free mechanisms.

"Caffeine–Acrylic Resin DLP-Manufactured Composite as a Modern Biomaterial" presented by Tomczak et al. [4] examines the innovative combination of caffeine with acrylic resin using digital light processing (DLP) for transdermal drug delivery. This study explores the mechanical properties and release capabilities of the composite, contributing significantly to the field of composite drug delivery systems.

The study titled "Dimensional Accuracy of Electron Beam Powder Bed Fusion with Ti-6Al-4V" by Bol et al. [5] delves into the precision of the electron beam powder bed fusion process, particularly focusing on Ti-6Al-4V specimens. It emphasizes the importance of understanding key process parameters to optimize AM for better dimensional accuracy.

"By Visualizing the Deformation with Mechanoluminescent Particles, Additive Manufacturing Offers a Practical Alternative to Stress and Strain Simulation", Einbergs et al. [6] introduces an innovative approach to stress analysis in mechanical components. This article describes the utilization of mechanoluminescent materials in 3D printing, offering a novel method for the real-time evaluation of complex forces.

"A Novel Feature-Based Manufacturability Assessment System for Evaluating Selective Laser Melting and Subtractive Manufacturing Injection Moulding Tool Inserts", El

Citation: Obeidi, M.A. Additive Manufacturing—Process Optimisation. *Designs* **2024**, *8*, 34. https://doi.org/10.3390/designs8020034

Received: 7 April 2024
Accepted: 8 April 2024
Published: 10 April 2024

Kashouty et al. [7] discusses the development of a feature-based manufacturability assessment system. This system integrates selective laser melting with subtractive manufacturing, focusing on the geometrical complexities of tool inserts.

In "A Structural and Thermal Comparative Review of 3D-Printed Wall Shapes", Bello et al. [8] the structural and thermal performance of various 3D-printed wall shapes in the concrete printing industry is analysed. This review utilizes simulation tools to study different wall types, contributing to a deeper understanding of 3D-printed structures in construction.

The article "Effects of Raster Angle on the Elasticity of 3D-Printed Polylactic Acid and Polyethylene Terephthalate Glycol", Albadrani et al. [9] explores the impact of the raster angle on the mechanical properties of 3D-printed materials, providing valuable insights for optimizing 3D printing parameters for enhanced product performance.

"Rheological Behaviour of ABS/Metal Composites with Improved Thermal Conductivity for Additive Manufacturing", Moritz et al. [10] investigates the properties of metal-reinforced polymer composites, particularly focusing on their thermal and mechanical characteristics. This study provides foundational knowledge for the development of composites with optimized processability and properties.

"Development and Performance Evaluation of Fibrous Pseudoplastic Quaternary Cement Systems for Aerial Additive Manufacturing", Dams et al. [11] presents a novel cementitious material developed for aerial additive manufacturing. It highlights the material's viability for in situ construction using UAVs, emphasizing its potential to revolutionize the construction industry.

Finally, "Enhancement of Fatigue Life of Polylactic Acid Components through Post-Printing Heat Treatment" presented by Jimenez et al. [12] addresses the improvement in fatigue resistance in 3D-printed components. This study demonstrates the potential of post-printing heat treatment to enhance the structural reliability of 3D-printed parts, especially in dynamic load scenarios.

Together, these articles provide a multifaceted view of AM, showcasing its versatility, adaptability, and transformative potential across various industries. From biomedical scaffolds to aerospace components, the breadth of AM's applications is vast, offering endless possibilities for innovation and optimization. As AM continues to evolve, its integration with emerging technologies like machine learning and bio-inspired design promises to unveil new horizons in manufacturing and design.

Funding: This research received no external funding.

Conflicts of Interest: The authors declare no conflicts of interest.

References

1. Egan, P.F. Design for Additive Manufacturing: Recent Innovations and Future Directions. *Designs* **2023**, *7*, 83. [CrossRef]
2. Surendran, R.; Pavithran, S.S.; Balachandran, A.; Vijayan, S.; Madanan, K.; Brabazon, D. A Review of the Current State of the Art of Polyether Ether Ketone (PEEK) Composite Based 3D-Printed Biomedical Scaffolds. *Designs* **2023**, *7*, 128. [CrossRef]
3. Sossou, G.; Demoly, F.; Gomes, S.; Montavon, G. An Assembly-Oriented Design Framework for Additive Manufacturing. *Designs* **2022**, *6*, 20. [CrossRef]
4. Tomczak, D.; Wichniarek, R.; Kuczko, W. Caffeine–Acrylic Resin DLP-Manufactured Composite as a Modern Biomaterial. *Designs* **2023**, *7*, 49. [CrossRef]
5. Bol, E.; Ramulu, M. Dimensional Accuracy of Electron Beam Powder Bed Fusion with Ti-6Al-4V. *Designs* **2023**, *7*, 53. [CrossRef]
6. Einbergs, E.; Spustaka, A.; Vitola, V.; Zolotarjovs, A. By Visualizing the Deformation with Mechanoluminescent Particles, Additive Manufacturing Offers a Practical Alternative to Stress and Strain Simulation. *Designs* **2023**, *7*, 54. [CrossRef]
7. El Kashouty, M.F.; Rennie, A.E.W.; Ghazy, M. A Novel Feature-Based Manufacturability Assessment System for Evaluating Selective Laser Melting and Subtractive Manufacturing Injection Moulding Tool Inserts. *Designs* **2023**, *7*, 68. [CrossRef]
8. Bello, N.D.; Memari, A.M. A Structural and Thermal Comparative Review of 3D-Printed Wall Shapes. *Designs* **2023**, *7*, 80. [CrossRef]
9. Albadrani, M.A. Effects of Raster Angle on the Elasticity of 3D-Printed Polylactic Acid and Polyethylene Terephthalate Glycol. *Designs* **2023**, *7*, 112. [CrossRef]
10. Moritz, V.F.; Prévost, H.; Crespo, J.S.; Ferreira, C.A.; Devine, D.M. Rheological Behaviour of ABS/Metal Composites with Improved Thermal Conductivity for Additive Manufacturing. *Designs* **2023**, *7*, 133. [CrossRef]

11. Dams, B.; Shepherd, P.; Ball, R.J. Development and Performance Evaluation of Fibrous Pseudoplastic Quaternary Cement Systems for Aerial Additive Manufacturing. *Designs* **2023**, *7*, 137. [CrossRef]
12. Jimenez-Martinez, M.; Varela-Soriano, J.; Carrera-Espinoza, R.; Torres-Cedillo, S.G.; Cortés-Pérez, J. Enhancement of Fatigue Life of Polylactic Acid Components through Post-Printing Heat Treatment. *Designs* **2024**, *8*, 7. [CrossRef]

designs

MDPI

Review

Design for Additive Manufacturing: Recent Innovations and Future Directions

Paul F. Egan

Department of Mechanical Engineering, Texas Tech University, Lubbock, TX 79409, USA; paul.egan@ttu.edu

Abstract: Design for additive manufacturing (DfAM) provides a necessary framework for using novel additive manufacturing (AM) technologies for engineering innovations. Recent AM advances include shaping nickel-based superalloys for lightweight aerospace applications, reducing environmental impacts with large-scale concrete printing, and personalizing food and medical devices for improved health. Although many new capabilities are enabled by AM, design advances are necessary to ensure the technology reaches its full potential. Here, DfAM research is reviewed in the context of Fabrication, Generation, and Assessment phases that bridge the gap between AM capabilities and design innovations. Materials, processes, and constraints are considered during fabrication steps to understand AM capabilities for building systems with specified properties and functions. Design generation steps include conceptualization, configuration, and optimization to drive the creation of high-performance AM designs. Assessment steps are necessary for validating, testing, and modeling systems for future iterations and improvements. These phases provide context for discussing innovations in aerospace, automotives, construction, food, medicine, and robotics while highlighting future opportunities for design services, bio-inspired design, fabrication robots, and machine learning. Overall, DfAM has positively impacted diverse engineering applications, and further research has great potential for driving new developments in design innovation.

Keywords: design; engineering; additive manufacturing; 3D printing; materials; processes; optimization; mechanics; modeling; applications

Citation: Egan, P.F. Design for Additive Manufacturing: Recent Innovations and Future Directions. *Designs* **2023**, *7*, 83. https://doi.org/10.3390/designs7040083

Academic Editor: Obeidi Muhannad

Received: 31 May 2023
Revised: 17 June 2023
Accepted: 26 June 2023
Published: 29 June 2023

1. Introduction

As additive manufacturing (AM) technologies continue to advance, there is a need for design methods to guide innovations that benefit from AM's capabilities for fabricating complex structures with novel materials [1]. Unfortunately, recent research suggests that AM technologies remain underutilized [2]. For designers to achieve AM innovations, they must do more than simply possess AM technologies—both the effective use of resources and information management for decision making are crucial. Design for additive manufacturing (DfAM) provides a framework that facilitates decision making with AM technologies. DfAM is a multifaceted field of study in which diverse topics such as creativity [3], bio-inspiration [4], materials [5], optimization [6], and validation [7] are all considered for enhancing AM design with integrated approaches. Advances in DfAM are necessary to keep up with the exponential increase in interest for AM applications [8], especially in fields such as medicine that benefit from on-demand design and manufacturing for personalized solutions [9]. Here, DfAM research is surveyed by considering perspectives from stages across the design process and discussed for diverse application areas that may foster AM design innovations.

An important aspect of DfAM is the establishment of a framework for learning and implementing relevant tools and techniques for a particular design application. A DfAM product development framework subdivided into the stages of process selection, functional redesign, and optimization has been demonstrated as an effective approach for both research and industry case studies [10]. These stages enable designers to reason about the entire life cycle of an AM product and iterate between steps in the design process. Another DfAM framework has focused on assembly by considering architecture minimization that relied on using AM's advantages for part consolidation and assembly-free mechanisms [11]. The framework provides the context of an 'AM-factory', which was illustrated with a gripper case study produced with automated design generation and lattice structures to improve mechanical efficiency. Lattices are common AM structures and have been investigated in a design–build–test framework that used an integrated computational generation and experimental validation strategy to tune lattices for biomedical applications [12]. The process resulted in the experimental validation of a design space where lattices were configured based on the specific physiological needs of patients. These DfAM endeavors provide a context for reviewing recent research in the framework proposed in Figure 1, in which Fabrication, Generation, and Assessment phases are considered during AM research and development.

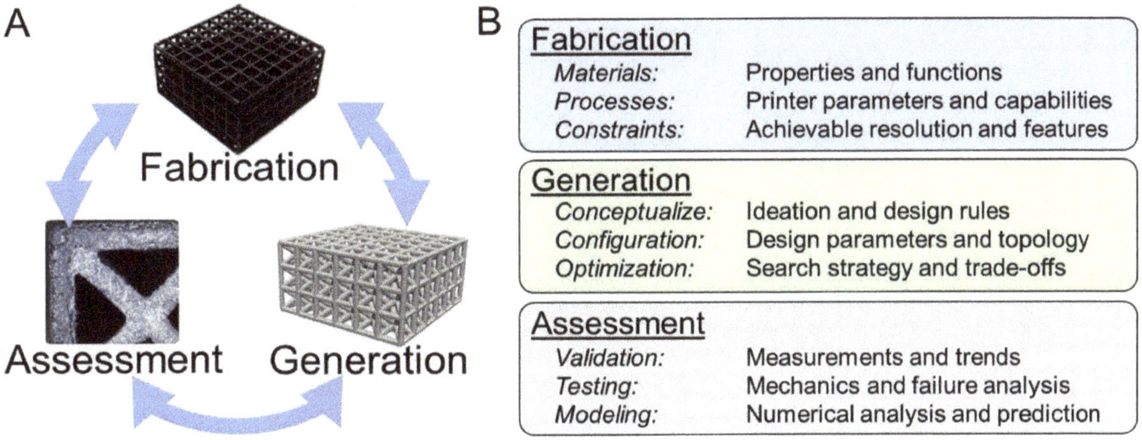

Figure 1. Design for additive manufacturing framework (**A**) for Fabrication, Generation, and Validation phases (**B**) with key steps to consider in each phase. Design example provided for porous nylon 11 lattices fabricated with powder bed fusion.

The Figure 1 framework demonstrates a case study with Nylon 11 material printed with powder bed fusion, which is a novel AM process that facilitates printing integrated parts and non-assembly mechanisms [13]. The Fabrication phase is demonstrated with a lattice design, in which minimum manufacturing constraints were measured during the Assessment phase using microscopy and mechanical testing to define a design space for configuring lattice structures in the Generation phase. Steps across phases may be iterated in a non-linear fashion to better understand AM capabilities that result in more efficient approaches for producing new designs, with specific steps highlighted in Figure 1.

Non-linearity for DfAM phases is necessary to consider since the specific sequence of steps and phases is contextually dependent on the current state of the art for an application. Depending on the application, it is potentially best to start by characterizing fabrication processes to develop new materials and manufacturing innovations, such as when novel biomaterials come to fruition and open a new design space for medical applications. In this case, the designer would begin with the Fabrication phase to determine the capabilities of a printing process in terms of materials and fabrication constraints that enable different design configurations using the new material. If the Generation phase were initiated prior to having this knowledge, it is highly likely a designer could explore infeasible solutions of the design space that would not result in manufacturable designs. In other cases, a designer may begin with generation steps that focus on innovations by proposing a novel structural configuration while using well-established AM technologies to fabricate the end product. Here, designers may begin directly with generating solutions according to rules and constraints that have been established as prior best practices and then move directly to the Assessment phase after printing.

Similar DfAM steps have been investigated for nylon powder printing using multijet fusion techniques. For instance, during fabrication, the differences in refresh rates of powder can affect the thermal, morphological, and mechanical characteristics of printed parts [14]. During design generation, parameter alterations for lattices may affect print accuracy and the reaction loads of structures [15]. During assessment, such as testing the mechanics of honeycomb structures, finite element (FE) models have been created that facilitate computational design with accurate predictions of structural performance [16]. Using these techniques can aid in the creation of novel systems using AM, such as multi-helical springs with experimentally validated stiffness predictions [17]. The directed design of AM springs has great potential for applications in multiple areas, including the aerospace and automotive sectors, by providing a higher stiffness for competing alternatives of a similar mass. By carefully considering the technology at hand and previous works across phases, it is possible for designers to recognize which steps and methods may facilitate the greatest opportunities for innovation.

The DfAM framework in Figure 1 has broad generalizability across AM technologies and applications. For instance, in the Fabrication phase for extrusion printing, anisotropic part mechanics are dependent on the base material selected and print orientation [18]. Such tuning extends to further processes for tailored thermoplastics and photopolymers, with a possibility to adjust parts mechanically by adding nanoparticles and fibers [19]. During the Generation phase, a configuration of efficient micro-architected structures has significantly improved heat exchanger effectiveness compared to traditionally manufactured designs [20]. The optimization of mechanical metamaterials has resulted in lightweight helical coils with advantageous structural rigidity and large deformation capabilities compared to equivalent coil springs of identical weight [21]. In the Assessment phase, experiments have measured the dimensional accuracy of metal powder bed fusion processes and found both micro- and millimeter scale deviations from the intended design [22]. Mechanoluminescent particles have also been embedded in prints to provide insights for mechanical failures of AM parts that open the possibility for the real-time evaluation of mechanical responses in designs [23]. These cases highlight a great potential for innovative AM designs to outperform designs from traditional technologies and a need for a careful consideration of DfAM methods and AM limitations to ensure the technology is used to its fullest potential.

In this review, DfAM research for innovations in diverse applications is surveyed in the context of Figure 1 phases. The review critically considers the integrated nature of DfAM steps, particularly across design phases and applications that are often non-obvious and necessitate multiple iterations, thereby contributing to the literature by filling gaps between individual research studies that focus on specifics. For instance, when designing for innovative applications, success requires considering AM materials and processes paired appropriately since not all printing processes use the same materials. These decisions in turn affect fabrication constraints and uncertainties in the manufacturing process that lead to printed parts responding differently than idealized models. Depending on the application area and current state of knowledge in the field, different design problems necessitate emphasis on different DfAM phases and steps to identify the greatest opportunities for innovation, which are difficult to identify when considering DfAM steps in isolation. Considering these phases and steps provides context for discussing innovative AM applications for diverse sectors including aerospace, automotives, construction, food, medicine, and robotics. The review concludes by considering future DfAM directions, highlighting them with challenges and directions for researchers to consider as both DfAM frameworks and AM technologies continue to advance. Reviewing these multifaceted tools and techniques across DfAM perspectives represents a critical advancement for better understanding how DfAM frameworks drive AM innovations.

2. Fabrication

Determining the fabrication capabilities of AM processes is an essential step in DfAM because it provides bottom-up knowledge for designers to form decisions. The capabilities of designs from fabrication depend primarily on the materials and deposition processes used to form three-dimensional parts. Each combination of a material and a process has constraints that dictate rules for minimum dimension size and features that further inform design generation.

2.1. Materials

Material selection is a crucial step in DfAM that influences the properties, functionality, and printability of AM parts. Common AM materials span from elastomers to polymers to metals. AM processes enable the creation of innovative designs by the directed deposition of novel materials such as superalloy metals [24]. For superalloys, AM reduces manufacturing steps and minimizes waste compared to traditional investment casting while also opening the possibility of fabricating hollow, foam-like, and lattice-based architectures. According to one study, nickel-based superalloys created with binder-jet processing do not introduce residual stress during fabrication due to the sintering process providing more consistent and controlled heating than traditional processes [25]. Fatigue-tested superalloy materials have been produced by comparing as-printed, as-sintered, and mechanically ground samples, as demonstrated in Figure 2A. Mechanically ground samples provided the highest engineering stress per number of cycles to failure (350 MPA at approximately 1 million cycles). The mechanical grinding resulted in fatigue performance that significantly surpassed that of a cast alloy. Further research in superalloys has resulted in a new class of alumina-forming superalloys with advantageous crack resistance and directional structures [26,27]. Such materials demonstrate that AM not only provides new capabilities for material deposition to form complex geometries but also provides enhanced mechanical properties compared to conventional manufacturing processes.

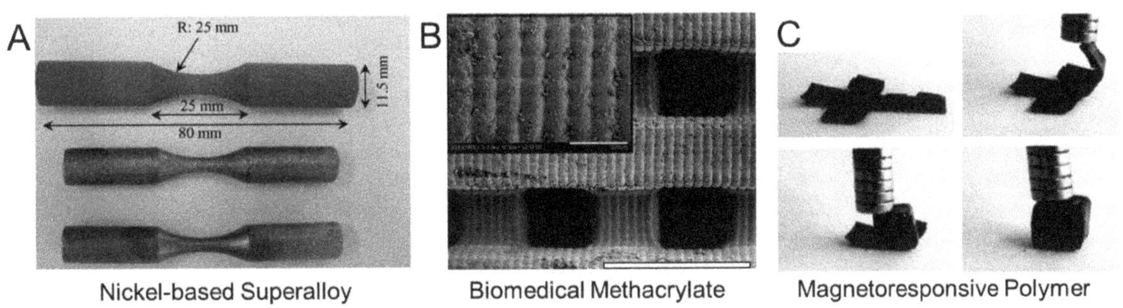

Figure 2. Highlighted material innovations for additive manufacturing including (**A**) nickel-based superalloy [25], (**B**) biomedical methacrylate (1 mm scale bar) [28], and (**C**) magnetoresponsive polymers [29]. Images adapted with permission.

AM is also providing new capabilities in medical applications in which DfAM enables personalization for patient-specific needs. Typical printing materials used in medicine include polylactic acid (PLA), acrylonitrile butadiene styrene (ABS), polyvinyl alcohol (PVA), thermoplastic polyurethane (TPU), polyether ether ketone (PEEK), biological tissues, carbon fiber, titanium, and nitinol [5]. One emerging application of AM materials in medicine is the tailoring of complex three-dimensional tissue scaffolds in which optimized geometry supports the curvature-driven growth of new tissues [30]. Biocompatible polymers are necessary for such applications, with methacrylates becoming a preferred option due to their advantageous stiffness and toughness, which are necessary to withstand forces of the human body [28]. Biocompatible methacrylates have been recently developed for printing with vat photopolymerization processes that produce complex structures with microscale resolutions [31], as demonstrated in Figure 2B. The methacrylate specimens had compressive moduli in the range of 10 to 19 MPa, with cytotoxicity results suggesting they are not toxic for porcine chondrocytes, thereby demonstrating their suitability for regenerative medicine.

AM is also enabling innovation via the fabrication of functional magnetic materials that are suitable for diverse applications in thermal generation, electronics, and small-scale robotics [32]. An advantage of AM is the creation of parts with intricate internal structures with complex geometries that enable the placement of magnetic materials strategically throughout a design. Such designs are difficult or impossible to replicate with traditional subtractive manufacturing processes. Magnetic photopolymerizable resin has been optimized by creating samples with varied proportions of solid loading for magnetic nanoparticles needed to maximize magnetic actuation forces by weight [33]. A high performance was achieved for samples with 30% solid proportions. Figure 2C demonstrates a printed magnetoresponsive polymer with 2% of its weight composed of magnetite nanoparticles [29]. The mechanical properties of the polymer were demonstrated with tailorable stiffness by varying urethane–acrylate resins and butyl acrylate, which functioned as a reactive diluent. The magnetic response was tailorable by altering the amount of Fe_3O_4 nanoparticles for up to 6% weight. Material tuning demonstrated capabilities for controlling movements of objects with magnetic responses, including rolling, translation, stretching, and folding/unfolding. These material examples demonstrate the novel capabilities and functions enabled by AM while also highlighting important considerations for designers when selecting AM processes based on material compatibility and needs.

2.2. Processes

There are diverse AM processes for fabricating designs including extrusion, photopolymerization, powder printing, and sheet lamination for polymers [19]. There are further processes to consider beyond polymer manufacturing in areas such as bioprinting, food printing, and metal melting/sintering. Each process has its own unique material libraries and capabilities. A brief description of these highlighted AM technologies will be provided as a necessary context for discussing their role in DfAM.

Fused deposition modeling (FDM) is an inexpensive and popular approach for extruding polymers that relies on melting filaments and depositing them layer by layer to construct a design [34]. FDM has resolution limits based on nozzle sizes that are typically 0.25 to 1.0 mm. The technology is capable of rapidly producing parts with a diverse material library including polymers of ABS, PLA, polyethylene terephthalate (PET), polycarbonate (PC), and nylon. Once a material is selected, process parameters for controlling orientation, infills, rasters, and layering all require consideration due to their influence on final part properties [35]. The Taguchi method is a common technique for designing experiments to systematically alter process parameters and measure their effects, which informs an optimum set of parameters for specified materials and applications [36]. By careful consideration of process parameters, designers may tune tensile strength, impact resistance, damping, and further properties for application-specific needs.

Further extrusion processes include those for bioprinting and food printing that rely on the direct deposition of materials using a nozzle [37,38]. Common biomaterials for bioprinting include polycaprolactone (PCL), tricalcium phosphate (TCP), and alginate [39]. Food printing commonly uses starch, protein, and gel materials [40]. Both bioprinting and food printing processes can use temperature control to improve printability by altering the rheological properties of the printed material. Further improvement in the printability of these soft materials is achievable by combining additives with materials to alter their consistency, such as thickeners for food.

Powder printing is another common AM approach that has several different implementations for fabricating designs. Direct metal laser sintering (DMLS) uses the energy from a laser to heat and join powder particles together layer by layer to produce parts with specified geometries. The process is suitable for metals such as gold, silver, stainless steel, and titanium that can produce jewelry or bone tissue engineering scaffolds with microscale features [41,42]. Selective laser melting (SLM) is another common powder process that melts metal powder that cools to form designed parts [43]. Selective laser sintering (SLS) has been used to process polystyrene parts [44], while mulijet fusion (MJF) can fuse polymer powders to form flexible designs such as nylon springs [45]. An advantage of powder processes is that unused power acts as support material during printing, which can enable the design of complex geometries and mechanisms difficult to fabricate with other AM processes.

AM fabrication from resin materials relies on photopolymerization to fabricate high-resolution parts with microscale resolutions. Stereolithography (SLA) uses a laser for curing resin, while digital light processing (DLP) uses a planar projection of light. In each case, liquid resin with photopolymers is exposed to ultraviolet light that promotes cross-linking to solidify material. Such processes have been used to produce ultralight and strong porous ceramics to form mechanically efficient lattice structures [46]. The printing process parameters of exposure time and printing angle can be altered during resin printing to tune the curing behavior and mechanical strength of printed samples [47]. DLP printing has been demonstrated for tissue scaffold applications in which its capabilities for microscale resolution create a biological niche with a suitable geometry to promote bone growth [48]. Overall, these diverse AM technologies and their respective process parameters present a complex space for designers to navigate and a need to identify the constraints and capabilities of each process to inform design decisions.

2.3. Constraints

Characterizing constraints for 3D printing processes is essential to provide a set of rules for designers to follow when selecting appropriate materials and processes for configuring parts [49]. Within a class of 3D printing processes, general capabilities, such as an extrusion printer's ability to produce overhangs, remain similar, which suggests that a set of general guidelines is useful for designers. When considering 3D-printed springs with material extrusion [50], guidelines were developed by printing springs with PLA material and assessing their print quality and capabilities. Determined guidelines suggest that square wire cross-sections, mono-directional in-fills, and thin layers are recommended for printing springs to ensure consistency and functionality.

Researchers have systematically studied printing constraints and created worksheets for designers to follow that promote the use of design decisions that improve printing success [51]. Figure 3 shows a highlighted scoring criteria in the Design for Additive Manufacturing Worksheet, in which designers score parts according to criteria in each column such as complexity, functionality, material removal, and unsupported features. Each criterion has multiple levels of assessment with associated scores in which designers choose one level that their design fits for each category. For instance, in the material removal category, a part with no support material scores the best, while a part with support material that is difficult to remove scores worst. Designers score the sum across rows with features that are more difficult to fabricate providing lower scores. Based on the total design score, designers are recommended that the part 'Needs redesign', should 'Consider redesign', has a 'Moderate likelihood of success', or has a 'Higher likelihood of success'. These guidelines were tested with FDM-printed parts designed by students that demonstrated that the use of the worksheet resulted in an 81% decrease in the rate of poorly designed parts. The study demonstrates the practical benefits of incorporating DfAM practices to train designers to use AM technologies properly and maximize their potential.

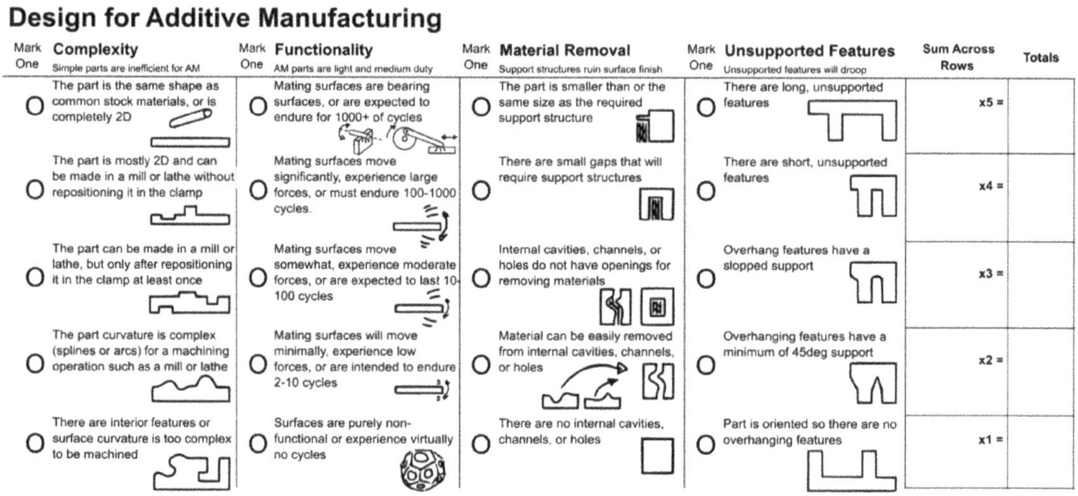

Figure 3. Highlighted scoring criteria from the Design for Additive Manufacturing Worksheet [51]. Image adapted with permission.

The sizes of clearances and tolerances are further constraints that designers must consider with AM parts, especially when creating working mechanisms. A study using a layered manufacturing process created revolute non-assembly mechanisms and demonstrated that the designed clearance of a mechanical joint critically affects its dynamic performance [52]. An algorithmic process was created to resize clearances to aid designers in configuring clearances that resulted in feasible designs. Non-assembly mechanisms have also been created with dissolvable support material with FDM processes [53]. In these studies, joint clearances that were too large were found to impair functionality and were a more important consideration than the length of linkage mechanisms. Powder printing processes are particularly useful for non-assembly mechanism creation since unused powder is present in areas between components of joints that, once removed, enables the movement of mechanisms requiring no further assembly. A recent study on nylon-powder-printed prosthetics found that relevant mechanisms operated best within a specified range of gap sizes from about 0.2 to 0.4 mm [13]. Further studies with geometric tolerances of laser powder bed fusion found that 15 mm diameter holes had mean printed diameters of 15.05 and 15.03 mm with standard deviations of 0.04 mm [54]. Based on these ranges, designers can specify suitable tolerances for parts based on empirical observations detailing the expected variation of 3D-printed parts. Such DfAM considerations are important to ensure that design efforts are focused on feasible designs that have a higher likelihood of working once printed.

3. Generation

Generation steps help designers use AM processes to create innovative designs based on material capabilities and constraints. Conceptualization is an early step in the design process in which designers use creative processes to form solutions to problems. AM enhances conceptualization by providing new capabilities compared to competing manufacturing technologies; however, AM also creates an exceedingly large and complex design space to navigate. Configuration steps aid designers in selecting a structural embodiment, such as a lattice architecture, that is further tunable via optimization that specifies parameter values within the design space for a specific application. By considering these DfAM steps, designers may leverage the unique facets of AM technologies while forming engineering solutions.

3.1. Conceptualization

During design conceptualization, it is important for designers to consider the advantages of AM, such as the generation of complex geometries, compared to its deficits, such as the difficulty in printing overhangs with some processes. Since AM has differing capabilities from traditional manufacturing methods, creativity is essential for designers to think outside the box to use AM to create novel designs that progress beyond the limitations of conventional technologies [55]. A design study comparing classrooms with and without formal DfAM training demonstrated that AM ideas were more elegant in terms of aesthetics but often less feasible [56]. Another study found that re-mixing AM designs, which is a process of adapting or recombining existing design elements into something new, promotes the creation of designs that are more often printed by online community members [57]. Since these designs modify already printable components, it promotes the diversity and quantity of ideas within the confines of DfAM constraints that ensures manufacturability. An experiment with 343 junior-level engineering students investigated effects from teaching three different design approaches: (1) no DfAM, (2) restrictive (considers limitations) DfAM, and (3) opportunistic (considers capabilities) and restrictive DfAM [58]. Each type of DfAM approach was found to improve the technical proficiency of AM designs but not creativity. These studies suggest that curriculum and delivery play an important role for influencing designers' AM capabilities with a need for effective approaches that promote diverse design generation while retaining practicality for printing.

Teaching multi-modal design heuristics has been demonstrated as a viable strategy for improving designer creativity for thinking beyond the traditional manufacturing mindset [59]. Heuristics are general rules that designers may follow to find viable solutions quickly. The multi-modal heuristics taught for AM included categories of part consolidation, customization, convey information, material, material distribution, embed–enclose, lightweight, and reconfiguration. Some specific heuristics included part consolidation to reduce assembly time, customization with geometry to the use case, and material distribution to absorb energy with small interconnected parts. The use of these multi-modal heuristics led to an increased creativity of concepts generated by individuals and teams while also stimulating more diverse design concepts. The timing of design heuristics provided to novices has also been investigated [60], and it was found that providing DfAM lessons improved the manufacturability of designs but did not affect overall quality. The novelty of designs decreased after being provided heuristics, which suggests a decrease in creativity. These studies demonstrate that improving creativity using DfAM is challenging, and further research in these areas could help discover improved means of increasing designer creativity while ensuring quality and viability of designs.

3.2. Configuration

Computational approaches are often used to aid human designers in the configuration of AM parts due to their inherent complexity. Parametric design enables the adjustment of definable values related to a design's geometry to rescale parts and mechanisms for specified applications. Scalable AM mechanisms have been created parametrically for revolute, prismatic, and spherical joints with cylindrical spur, spiral bevel, and straight bevel gears [61]. Spiral gear mechanisms have been printed using a stereolithography apparatus machine that demonstrated a successful rescaling and functioning of the mechanism. Parametric design can also benefit from biomimetic approaches, such as approaches for modeling and optimizing L-systems inspired by plant growth algorithms [62]. Since biological systems are typically optimized via evolution for efficiency, mimicking such structures often leads to mechanically efficient systems. Honeycombs inspired by nature are common AM structures due to their high relative mechanical properties per weight. Mechanical testing has demonstrated that increases in material distribution at honeycomb nodes increases both stiffness and absorbed energy but may weaken other aspects of the structure if material volume is further increased [16]. The design of more complex architected structures, such as regular truss lattice cells, is another AM configuration strategy that enables mechanical efficiency [63] (Figure 4). Different configurations of lattices provide trade-offs in performance, such as those among strength, stiffness, and energy absorption, which designers may manipulate to tailor structures with properties for specified applications.

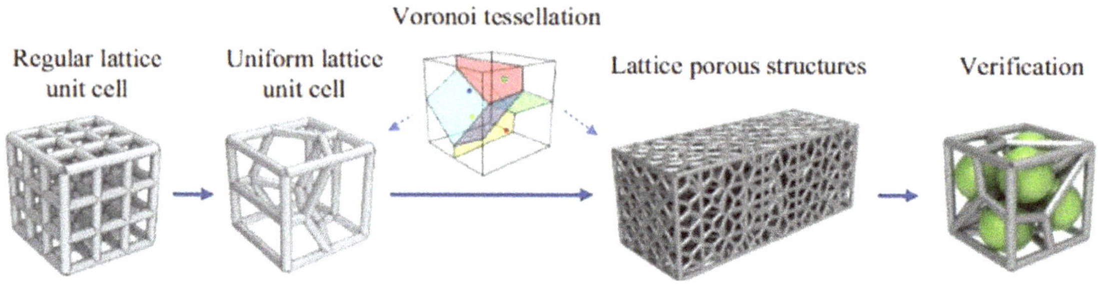

Figure 4. Parametric design of porous Voronoi lattice structures [64]. Image adapted with permission.

Voronoi-based lattice structures have been designed parametrically to create porous structures with properties that are globally controllable while locally uniform [64]. In the process, individual unit cells are divided throughout the design space, seed points and beam radii are determined for unit cells, and a Voronoi tessellation is used to distribute seed points uniformly. The seed points are tessellated globally with edges cylindered according to their corresponding beam radius values. Such structures have high stability that ensures that spatial deviations in porosity and surface area are small. Voronoi models are suitable for applications that need mechanical efficiency and have geometry-dependent needs, such as tissue growth based on a scaffold's shape [65]. Further strategies for configuring AM structures include functional gradients that gradually change parametric values spatially through a lattice [66], altering topologies of unit cells for transitioning properties throughout a lattice [67], or using multiple materials placed within a lattice to generate anisotropic properties [68]. Due to the diversity of strategies and decisions at local and global levels during design generation, optimization is often necessary to maximize the potential of configuration strategies.

3.3. *Optimization*

Optimization is used in DfAM for improving both the manufacturing process and the performance of final parts. One study investigating anisotropic lattice mechanical properties and support material removal employed a snap-fit configuration strategy to enable more efficient printing of structures while resulting in a 100% increase in strength and energy absorption [69]. The approach demonstrated that the overall configuration strategy plays an important role in final achievable properties and performance. Once a configuration strategy is determined, computational approaches are often employed to refine designs according to specified objectives and constraints. For instance, one successful approach paired a lattice unit-cell library with parametric design and topology optimization to obtain lightweight structures [70]. Topology optimization is a process for optimizing the material layout within a given design space with reference to loads, boundary conditions, and constraints. The combined approach led to designs with improved stiffness while enabling the rapid generation of different topologies to find optimized stiffnesses.

Topology optimization has been used for medical AM applications to design tissue scaffolds and customizable shoes, thus demonstrating the adaptability of the approach for varied systems [71]. The core algorithm used was termed as an automatic complex topology lightweight structure generation method. The algorithm uses a mesh generation algorithm to produce a node distribution inside an object representing a boundary surface of a targeted complex structure. Low-weight trusses are generated using the distribution of nodes. Radii are then adjusted based on an FE analysis with an optimization algorithm to produce a lightweight truss structure. The algorithm pseudocode is provided in Figure 5A with shoe design results for the loading of the truss structure without optimization and with optimization provided in Figures 5B and 5C, respectively. The results of the topology optimization demonstrate a more consistent level of loading throughout the shoe and a lower peak load, thus improving the pressure distribution for patients.

A

1	Prepare the boundary edge (for 2D case) or surface (for 3D case) of the target lightweight truss structure; assign the maximum stress σ_c that is allowed in the truss structure
2	Assign target node distances inside the boundary object
3	Assign nodes inside the object using the PIMesh
4	Generate a raw lightweight truss structure based on the node distributions. Let N be the number of the bars of the truss structure, and $N_{\text{updated}} = N$
5	While $\sigma_{\max} > \sigma_c$ or $N_{\text{updated}} > 0.05N$ (optimize the lightweight truss structure)
5.1	Run finite element analysis to obtain the von Mises stress distributions $\boldsymbol{\sigma}_v$ of the structure, and let $\sigma_{\max} = \max(\boldsymbol{\sigma}_v)$
5.2	Update the diameter of the bars of the truss structure according to the von Mises stress distributions, and N_{updated} is the number of the bars with radius changed
6	Generate the 3D models of the optimized lightweight truss structures

B

C

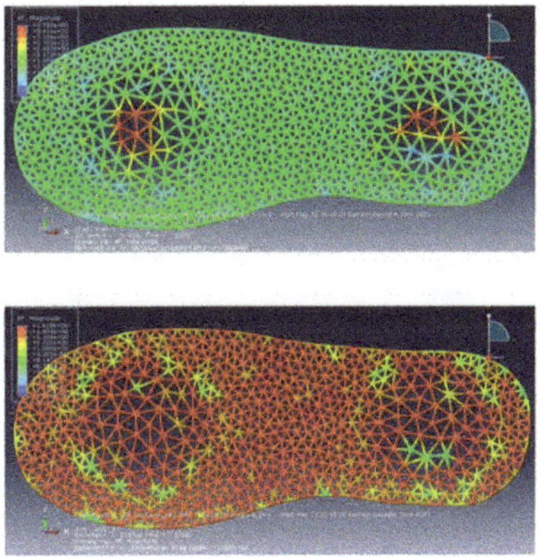

Figure 5. Topology optimization. (**A**) Algorithm for 3D-printed shoes with reaction forces of (**B**) a non-optimized shoe (max force = 1.88 N) compared to (**C**) an optimized shoe (max force = 0.94 N) [71]. Image adapted with permission.

It is possible to place constraints on structural features, such as overhang angles, to reduce the amount of support material necessary to fabricate a topology-optimized print. Build orientation and topology have been optimized together to accomplish these goals by using direction gradients to control the overhang angle within the design domain to reduce internal supports. A second density-based global constraint was used to control design domain boundaries for a reduction in external supports [72]. Another study used truss topology optimization with anisotropic struts that demonstrated improved material properties for trusses with anisotropic struts over struts optimized with isotropic materials [73]. The approach simultaneously optimized the struts along with the volume fraction of fibers or holes as reinforcements. Further steps included accommodating material symmetries, penalizing size variables, and improving manufacturability by introducing no-cut constraints to ensure symmetry across planes and faces of unit cells. Design and optimization has also been conducted on hybrid structures with solid and lattice portions [6]. Empirical validation with simulation results demonstrated that hybrid structures achieved higher stiffness, yield strength, and critical buckling compared to pure lattice or solid structures. Important last steps in optimization from these studies are validation and controlled comparisons. Validation with fabrication and experiments is necessary to confirm the accuracy of predictions from optimization, while controlled comparisons to different algorithms and cases enables designers to identify the most favorable design methods for a particular domain.

4. Assessment

Once generated designs are fabricated, it is necessary to assess their capabilities to determine if they perform as intended. A first step is the validation of printed designs to determine if features are printed accurately, especially those that may affect engineering performance. Testing, such as mechanical characterization, is then conducted to measure the properties of printed parts. Modeling steps can incorporate validation and testing results to account for differences between theorized performance and actual performance. Depending on the application and best practices, these steps may occur in a non-linear and iterative fashion. For instance, when using well-established FDM processes for macroscale designs with simple geometries, modeling and computational assessment may be considered prior to testing and validation to confirm predictions. However, when considering lattices that are printed near the fabrication limits of AM processes, such as those with microscale struts, designs may not print at all or print with highly different consistencies than predicted by ideal cases in modeling. Therefore, validation is necessary to determine achieved features prior to the creation of accurate modeling techniques that incorporate deviations incurred by printing processes. If validation is not conducted prior to evaluating the mechanics of a design, designers may spend excessive amounts of time modeling and testing designs that are impossible to fabricate or printed with very different structures and performances than expected. In all cases, by considering these steps and iterating appropriately, designers may review their designs and improve them according to the information obtained during each step.

4.1. Validation

Validation is necessary to determine if parts are printed accurately and function as intended. Discrepancies emerge in the fidelity of final parts due to inherent limitations in the printing process. For instance, in FDM printing, globules of material form due to the inconsistent melting of filaments, which, in turn, affects mechanical functioning [74]. Subjective assessments have been conducted for FDM surfaces that had 93 observers rate 107 images of printed surfaces prepared by nine different types of ABS filaments for three different printers [75]. Their approach quantified metrics for monitoring printed parts for use in automated systems to either improve processing parameters during printing or determine early part failures. During the laser sintering of polyamide-12, research has found inconsistencies in printed surfaces from powder properties, processing parameters, and surface orientation. These have been investigated using contact profilometry, focus variation, and micro-CT techniques as a function of applied energy density, XY location, orientation, and percentage porosity [76]. The results demonstrated that the roughness profile for top/bottom surfaces were distinct, with top surfaces having peaks of greater amplitude. Such inconsistencies can inform designers about where cracks may occur that could lead to part failure. Fidelity has also been studied for bioink printed structures in which extrusion can lead to inconsistencies in printed diameters of lattices and the relative placement of material [77]. Additionally, bioprinted structures may become distorted due to the deformation of soft materials used for printing.

The effects of design and processing strategies have been studied for lattices fabricated with DLP printing. Microscopy measurements have determined that structures with higher relative density and more beams per unit cell tended to print with more material than expected compared to other designs [7]. Mechanical testing demonstrated that the added material improved mechanical properties. However, material added by altering process parameters led to more mechanically efficient structures than those with increased relative density from designed topology alterations. Figure 6 demonstrates the microscopy process used to validate each structure's accuracy by measuring beam diameters, which was affected by the processing parameters of print orientation and exposure time. Print orientation affects beams due to layers needing to bridge large gaps or build on top of previously placed materials. Beams were printed with smaller diameters than expected on average, and fusion of pores tended to occur towards the center of the lattice. Such inconsistencies occur in DLP printing due to pixel size, stage motion, optical focus, and resin properties, which has led to new methods proposed as corrective factors. The grayscale manipulation of pixels has been demonstrated for smoothing discontinuities in surfaces with manipulations to improve accuracy informed by a reaction–diffusion simulation to predict final cured shape [78]. Compensation methods have been used to alter beam diameters throughout a lattice design to correct for spatially dependent inconsistencies such as beams printing larger than expected towards the center of a lattice [79]. These methods improve upon the already highly accurate DLP process to provide prints that better match a designer's intentions with more consistent, predictable performance.

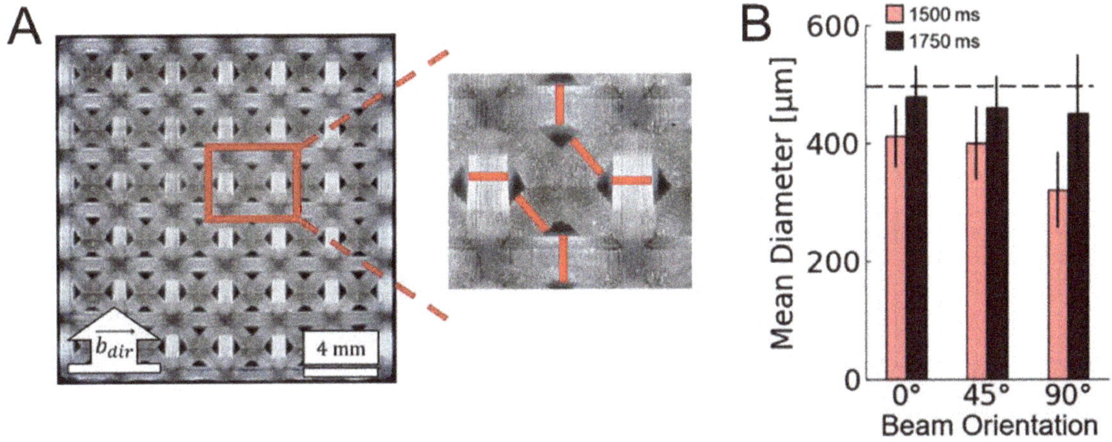

Figure 6. Validation of BCC lattice using (**A**) microscopy to (**B**) measure mean diameters based on beam orientation [7]. Images adapted with permission.

Inconsistencies in printing have also been observed for metal lattices produced with SLM [80]. Beams were printed larger than intended with material agglomeration occurring at corners due to overmelting, which resulted in fillet-like features. Octet topologies were more sensitive to inconsistencies than tetrahedron topologies, which demonstrated a dependency of accuracy on the configured design. Due to inherent limitations in metal additive manufacturing, methods have been developed for quantifying the effect of manufacturing defects using the automated analysis of microscope images [81]. The results found that parts were generally printed larger than expected and that increases in the sizes of nodes/beams increased stiffness. When designs were scaled smaller, mechanical properties increased due to the proportional increase in material per volume. Uncertainty quantification is another validation technique used to characterize the inconsistencies in printed lattices that uses data to inform design decisions [82]. Error propagation for metal printing has been determined in tandem with multiphysics simulations to better predict how microscale defects may affect global functioning, which can inform design decisions. Such methods are essential for designers to improve AM outcomes by understanding how differences in ideal models and tested outcomes affect final design performance.

4.2. Testing

Testing is a key assessment step that helps validate findings while also providing empirical measurements of mechanical behavior for creating accurate models to predict performance. Mechanical testing in compression is often used for measuring lattice properties and provides information regarding the effective elastic modulus, yield strength, and ultimate strength. The effective elastic modulus is the slope of the stress–strain curve, which accounts for the relative density of the lattice and its topology that is proportional to the elastic modulus of the base material used to construct the lattice. For SLA printed lattices with a base material elastic modulus of about 1670 MPa, the effective elastic moduli of lattices with four different beam-based topologies ranged from 100 to 260 MPa [83]. Designs tended to have higher effective elastic moduli when more beams were aligned with the loading direction. Mechanics can also fluctuate due to manufacturing inconsistencies for the same design. Studies of DLP-printed lattices that tested 30 prints of the same design demonstrated a distribution of density and mechanical property measurements that occur due to fabrication inconsistencies [84]. The study investigated uncertainty and reliability for lattice mechanics and found that lattice failure due to yielding was highly affected by fluctuations in beam diameter.

Further studies in lattice mechanics have investigated energy absorption for multi-material honeycomb structures printed with ABS and TPU by measuring the area under the force–displacement curve [85]. The study found that adding material bands of TPU to ABS hexagonal honeycombs resulted in a linear decrease in energy absorption from 15.1 to 2.9 kN·mm, which was linearly proportional to the amount of TPU added for out-of-plane testing. Out-of-plane failure responses for the two different designs were similar and demonstrated a deformation of the flexible TPU structure prior to the deformation of the stiff ABS material. Since the TPU deforms first and is highly recoverable, the system is reusable for low displacement cases yet retains high energy absorption for failure scenarios, such as crashing, that benefit from the stiff ABS material with more overall energy absorption. For in-plane testing, failure mechanisms of square and hexagonal honeycombs demonstrated that more consistent failures occurred from hexagon designs, while square designs resulted in discrete collapses of unit cell layers. These results provide quantifications for designers to tailor designs while also highlighting the differences in behavior during failure that inform decisions for selecting one type of design over another.

Studies on titanium SLM lattice structures have conducted extensive measurements of the deformation and failure behavior of diverse lattice topologies [86]. The study sought to understand the influences of lattice cell topology, cell size, and unit cell count. By mechanically testing diverse lattices, it was found that 10^3 unit cells and higher are necessary for the convergence of mechanical properties. During compression, different states of lattice behavior were identified from initial to elastic preload, to initial failure, to progressed failure, and to gross failure, the results of which are highlighted in Figure 7. The pattern of unit cells failing differed based on the design's topology. BCC structures had a comparatively large compliance followed with deflection that occurred due to strut bending near the lattice joints, with subsequent failure that leads to unit cell collapses. BCC-S3 lattices had a horizontal layer collapse with diagonal shear failure, while BCC-S2 lattices had a combined collapse of diagonal and horizontal layers. FCCZ structures had struts buckling vertically, followed by a fracture that led to unit cell collapse, followed by a diagonal layer collapse in two directions for the specimen's lower half. Trends from the study suggest that the Maxwell number and the alignment of struts along the loading direction affect lattice failure modes. Topologies with high lateral stiffness were observed to have horizontal layer crushing, which suggests they are appropriate for energy absorbing applications.

Further studies have investigated 30 strut-based lattices with cubic structures printed with SLS using polymeric materials [87]. Lattices were constructed by combining unit cells with topologies for cubic, diagonal, octahedron, and V-octet topologies. FE simulations were conducted to help interpret experimental results for effective stiffness, yield strength, and buckling strength for uniaxial, shear, and hydrostatic loadings. General results concluded that the modes of deformation differed for various loading conditions for stretching-dominated, bending-dominated, and mixed types of structures. Stretching-dominated structures with triangulated micro-architectures within unit cells provided greater stiffness and strength per unit weight than bending-dominated structures. The study demonstrated that topology and relative density also played a major role in mechanics, which suggests there are many routes for manipulating designs and conditions to achieve a desirable set of mechanical properties.

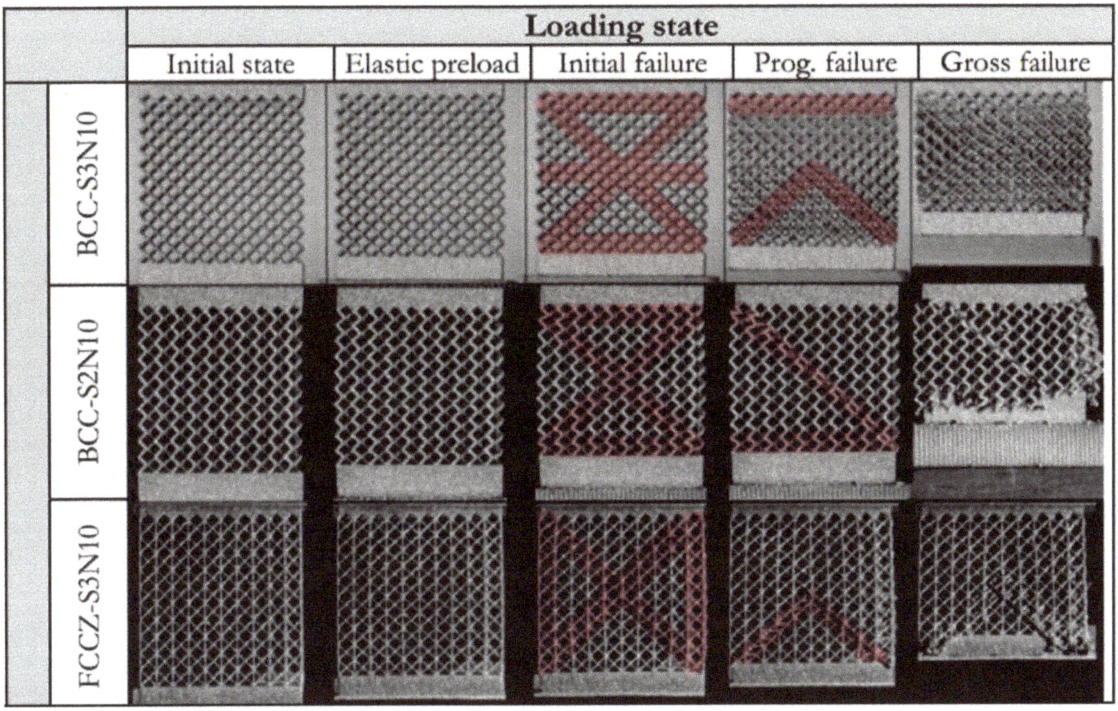

Figure 7. Lattices tested in compression with failure modes highlighted [86]. Image adapted with permission.

Diverse topologies for metal structures have also been studied from SLM processes with Inconel, which provides a high resistance to oxidization, creep, and a loss of mechanical properties at high temperatures [88]. The lattices demonstrated stable crushing behavior up to densification with exceptional ductility and high damage tolerance. Several observations were summarized regarding failure modes for the lattices. For all topologies and cell sizes, the behavior was apparently linear for strains of 2%. When strain was increased to 4%, BCC and FCC topologies remained linear while BCCZ and FCCZ topologies demonstrated a local plastic collapse for specific unit cells. The collapse had bands associated with maximum shear stresses aligned 45° to the loading direction. Topologies with Z-struts also had catastrophic failures followed by densification, with cyclic behavior of plateau stresses as local collapses occurred. For the FCC and BCC topologies, further strains resulted in mostly homogenous deformation behavior throughout the lattice. Additionally, the degree of bending and stretching-dominated behavior was possible to control by modifying the local cell structure, with Z-strut addition driving structures to have a greater degree of stretching-dominated behavior that was increased with larger cell sizes. Overall, mechanical testing demonstrates the diverse behaviors exhibited by AM structures that necessitate designers to create and validate models to accurately predict outcomes.

4.3. Modeling

Modeling is the process of creating a design and predicting its performance, which occurs throughout different periods in the design process as information is gained. Although a digital design is necessary to create prior to printing, proper modeling of its mechanics requires validation and testing to create an accurate digital geometry to determine how printing defects play a role in mechanical behavior. The most common approach for modeling mechanics of AM lattices is the finite element method (FEM), which numerically solves differential equations. FEM subdivides a large system into smaller simpler parts via discretization and mesh creation. Titanium structures fabricated with PBF processes have been represented with FEM for 50% to 90% porosity designs that found a high agreement with the Gibson–Ashby model [89]. The Gibson–Ashby model predicts mechanical properties, including elastic modulus and yield strength, from the relative density of a structure using an exponential scaling law and empirically informed constants. Scaling laws have been used to predict the compressive mechanical properties of BCC lattice structures with 117 design variations, first with FEM and then using mathematical equations to reduce computational effort [90]. The study also found that increased strut angles lead to increased relative stresses and elastic moduli for the stretch-dominated behavior of lattices.

FEM approaches have demonstrated the importance of modeling as-manufactured lattice geometries that differ from the idealized design case [91]. Numerical simulations from micro-CT assessments have provided a means for including manufacturing defects in modeling to determine their effect on buckling. The study found that analytical and linear numerical assessments for buckling consistently over-estimated effective buckling performance, which suggests the need for corrective factors to account for reduced performance from ideal cases. The modeling approach considered the boundary conditions, unloaded response, and first buckling mode of struts, while evaluating the ratio of the buckling strength to the slenderness ratio. Struts with a greater slenderness ratio had a higher tendency to buckle and fail. Numerical simulations in which non-linear geometry was considered with a non-linear material model provided higher consistency with experimental results compared to linear numerical methods.

Further modeling methods are necessary to consider as AM capabilities continue to advance and outpace traditional evaluation methods. For instance, researchers have modeled multi-material AM structures with complex geometries that used smooth material transitions with a semi-analytical unit cell decomposition strategy [92]. The strategy splits the complex lattice into units of struts and connectors with further interpolation of discrete material property values using a multiquadric radial basis function network. When uses of the model are compared to functions for traditional material distributions, the approach has advantages for handling an arbitrary number of base materials. Case studies have demonstrated consistency with theoretical material composition values. For lattices in biomedical applications FEM is possible to use in tandem with lattice and biomechanical models to determine lattice behavior for cases such as spine fusion [93]. These models can be combined with tissue growth simulations to determine trade-offs between lattice mechanical and biological performance [94]. Generally, these are opposing trade-offs due to the need for high relative density to improve stiffness, with low relative density to improve tissue growth density and nutrient transport. Due to the complexity in trade-offs, simplified models are useful to characterize trends and identify high-performing candidate designs prior to using more expensive modeling and optimization techniques to finalize designs for specific applications.

5. Applications

There are diverse engineering application areas enhanced by AM's capabilities for novel materials, complex optimized geometries, and rapid design iterations. The use of DfAM enables the creation of efficient structures for load-bearing applications relevant for the aerospace, automotive, and construction industries in which minimizing weight is crucial. Application-specific tailoring is particularly beneficial in food and medical applications in which design performance is dependent on each person's individual needs. Novel AM materials, such as the combination of soft and magnetoresponsive polymers, enable the creation of new engineered systems such as mobile tensegrity robots. Here, the state of the art in these application areas (Figure 8) is briefly reviewed while highlighting how DfAM may benefit future design innovations. A table is provided at the end of the section that highlights the relevant materials and processes for each application area considered.

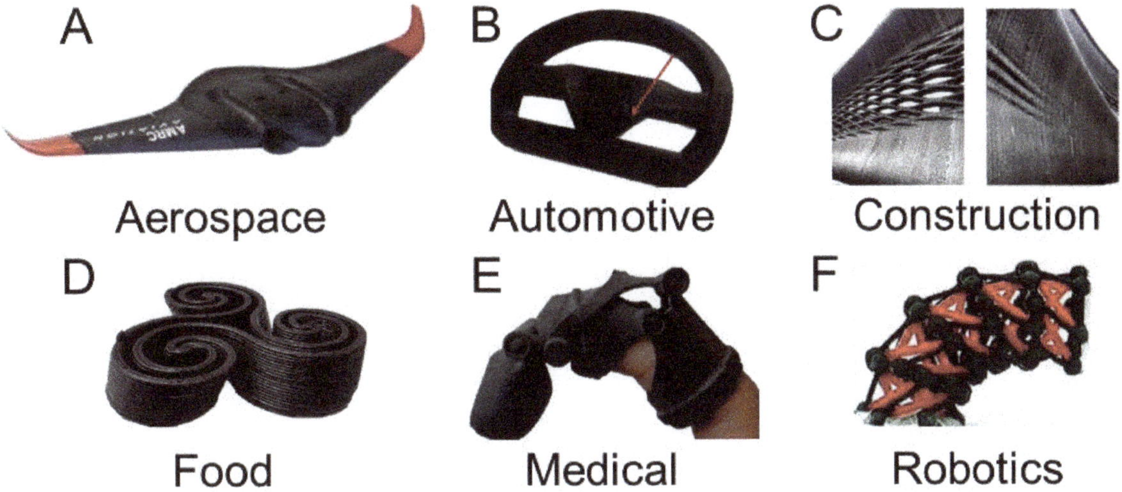

Figure 8. Representative innovative AM applications in diverse fields demonstrating (**A**) an unmanned aerial vehicle [95], (**B**) a lightweight steering wheel [96], (**C**) bridge construction [97], (**D**) a chocolate printed geometry [98], (**E**) a finger prosthetic [13], and (**F**) a soft tensegrity robot [99]. Images adapted with permission.

5.1. Aerospace

AM technologies have high relevance in aerospace applications since they provide high mechanical efficiency for high-strength low-weight systems favorable for flight. FDM technology has been applied for printing wings and/or fuselages for small remote control (RC) and unmanned aerial vehicles (UAVs) due to its low machine/material cost and high print speed/quality [100]. Parts for UAVs were assembled with the aid of glue and carbon rods for positioning, while support material was removed with pliers. Printed planes performed well in thermal soaring and high-speed glides but required further work for integrating airbrakes and reducing pitching-up effects. Fixed-wing UAVs have been successfully built from ABS materials that required less than 24 h to print all parts for the airframe (Figure 8A) [101]. Studies on process parameters recommended that layers with long contours aligned with the loading direction provided the greatest mechanical performance [102]. When considering AM in the aircraft industry, fuzzy systematic approaches have identified five critical factors for its success: (1) cost effectiveness, (2) special demand capabilities, (3) part printability, (4) a lack of manufacturing technologies, and (5) the size of the local maintenance market [103]. These considerations highlight that, beyond the design itself, logistics and market demands play an important role in the success of AM technologies.

AM has also been investigated for space applications, including the use of microsatellites. Microsatellites perform functions of earth observation, service, and on-orbit inspection and, due to their low volume production, may benefit from customized AM designs [104]. AM is suitable for creating heat shields for microsatellites using SLM processes, and many companies are fabricating relevant parts in rocket engines, solar panel supports, and valves suitable for use in space. AM may be suitable for Moon or Mars colonization due to its capabilities for using localized resources for construction without the need for space transport [105]. For instance, suitable sulfur concretes for Martian habitats have been created using microwave processes combined with FDM for casting. A comparison of suitable technologies between AM and casting methods for lunar regolith simulants has demonstrated ranges of compressive strengths between 2 to 31 MPa [106]. These trade-offs highlight the need for DfAM considerations for these applications to ensure that the optimal processing approaches are realized to reach the desired design capabilities.

5.2. Automotive

The automotive sector is another growing application area of AM, which is contributing via rapid prototyping, tooling creation, and the production of finished components [107]. Using AM, designers can create simple interior elements for a dashboard or fabricate scale models of entire cars. The technology saves time during prototyping by significantly lowering costs for manufacturing, leading to improved agility and reduced prices for the company. FDM printing using PLA materials has demonstrated success for rapid prototyping deep drawing tools for automobile parts manufacturing [108]. Several printers, material compositions, and processing parameters were varied to fabricate the drawing tools. Tool wear was measured using digital image correlation principles, and printing parameters and materials capable of safely producing a minimum batch of 100 parts were found. AM has also been used to improve sustainability in the automotive sector by promoting a circular economy for scrap metal [109]. A sequence of metallurgical operations was proposed to carry out circular economy component manufacturing that included steps for (1) milling, (2) physical–chemical treatment, (3) 3D printing, and (4) mechanical tests for validation. Prior to the establishment of the circular economy, further research is necessary for empirical validation with equipment and an assessment of financial feasibility.

Lattice structures are advantageous for automobiles and have been considered for use in electric vehicles to protect the battery pack against impact loading [110]. Lattices with high energy absorption can improve a vehicle's crashworthiness, which is a measure of a vehicle's ability to absorb energy from a collision. One proposed bio-inspired concept used a multi-layer approach to protect the battery by combining aluminum and steel parts that provide differing structures and functions for crash impacts throughout a hierarchical system. Topology optimization has been applied for AM parts in cars for weight reduction, which was demonstrated by redesigning a suspension arm [111]. A static analysis of suspension arm models before and after optimization demonstrated that topology optimization reduces stress and weight, with results dependent on a lattice's unit cell type. Weight reduction has been conducted by altering the infill of FDM-printed parts for automobiles, which is demonstrated by an optimized steering wheel example in Figure 8B [96]. During optimization, twelve different infill configurations were experimentally tested under tensile and flexural loading, which revealed that their elastic moduli ranged from 0.74 to 1.80 GPa, while their yield strength ranged from 15.3 to 30.7 MPa. These results demonstrate the wide-ranging performance of AM parts and the need for DfAM to consider and evaluate variations in designs to determine optimal configurations for applications.

5.3. Construction

AM is growing in prominence in construction applications due to advancements in printable concrete materials that enable automated fabrication for large structures that would otherwise require extensive human labor. Researchers have created complex walls fabricated from two polyurethane foams encased in a third wall of concrete [112]. The concrete consisted of CEM III cement, limestone filler, sand, gravel, water, and a set accelerator. The concrete was directly pumped to flow easily without any need for vibration. A robotic printer was used for construction to produce a finished house with 95 m^2 surface area built on-site that housed one family in Nantes, France. Another study using a self-developed printer that constructed structures in Guangzhou, China processed materials of C25 ready-mixed concrete with 5 to 15 mm coarse aggregate [113]. Large-scale construction has also been facilitated for on-site printing in Dresden, Germany using cement materials [114]. The concept of CONPrint3D was proposed, which adapts concrete 3D printing to current trends in architecture and structural design, provides a maximal use of common construction machinery, uses concrete compositions aligned with existing standards, and has printheads for construction with suitable surface quality and precision. The printer has demonstrated successful construction for a variety of materials and designs.

Concrete bridges have been 3D printed with certification for public use in Eindhoven [115], which went through a sequence of testing to determine safe mechanics prior to use. The construction of 3D-printed bridges has also been successfully completed and tested using metal materials with a wire and arc additive manufacturing method (Figure 8C) [97]. The bridge has a 10.5 m span, is suitable for walking and was validated by testing, analysis, and verification steps. Experiments were conducted to first test the handrails, followed by the substructures with handrails, and finally the completed bridge with its deck welded to its serviceability limit. FE simulations were used to provide further insights for the structural response, load-bearing capacity, and long-term health of the printed structure. In further uses of concrete, digital design-to-manufacture processes have been used to create a post-tensioned concrete girder with a novel topology and shape optimization procedure [116]. Experimental methods were used to investigate the girder's load-carrying capacity, which demonstrated that the optimized printed structure had a significant material reduction. Material reduction is favorable in construction applications due to the decrease in cost and environmental impact.

Metal printing is desirable for construction due to its high quality and resource efficiency [117]. Sustainability is particularly important for construction since it is responsible for a large share of the world's carbon footprint. The high strength of metals, combined with DfAM methods such as topology optimization, enable a reduction in material use compared to traditionally manufactured parts while also facilitating rapid construction and customization. Other materials, such as cob and concrete, have also been assessed for their environmental impact. A comparison between the materials found trade-offs: Concrete had a higher overall environmental impact when considering factors such as global warming potential but less impact on land use and resource scarcity [118]. Further studies in DfAM for environmental impact are essential to optimize sustainability in construction, particularly in determining the impact of the materials themselves, the energy used for processing, and further logistical considerations such as transport.

5.4. Food

Food printing is an emerging area in AM that enables the directed deposition of foods to create appealing shapes with personalized nutritional profiles. The most common type of printing for food is extrusion, which is suitable for soft materials. A DfAM study on food printing created custom food inks by combining pureed pumpkin with corn starch or guar gum materials to alter the material's rheological properties [119]. The results found that a formulation with 4% guar gum added by weight provided the most accurate print for a three-dimensional squirrel design. Further trade-offs in printability, texture, and sensory properties have been investigated for mashed potatoes fortified with protein or lipid materials to improve nutrition and/or taste [120]. The results found that added butter improved the sensory properties of taste, mouthfeel, smell, and visual appeal, whereas added pea protein reduced sensory appeal while reducing calories and increasing protein. The use of a non-conventional cricket powder improved the protein content of the food while decreasing sensory appeal. 3D food printing is extendible to diverse foods, such as mimicking the mechanical properties of apple tissue with plant-based ingredients to create innovative cereal-based snacks [121]. Genetic algorithms and response surfaces have been used to optimize the printing of a chicken gel [122], which demonstrates the use of DfAM principles to create novel foods.

A key advantage for printing foods over conventional manufacturing processes is the possibility to create complex shapes with geometries that appeal to consumers [98], as demonstrated with the chocolate print in Figure 8D. The study determined fabrication constraints for chocolate and marzipan printing materials and assessed their capabilities for printing complex features, such as overhangs. The study created designs of varied complexity that were rated by participants who preferred the most complex shapes enabled by AM. The fabrication of appealing shapes can promote healthy eating practices, such as creating illusions with foods to look larger while retaining a fixed amount of calories [123]. The improved aesthetics of food can also increase the adoption of healthy foods that consumers are reluctant to eat otherwise [124], especially when combined with further strategies such as serving foods in facilitating environments or social structures. These strategies highlight the need to consider DfAM principles beyond just improving printability since social factors also play a key role in the success of AM designs.

5.5. Medical

AM technologies are well suited for medical application due to their customizable geometry that enables the design and fabrication of parts specifically tailored for patients. Prosthetics are a prominent application for AM that could benefit the approximately 3 million people across the globe that have upper-limb amputations [125]. Flexibility for prosthetic hands has been achieved by fabricating soft joints and supporting body-powered movements with non-elastic cables for flexion. Powder bed fusion has been used to create components for integrated prosthetics that include lattices for mechanical efficiency, springs for energy absorption, and non-assembly mechanisms to support movement [13]. An example finger prosthetic is provided in Figure 8E, which was constructed via a series of design, fabrication, and assessment steps to determine printable dimensions and tolerances for constructing functional mechanisms. Orthopedic solutions are another area for AM designs to aid patients, especially by computational engineering processes to configure structures for unique patient needs [126]. Such solutions reduce the labor requirements for creating customized solutions while also improving fit and compliance for patients. There is also a possibility for extending AM designs to orthopedic footwear, which is a prominent issue for patients with diabetes [127]. Medical solutions can benefit from the ubiquity of lattice structures in AM [128], which enable tailorable and anisotropic solutions for functioning in relation to the uneven forces and complex geometries of the human body.

Regenerative medicine applications benefit from AM designs for tissue scaffolds that are mechanically efficient and tailorable. Tissue scaffolds, such as those for bone, have mechanical and biological trade-offs that are often in conflict. Scaffolds have been created using DLP printing processes to produce hierarchical structures using truss-based unit cells for mechanical efficiency while providing large voids to support the biological growth of blood vessels [129]. The structures were tuned using a combination of computational design approaches to generate hierarchies and finite element analysis to evaluate mechanical stiffness. Further computational design approaches for tissue scaffolds have used voxel-based simulations to predict tissue growth in unit cell structures that were validated with in vitro biological experiments [130]. Lattice structures enable the creation of tunable geometries to adjust surface area to provide more places for tissue seeding while providing curvature that is necessary for three-dimensional growth. Open-source libraries of tissue scaffolds have also been created using lattice structures to facilitate multi-scale and multi-material AM designs [131]. The libraries benefit from enabling systematic parameter variations to generate solutions and mapping to compare design trade-offs. Advances in fabrication, such as dual extrusion printing, have enabled the regeneration of tissues with high aesthetics and shape retention [132]. In vitro and in vivo trials were conducted that demonstrated complex bioprinted scaffolds could facilitate nipple–areola reconstruction. These solutions highlight the practicality for AM solutions in medicine, in which continued design research can enable the creation of algorithms for tuning structures for patient-specific needs.

5.6. Robotics

AM has become a key technology in the fabrication of robotics, particularly for soft robots, by providing capabilities for printing materials with large deformations and varied functionalities [133]. Soft robots are a new generation of robots that may cooperate with humans or traverse constrained environments, such as steering through narrow environments. The primary materials for soft robots are fluids, gels, and functional polymers that were traditionally created with molds and are now being replaced with faster and more reliable AM technologies. For instance, AM-fabricated hydrogel actuators have been printed for use in jellyfish soft robots [134]. Researchers conducted compression testing to evaluate the gel actuators with normalized trade-offs between length, contraction, pressure, and volume as key design criteria. Tensegrity structures, which are a combination of stiff struts and flexible tendons, have been used to create soft robots that function with the aid of magnetic forces for actuation (Figure 8F) [99]. The robots were created using smart materials with no need for further assembly. Printers with dual print heads created a sacrificial mold for the tendon material fabricated with PVA while struts were printed with PLA. After printing, polymeric smart materials were injected into the sacrificial mold. The robot was able to conduct various transformations in response to torsional, compression, and shearing stresses with designers able to predict outcomes using simulations informed by experiments.

Robotic applications are well suited for medicine, in which tactile sensing has been enabled by AM-printed soft pressure sensors [135]. Several materials including TPU, conductive PLA composites, and graphite ink were used to develop five different variations of sensors fabricated with a modified 3D printer capable of outputting diverse inks, pastes, and polymer materials. The highest performing touch sensor formed with silver paint and soft rubber exhibited a stable response with a sensitivity of 0.00348 kPa^{-1} for pressures less than 10 kPa and a pressure of 0.00134 kPa^{-1} at higher pressures. Surgical robots have been created for patient-specific applications, which is advantageous for operations such as treating lesions of different sizes and shapes for the removal of deep intracranial brain tumors [136]. The customized AM robots can reach a surgical site by avoiding or minimizing damage to critical brain structures. Although the technology for customizable surgical robots is promising, there is a need for further research to improve sterilizability, biocompatibility, and stiffness. Multi-material robots in medicine have also been proposed for on-demand drug delivery that benefits from the design freedom offered by AM [137]. Digital design optimization has been used to create a robot divided into parts that mimicked the movements of an inchworm with modeling supported with multiphysics software to inform design decisions. The robot exhibited linear and turning locomotion powered by magnets, which demonstrated its capabilities for traversing the inside of the human body, such as traveling through the lungs. These works highlight the potential for DfAM to further improve capabilities for robotics, in which new advances could considerably improve robotic functionality in healthcare.

5.7. Applications Summary

Application areas commonly use different additive manufacturing processes and materials according to their specific needs. Table 1 summarizes the materials and processes for each surveyed application that provides designers with starting points to consider when approaching DfAM in these areas. The table includes material used directly during the AM process and, for some cases, such as Martian habitats, materials that are used to form final parts from AM casts. By referring to the table, designers may recognize the current state of the art and gaps in which the consideration of further applications, materials, and processes open opportunities for innovation.

Table 1. Highlighted materials and processes considered for each application area.

Area	Applications	Materials	Processes	References
Aerospace	Aircraft, UAVs, air ducts, wings, turbocharger blades, spare parts, prototyping, and rapid tooling	PLA, ABS, PET, nylon, carbon fiber, steel, titanium, and custom resins	FDM, SLS, DMLS, and SLA	[95,100–103]
Aerospace	Microsatellites, heat shields, magnetic shields, and spacecrafts	PEEK, PEKK, PEI, polyamide, carbon fiber, $AlSi_{10}Mg$, and molybdenum	FDM, SLM, and DMLS	[104]
Aerospace	Lunar and Martian habitats	Sulfur, magnetite aggregate, silica sand, regolith, and geopolymer composites	Extrusion, powder jetting/fusion, and vat polymerization	[105,106]
Automotive	Steering wheel, suspension arm, interior elements, drawing tools, and replacement parts	PLA, ABS, metal powders, and composites	FDM and metal powder printing	[96,107–111]
Construction	Bridges, walls, houses, girders, joints, and stiffeners	Concrete, cob, polymer-foam, glass fibers, geopolymers, and metals	Large-scale extrusion, robotic printing, and wire and arc	[97,112–118]
Food	Health treatment, weight loss, consumer appeal, and sustainable food production	Cereals, fruits, vegetables, chocolate, marzipan, meats, dairy, gels, and insects	Extrusion and powder binding	[98,119–124]
Medical	Prosthetics, medical devices, footwear, implants, orthopedics, and prototyping	PLA, TPU, PETG, carbon fiber, nylon, and stainless steel	FDM, SLA, and PBF	[13,125,126,128]
Medical	Tissue scaffolds and interbody spinal cages	PLA, PCL, PEEK, TCP, methacrylates, titanium, biomaterials, and living cells	Extrusion, SLA, DLP, and SLS	[129–132]
Robotics	Soft robotics and actuators	ABS, silicone, polyurethane, hydrogels, nylon, varied elastomers, smart composites	FDM, SLA, inkjet, SLS, Multi-material processes, light-scanning gel printer	[99,133,134,137]
Robotics	Surgical robots and robotic prosthetics	PLA, ABS, PVA, TPU, methacrylates, polymer resins, multi-material, and metal paste	FDM, SLS, SLA, inkjet, and multijet	[135,136]

6. Outlook

As AM continues to provide innovations across diverse industry sectors, there are also many new opportunities in research and development for applying the technology. Figure 9 highlights four areas of interest, including innovation services, bio-inspired design, robotic fabrication, and machine learning, that are relevant to the advancement of AM using DfAM frameworks.

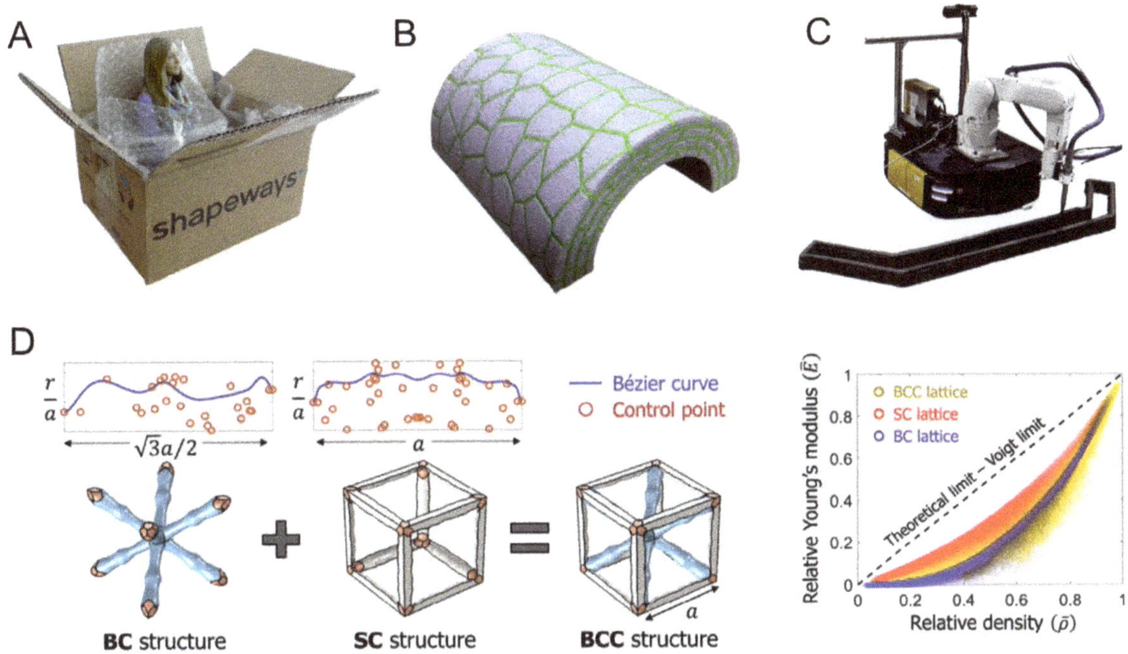

Figure 9. Future outlook in highlighted areas for DfAM, including emerging (**A**) innovation services [138], (**B**) bio-inspired design [139], (**C**) robotic fabrication [140], and (**D**) machine learning approaches [141]. Images adapted with permission.

6.1. Innovation Services

The emergence of AM technologies is producing new services and maker spaces that encourage widespread innovation by lowering the barrier required to design and fabricate parts. Outsourced 3D printing services are companies that receive digital designs primarily via online orders from consumers and then fabricate and deliver prints, as demonstrated in Figure 9A for a customized figurine [138]. A reason consumers prefer outsourced printing services over at-home fabrication is due to the higher costs associated with printers that fabricate with multiple materials and provide high-quality finishes for personal use. Makerspaces, in contrast, provide designers access to shared equipment and materials at a low cost that encourages more hands-on innovation. Makerspaces provide environments for these designers to create and fabricate who otherwise would not have the resources to do so, thereby democratizing innovation [142]. Makerspaces can combine resources in new ways and, by providing open access, facilitate the creation of designs that may otherwise never be realized. Although a bottom-up approach is often employed in the organization of makerspaces, a more structured approach can improve the commercial viability of designs. One interesting example of AM for commercial products is the creation of nail-art technology to produce diverse patterns with varying aesthetic appeal for consumers [143]. The technology provides precision control with a resolution below 50 microns using various colors that are printable on demand. Such technologies and approaches have viability in other sectors, such as when overlapping with aesthetics for the controlled release of drugs, which could improve the appeal of vitamins for children.

Innovations enabled by AM also have great potential to influence sustainability globally and play a large role in the advancement of Industry 4.0 by interfacing with the Internet of Things. As AM replaces traditional manufacturing processes in obvious metrics such as material efficiency and mechanics, there is a need to consider less obvious metrics such as AM's impact on sustainability [144]. AM is sustainable with less processing steps to reduce energy use, waste, and emissions while also requiring little post-processing. The technology is also sustainable societally by enabling automation in developed countries and lowering the barrier for skills use in under-developed countries. AM's ability to support mobile phone production has been examined to determine its effects on sustainable supply chains by considering the interactions between printer availability, consumer attitudes, and entry to market [145]. The effects of AM technologies for small firms were investigated and suggest that there is an unmet need for social sustainability that could be driven by AM to form new business models. Industry 4.0 is poised to automate manufacturing and data across products, which can help AM technologies become interconnected and respond to customer requirements more efficiently with smart manufacturing [146]. AM can support Industry 4.0 directly with applications or indirectly with processes in which interactive experiences with consumers, such as the use of augmented reality, can result in the tangible construction of high-value designs. Smart materials are also possible to construct with AM for integration, which is expected to gain wider prominence as Industry 4.0 technologies advance [147]. Application areas that could benefit from further AM research include beacon technology for signals with unique identifiers, cyber-physical systems for maximizing product fabrication rates, and big-data-driven manufacturing that informs decision making across a product's entire lifecycle.

6.2. Bio-Inspired Design

Bio-inspired design relies on mimicking structures, behaviors, and/or functions of natural systems to improve engineering systems, often by increasing efficiency. Protheses have been improved by bio-inspired workflows that incorporated steps of 3D imaging, modeling, and optimization to design and fabricate a transtibial prosthesis [148]. The designs used a combination of FE analysis with topology optimization to create foot geometries printed with nylon materials. When compared to traditional prosthetics, these designs lowered the average cost by 95%, the weight by 55%, and the production time by 95%. Novel viscoelastic dampers with mechanical interlocks have also been developed using additive manufacturing with bio-inspired principles [149]. The dampers were designed with a jigsaw-like mechanism informed by various natural systems including turtle shells, skull suture joints, and frost crystal fractals. The device had a steel hard phase and a TPU soft phase, which were configured using experimental and computational methods for testing and validation. DfAM principles were used to investigate loading rates and patterns of specimens that found that density greatly affects the design's mechanical response.

A bimaterial structure that mimics nacres' multilayer structure has been developed by observing natural nacreous shells and mapping features, such as their dome-shaped structures (Figure 9B) [139]. The design was generated with Voronoi tessellation and printed using ABS plastic for impact resistance, while softer TPU material was selected for bonding. The design's performance under impulse loads was assessed numerically and demonstrated that cohesive and adhesive bonds within the nacre mitigated energy that reduced damage to the composite. Bio-inspiration has also been used to mimic plant cell morphology, which facilitates water and mineral transport to enable powder removal from printed parts such as lattices [150]. Powder removal is difficult in lattice structures due to their complex network of interconnected pores, which can be improved by adjusting unit cells' centers and the size of ventilation holes that promote material removal. Bio-inspired femoral stems have benefitted from an efficient lattice design [151], which reduced stress shielding by 28%, thereby promoting better long-term outcomes for bone health.

Further cases in bio-inspired design have used honeycomb structures to achieve efficient mechanics that mimic natural structures at micro- and macroscales [152]. Hierarchy was introduced by altering hole positions, orientations, and shapes, which resulted in tunable mechanics. Mechanical testing results demonstrated that hierarchical honeycombs with circular holes performed best compared to those with square or hexagonal holes for stiffness, strength, and energy absorption metrics. Zero Poisson's ratio structures have also been investigated with combined soft and stiff unit cells for controlled deformation patterns during compression [153]. When novel bio-inspired designs were compared to auxetic materials, they possessed an improved ability to remain stable while simultaneously providing high energy absorption. Bio-inspired principles of lattice designs have also been applied towards scooter decks for enhanced performance using honeycombs [154], as well as for water sports boards [155], which demonstrates the versatility for such structures across application areas. Future research has great potential for uncovering useful structures and functions in natural systems, in which rapid testing and creation can occur with AM to further improve the performance of engineered systems.

6.3. Robotic Fabrication

Robotic fabricators are systems that aim to move beyond the limitations of current AM processes by incorporating robotic movements and control during fabrication. One effort combined a six-axis robotic arm system with hybrid extrusion–photopolymerization to enable the fabrication of layerless lattices [156]. Octet lattices were printed with processing phases including nozzle approach, pre-extrusion, extrusion, post-curing, nozzle removal, and nozzle travel, which were facilitated by the robotic arm. The failure mechanisms of the hybrid printed lattice compared to lattices produced using DLP resulted in a higher maximum load achieved with a smoother load–displacement response. Parametric programming for robots has been used to produce large-scale steel constructions [157]. Programmed robots have capabilities for printing with steel, concrete, or clay, and the parametric programming approach facilitates automatic calculations of z-coordinates, requires small memory capacities, and enables the efficient scaling of objects. Robotic fabrication has been improved with error modeling and optimization via analytical derivation and experimentation [158]. The process is necessary to minimize the maximum trajectory execution error in a computationally efficient manner to support the six degrees of freedom of motion necessary to print complex objects.

Mobility is essential for fabrication robots so that they may print structures on large scales. One mobile robotic printing system was built using components of a mobile base, a six-degree freedom manipulator, a 1 cm nozzle, and a hose/pump system to enable material flow (Figure 9C) [159]. The robot was able to construct a structure of approximately 200 cm by 45 cm by 10 cm, which was much larger than the 87 cm reach of the robotic arm used to deposit the material. Ten layers of printed material were deposited in about nine minutes using a nozzle speed of 10 cm/s. Large-scale construction is also possible by coordinating the efforts of multiple collaborative robots [160]. The use of multiple robots was coordinated by the optimized scheduling of tasks that begin with a segmentation process to sub-divide a design into smaller pieces assigned to each robot. Since interference between robot paths can lead to collisions, an approach was developed that divides printing areas into safe and interference zones, with only one robot acting in the interference zone at a time. There are many research opportunities and challenges for further improvements in robotic fabrication, with context, requirements, materials, and mechanics identified as key criteria to consider [161]. By further developing robotic fabrication technologies it is possible to repair, renovate, and retrofit constructions with novel AM designs, such as curved walls and intricate structures that enable high aesthetics and function in future applications.

6.4. Machine Learning

Machine learning support for AM is gaining prominence for improving processes and design using algorithms and statistical analyses [162]. Recent applications in machine learning for AM have incorporated reinforced, unsupervised, and supervised learning. Supervised learning, which requires training, is suitable for regression and classification tasks that enable closed loop control and defect detection. Unsupervised learning, which identifies patterns within datasets, can provide capabilities for clustering and principal component analysis. Reinforcement learning is suitable for improved AM by sequentially forming decisions based on reward signals. All types of learning have their own inconsistencies and biases that designers must consider.

Machine learning has been applied in areas such as medical diagnoses, the processing of images, and associated learning while also optimizing AM processing parameters [163]. Incorporating machine learning in DfAM is advantageous as there is little upfront work required for establishing equations, and the approach is well suited for managing data of highly complex design problems. Multi-material process parameters have been optimized using artificial neural networks for multi-material FDM printing that were experimentally validated via mechanical testing [164]. The artificial neural network model was trained with data from experimental designs and used in tandem with a genetic algorithm to determine optimized parameters that increased tensile strength. Machine learning has also been used to detect defects in real time using a convolutional neural network [165]. The method was built from image classification with computer vision to check the quality of parts using an integrated camera. The approach represents an initial step for improving the consistency of printed parts using real-time parameter adjustments, which is one of the largest open problems for improving AM outcomes.

Machine learning can improve the design of AM parts, such as architected materials for improved mechanical efficiency [141] (Figure 9D). A process was used to propose unit cells of different topologies, superimpose them, and assess their mechanical properties such as Young's modulus relative to a theoretical limit based on the Gibson–Ashby model. Beam shapes were optimized while retaining a fixed relative density of lattices. Neural networks were trained to understand the relationship among high-dimensional design inputs and mechanical outputs that resulted in modulus and density predictions at a much faster rate than FE analysis. Artificial neural networks have also been used for patient-specific optimization for spinal disc applications [166]. The approach was combined with topology optimization to size lattice components for fast convergence that was facilitated with design space reduction by optimizing unit cell distributions with a predefined grid. Machine learning has also been used to design multi-material AM tissue mimics using neural networks to consider complicated combinations of design parameters [167]. Further AM applications in medicine have used machine learning to produce pharmaceutical products fabricated via hot melt extrusion and FDM techniques that necessitated an integrated computational and experimental approach to generate and understand data [168]. These cases highlight the need to integrate steps across all phases of DfAM to achieve innovations while also demonstrating how new advances in machine learning can greatly enhance AM outcomes.

7. Conclusions

This review provided a recent survey of DfAM advances in the context of Fabrication, Generation, and Assessment phases, which are the foundation for engineering design innovation. Each of these phases plays a pivotal role in the design process for maximizing the use of AM technology by implementing novel materials, optimizing designed structures, or providing validated models to support improvement in future iterations. Engineering innovations were highlighted in diverse areas, including the creation of Martian concrete for in situ construction, texture modified foods for personalized nutrition, a sustainable circular model for automotive scrap, and soft tensegrity robots with walking capabilities. The reviewed DfAM phases facilitated a discussion for future research opportunities, such as the creation of innovation services that democratize design, bio-inspired design methods for high-performance structures, robots with AM capabilities to fabricate large-scale designs, and machine learning for design optimization. The culmination of continued research in these areas highlight promising opportunities in DfAM, in which further advancements have great promise for driving wide-spread engineering innovation.

Key implications and considerations for future work are as follows:

- **Fabrication:** There are diverse materials and printing processes currently available, with research innovations leading to new material capabilities such as strong superalloys, biocompatibility, and magnetofunctional materials. Many of these materials are limited by suitable printing processes and fabrication constraints. Future work could consider expanding the available AM design space by improving printing time and accuracy across scales as new printable materials emerge.

- **Generation:** Due to the complex AM design space, it is challenging to configure optimal designs tailored for specific applications. Human designers are necessary to generate innovative solutions but have difficulty overcoming biases and barriers, while computational design requires a well-defined search space that limits innovation. Future work could consider intelligent computational methods that work with, or mimic, humans while incorporating mechanisms to reduce biases during searches.
- **Assessment:** There is often a mismatch between the ideal digital design and the as-fabricated design that can affect dimensional accuracy and mechanical performance, especially near the resolution limits of a printing process in which innovation occurs. Future work is necessary to better characterize how different printing processes affect design accuracy and mechanics, in addition to understanding the complex failure mechanics of AM parts based on their topological configurations.
- **Applications:** AM innovations are prevalent across numerous application areas that benefit from the diversity of materials and processes available, especially regarding polymers and metal printing. Soft material printing, such as that for food and tissue engineering, are emerging areas with a need for applying well-established DfAM principles to characterize new design opportunities. Across all domains, there is a need for advancing DfAM to more efficiently tailor and optimize designs to fully leverage AM's capabilities for customization.
- **Outlook:** Emerging areas in DfAM include considerations of on-demand printing services and using the broad capabilities of DfAM to enable new functionality via bio-inspiration. Robotic printing processes are enabling large-scale design and more efficient printing, while machine learning capabilities provide promise for automated design tailoring and integration across DfAM phases. Further advances in these fields provide great opportunities for researchers to impact AM and drive new innovations.

Funding: This research received no external funding.

Data Availability Statement: No new data were generated for the creation of this review.

Conflicts of Interest: The author declares no conflict of interest.

References

1. Thompson, M.K.; Moroni, G.; Vaneker, T.; Fadel, G.; Campbell, R.I.; Gibson, I.; Bernard, A.; Schulz, J.; Graf, P.; Ahuja, B.; et al. Design for Additive Manufacturing: Trends, opportunities, considerations, and constraints. *CIRP Ann. Manuf. Technol.* **2016**, *65*, 737–760. [CrossRef]
2. Candi, M.; Beltagui, A. Effective use of 3D printing in the innovation process. *Technovation* **2019**, *80*, 63–73. [CrossRef]
3. Lindwall, A.; Dordlofva, C.; Öhrwall Rönnbäck, A.; Törlind, P. Innovation in a box: Exploring creativity in design for additive manufacturing in a regulated industry. *J. Eng. Des.* **2022**, *33*, 567–586. [CrossRef]
4. Li, S.; Xin, Y.; Yu, Y.; Wang, Y. Design for additive manufacturing from a force-flow perspective. *Mater. Des.* **2021**, *204*, 109664. [CrossRef]
5. Haleem, A.; Javaid, M. 3D printed medical parts with different materials using additive manufacturing. *Clin. Epidemiol. Glob. Health* **2020**, *8*, 215–223. [CrossRef]
6. Dong, G.; Tang, Y.; Li, D.; Zhao, Y.F. Design and optimization of solid lattice hybrid structures fabricated by additive manufacturing. *Addit. Manuf.* **2020**, *33*, 101116. [CrossRef]
7. Egan, P.F.; Khatri, N.R.; Parab, M.A.; Arefin, A.M. Mechanics of 3D-Printed Polymer Lattices with Varied Design and Processing Strategies. *Polymers* **2022**, *14*, 5515. [CrossRef]
8. Egan, P.F. Special Issue Editorial: Applications of 3D Printing for Polymers. *Polymers* **2023**, *15*, 1638. [CrossRef]
9. Al-Dulimi, Z.; Wallis, M.; Tan, D.K.; Maniruzzaman, M.; Nokhodchi, A. 3D printing technology as innovative solutions for biomedical applications. *Drug Discov. Today* **2020**, *26*, 360–383. [CrossRef]
10. Vaneker, T.; Bernard, A.; Moroni, G.; Gibson, I.; Zhang, Y. Design for additive manufacturing: Framework and methodology. *CIRP Ann.* **2020**, *69*, 578–599. [CrossRef]
11. Sossou, G.; Demoly, F.; Gomes, S.; Montavon, G. An Assembly-Oriented Design Framework for Additive Manufacturing. *Designs* **2022**, *6*, 20. [CrossRef]

12. Egan, P.F.; Bauer, I.; Shea, K.; Ferguson, S.J. Mechanics of Three-Dimensional Printed Lattices for Biomedical Devices. *J. Mech. Des.* **2019**, *141*, 031703. [CrossRef]
13. Khatri, N.; Smith, J.A.; Egan, P.F. Empirical characterization of lattice, spring, and non-assembly mechanisms fabricated with nylon powder printing. In Proceedings of the ASME IDETC Design Automation Conference, St. Louis, MO, USA, 14–17 August 2022.
14. Pandelidi, C.; Lee, K.P.M.; Kajtaz, M. Effects of polyamide-11 powder refresh ratios in multi-jet fusion: A comparison of new and used powder. *Addit. Manuf.* **2021**, *40*, 101933. [CrossRef]
15. Porter, D.A.; Di Prima, M.A.; Badhe, Y.; Parikh, A.R. Nylon lattice design parameter effects on additively manufactured structural performance. *J. Mech. Behav. Biomed. Mater.* **2022**, *125*, 104869. [CrossRef]
16. Nazir, A.; Arshad, A.B.; Lin, S.-C.; Jeng, J.-Y. Mechanical Performance of Lightweight-Designed Honeycomb Structures Fabricated Using Multijet Fusion Additive Manufacturing Technology. *3D Print. Addit. Manuf.* **2022**, *9*, 311–325. [CrossRef]
17. Arshad, A.B.; Nazir, A.; Jeng, J.-Y. Design and performance evaluation of multi-helical springs fabricated by Multi Jet Fusion additive manufacturing technology. *Int. J. Adv. Manuf. Technol.* **2022**, *118*, 195–206. [CrossRef]
18. Somireddy, M.; Czekanski, A. Anisotropic material behavior of 3D printed composite structures–Material extrusion additive manufacturing. *Mater. Des.* **2020**, *195*, 108953. [CrossRef]
19. Park, S.; Shou, W.; Makatura, L.; Matusik, W.; Fu, K.K. 3D printing of polymer composites: Materials, processes, and applications. *Matter* **2022**, *5*, 43–76. [CrossRef]
20. Dixit, T.; Al-Hajri, E.; Paul, M.C.; Nithiarasu, P.; Kumar, S. High performance, microarchitected, compact heat exchanger enabled by 3D printing. *Appl. Therm. Eng.* **2022**, *210*, 118339. [CrossRef]
21. Cimolai, G.; Dayyani, I.; Qin, Q. Multi-objective shape optimization of large strain 3D helical structures for mechanical metamaterials. *Mater. Des.* **2022**, *215*, 110444. [CrossRef]
22. Bol, E.; Ramulu, M. Dimensional Accuracy of Electron Beam Powder Bed Fusion with Ti-6Al-4V. *Designs* **2023**, *7*, 53. [CrossRef]
23. Einbergs, E.; Spustaka, A.; Vitola, V.; Zolotarjovs, A. By visualizing the deformation with mechanoluminescent particles, additive manufacturing offers a practical alternative to stress and strain simulation. *Designs* **2023**, *7*, 54. [CrossRef]
24. Panwisawas, C.; Tang, Y.T.; Reed, R.C. Metal 3D printing as a disruptive technology for superalloys. *Nat. Commun.* **2020**, *11*, 2327. [CrossRef] [PubMed]
25. Mostafaei, A.; Neelapu, S.H.V.R.; Kisailus, C.; Nath, L.M.; Jacobs, T.D.; Chmielus, M. Characterizing surface finish and fatigue behavior in binder-jet 3D-printed nickel-based superalloy 625. *Addit. Manuf.* **2018**, *24*, 200–209. [CrossRef]
26. Ghoussoub, J.N.; Klupś, P.; Dick-Cleland, W.J.; Rankin, K.E.; Utada, S.; Bagot, P.A.; McCartney, D.G.; Tang, Y.T.; Reed, R.C. A new class of alumina-forming superalloy for 3D printing. *Addit. Manuf.* **2022**, *52*, 102608. [CrossRef]
27. Fortuna, S.; Gurianov, D.; Kalashnikov, K.; Chumaevskii, A.; Mironov, Y.P.; Kolubaev, E. Directional solidification of a nickel-based superalloy product structure fabricated on stainless steel substrate by electron beam additive manufacturing. *Met. Mater. Trans. A* **2021**, *52*, 857–870. [CrossRef]
28. Risangud, N.; Jiraborvornpongsa, N.; Pasee, S.; Kaewkong, P.; Kunkit, N.; Sungkhaphan, P.; Janvikul, W. Poly (ester-co-glycidyl methacrylate) for digital light processing in biomedical applications. *J. Appl. Polym. Sci.* **2021**, *138*, 51391. [CrossRef]
29. Lantean, S.; Barrera, G.; Pirri, C.F.; Tiberto, P.; Sangermano, M.; Roppolo, I.; Rizza, G. 3D printing of magnetoresponsive polymeric materials with tunable mechanical and magnetic properties by digital light processing. *Adv. Mater. Technol.* **2019**, *4*, 1900505. [CrossRef]
30. Egan, P.; Wang, X.; Greutert, H.; Shea, K.; Wuertz-Kozak, K.; Ferguson, S. Mechanical and Biological Characterization of 3D Printed Lattices. *3D Print. Addit. Manuf.* **2019**, *6*, 73–81. [CrossRef]
31. Steyrer, B.; Neubauer, P.; Liska, R.; Stampfl, J. Visible light photoinitiator for 3D-printing of tough methacrylate resins. *Materials* **2017**, *10*, 1445. [CrossRef]
32. Zhang, C.; Li, X.; Jiang, L.; Tang, D.; Xu, H.; Zhao, P.; Fu, J.; Zhou, Q.; Chen, Y. 3D printing of functional magnetic materials: From design to applications. *Adv. Funct. Mater.* **2021**, *31*, 2102777. [CrossRef]
33. Shao, G.; Ware, H.O.T.; Li, L.; Sun, C. Rapid 3D printing magnetically active microstructures with high solid loading. *Adv. Eng. Mater.* **2020**, *22*, 1900911. [CrossRef]
34. Samykano, M.; Selvamani, S.; Kadirgama, K.; Ngui, W.; Kanagaraj, G.; Sudhakar, K. Mechanical property of FDM printed ABS: Influence of printing parameters. *Int. J. Adv. Manuf. Technol.* **2019**, *102*, 2779–2796. [CrossRef]
35. Sheoran, A.J.; Kumar, H. Fused Deposition modeling process parameters optimization and effect on mechanical properties and part quality: Review and reflection on present research. *Mater. Today: Proc.* **2020**, *21*, 1659–1672.
36. Kam, M.; İpekçi, A.; Şengül, Ö. Investigation of the effect of FDM process parameters on mechanical properties of 3D printed PA12 samples using Taguchi method. *J. Thermoplast. Compos. Mater.* **2023**, *36*, 307–325. [CrossRef]
37. Zaeri, A.; Zgeib, R.; Cao, K.; Zhang, F.; Chang, R.C. Numerical analysis on the effects of microfluidic-based bioprinting parameters on the microfiber geometrical outcomes. *Sci. Rep.* **2022**, *12*, 3364. [CrossRef]
38. Pant, A.; Lee, A.Y.; Karyappa, R.; Lee, C.P.; An, J.; Hashimoto, M.; Tan, U.X.; Wong, G.; Chua, C.K.; Zhang, Y. 3D food printing of fresh vegetables using food hydrocolloids for dysphagic patients. *Food Hydrocoll.* **2021**, *114*, 106546. [CrossRef]
39. Schwab, A.; Levato, R.; D'Este, M.; Piluso, S.; Eglin, D.; Malda, J. Printability and shape fidelity of bioinks in 3D bioprinting. *Chem. Rev.* **2020**, *120*, 11028–11055. [CrossRef]
40. Pérez, B.; Nykvist, H.; Brøgger, A.F.; Larsen, M.B.; Falkeborg, M.F. Impact of macronutrients printability and 3D-printer parameters on 3D-food printing: A review. *Food Chem.* **2019**, *287*, 249–257. [CrossRef]

41. Korium, M.S.; Roozbahani, H.; Alizadeh, M.; Perepelkina, S.; Handroos, H. Direct metal laser sintering of precious metals for jewelry applications: Process parameter selection and microstructure analysis. *IEEE Access* **2021**, *9*, 126530–126540. [CrossRef]
42. Choi, S.; Kim, J.-W.; Lee, S.; Yoon, W.Y.; Han, Y.; Kim, K.-J.; Rhie, J.-W.; Suh, T.-S.; Lee, K.-D. Mechanical and biocompatibility properties of sintered titanium powder for mimetic 3D-printed bone scaffolds. *ACS Omega* **2022**, *7*, 10340–10346. [CrossRef] [PubMed]
43. Wei, X.; Li, X.; Wen, S.; Zheng, Y.; Tian, Y. Channel design for 3D models with applications in powder-bed additive manufacturing. *Rapid Prototyp. J.* **2019**, *25*, 1536–1544. [CrossRef]
44. Zeng, Z.; Deng, X.; Cui, J.; Jiang, H.; Yan, S.; Peng, B. Improvement on selective laser sintering and post-processing of polystyrene. *Polymers* **2019**, *11*, 956. [CrossRef] [PubMed]
45. Nazir, A.; Ali, M.; Hsieh, C.-H.; Jeng, J.-Y. Investigation of stiffness and energy absorption of variable dimension helical springs fabricated using multijet fusion technology. *Int. J. Adv. Manuf. Technol.* **2020**, *110*, 2591–2602. [CrossRef]
46. Fu, Y.; Chen, Z.; Xu, G.; Wei, Y.; Lao, C. Preparation and stereolithography 3D printing of ultralight and ultrastrong ZrOC porous ceramics. *J. Alloy Compd.* **2019**, *789*, 867–873. [CrossRef]
47. Shen, M.; Zhao, W.; Xing, B.; Sing, Y.; Gao, S.; Wang, C.; Zhao, Z. Effects of exposure time and printing angle on the curing characteristics and flexural strength of ceramic samples fabricated via digital light processing. *Ceram. Int.* **2020**, *46*, 24379–24384. [CrossRef]
48. Song, P.; Li, M.; Zhang, B.; Gui, X.; Han, Y.; Wang, L.; Zhou, W.; Guo, L.; Zhang, Z.; Li, Z.; et al. DLP fabricating of precision GelMA/HAp porous composite scaffold for bone tissue engineering application. *Compos. Part B: Eng.* **2022**, *244*, 110163. [CrossRef]
49. Huang, J.; Chen, Q.; Jiang, H.; Zou, B.; Li, L.; Liu, J.; Yu, H. A survey of design methods for material extrusion polymer 3D printing. *Virtual Phys. Prototyp.* **2020**, *15*, 148–162. [CrossRef]
50. Enea, S.; Moon, S.K. Guidelines for 3D printed springs using material extrusion. *Rapid Prototyp. J.* **2021**, *28*, 409–427. [CrossRef]
51. Booth, J.W.; Alperovich, J.; Chawla, P.; Ma, J.; Reid, T.N.; Ramani, K. The design for additive manufacturing worksheet. *J. Mech. Des.* **2017**, *139*, 100904. [CrossRef]
52. Wei, X.; Tian, Y.; Joneja, A. A study on revolute joints in 3D-printed non-assembly mechanisms. *Rapid Prototyp. J.* **2016**, *22*, 901–933. [CrossRef]
53. Schaechtl, P.; Hallmann, M.; Schleich, B.; Wartzack, S. Tolerance analysis of additively manufactured non-assembly mechanisms considering joint clearance. *Procedia CIRP* **2020**, *92*, 27–32. [CrossRef]
54. Rupal, B.S.; Anwer, N.; Secanell, M.; Qureshi, A.J. Geometric tolerance characterization of laser powder bed fusion processes based on skin model shapes. *Procedia CIRP* **2020**, *92*, 169–174. [CrossRef]
55. Prabhu, R.; Bracken, J.; Armstrong, C.B.; Jablokow, K.; Simpson, T.W.; Meisel, N.A. Additive creativity: Investigating the use of design for additive manufacturing to encourage creativity in the engineering design industry. *Int. J. Des. Creat. Innov.* **2020**, *8*, 198–222. [CrossRef]
56. Sinha, S.; Chen, H.-E.; Meisel, N.A.; Miller, S.R. Does Designing for Additive Manufacturing Help Us Be More Creative? An Exploration in Engineering Design Education. In Proceedings of the International Design Engineering Technical Conferences and Computers and Information in Engineering Conference, Cleveland, OH, USA, 6–9 August 2017; p. V003T04A014.
57. Friesike, S.; Flath, C.M.; Wirth, M.; Thiesse, F. Creativity and productivity in product design for additive manufacturing: Mechanisms and platform outcomes of remixing. *J. Oper. Manag.* **2019**, *65*, 735–752. [CrossRef]
58. Prabhu, R.; Miller, S.R.; Simpson, T.W.; Meisel, N.A. Teaching design freedom: Understanding the effects of variations in design for additive manufacturing education on students' creativity. *J. Mech. Des.* **2020**, *142*, 094501. [CrossRef]
59. Blösch-Paidosh, A.; Shea, K. Enhancing Creative Redesign Through Multimodal Design Heuristics for Additive Manufacturing. *J. Mech. Des.* **2021**, *143*, 102003. [CrossRef]
60. Schauer, A.M.; Fillingim, K.B.; Fu, K. Impact of timing in the design process on students' application of design for additive manufacturing heuristics. *J. Mech. Des.* **2022**, *144*, 062301. [CrossRef]
61. Li, X.; Zhao, J.; He, R.; Tian, Y.; Wei, X. Parametric design of scalable mechanisms for additive manufacturing. *J. Mech. Des.* **2018**, *140*, 022302. [CrossRef]
62. Al Khalil, M.; Belkebir, H.; Lebaal, N.; Demoly, F.; Roth, S. A Biomimetic Design Method for 3D-Printed Lightweight Structures Using L-Systems and Parametric Optimization. *Appl. Sci.* **2022**, *12*, 5530. [CrossRef]
63. Li, C.; Lei, H.; Zhang, Z.; Zhang, X.; Zhou, H.; Wang, P.; Fang, D. Architecture design of periodic truss-lattice cellsfor additive manufacturing. *Addit. Manuf.* **2020**, *34*, 101172.
64. Lei, H.-Y.; Li, J.-R.; Xu, Z.-J.; Wang, Q.-H. Parametric design of Voronoi-based lattice porous structures. *Mater. Des.* **2020**, *191*, 108607. [CrossRef]
65. Fantini, M.; Curto, M.; De Crescenzio, F. A method to design biomimetic scaffolds for bone tissue engineering based on Voronoi lattices. *Virtual Phys. Prototyp.* **2016**, *11*, 77–90. [CrossRef]
66. Hindy, A.; Farahmand, F.; Pourdanesh, F.; Torshabi, M.; Al Janabi, A.H.; Rasoulianboroujeni, M.; Tayebi, L.; Tabatabaei, F.S. Synthesis and characterization of 3D-printed functionally graded porous titanium alloy. *J. Mater. Sci.* **2020**, *55*, 9082–9094. [CrossRef]
67. Letov, N.; Zhao, Y.F. Beam-based lattice topology transition with function representation. *J. Mech. Des.* **2023**, *145*, 011704. [CrossRef]

68. Mukhopadhyay, T.; Naskar, S.; Adhikari, S. Anisotropy tailoring in geometrically isotropic multi-material lattices. *Extrem. Mech. Lett.* **2020**, *40*, 100934. [CrossRef]
69. Liu, W.; Song, H.; Huang, C. Maximizing mechanical properties and minimizing support material of PolyJet fabricated 3D lattice structures. *Addit. Manuf.* **2020**, *35*, 101257. [CrossRef]
70. Liu, Y.; Zhuo, S.; Xiao, Y.; Zheng, G.; Dong, G.; Zhao, Y.F. Rapid modeling and design optimization of multi-topology lattice structure based on unit-cell library. *J. Mech. Des.* **2020**, *142*, 091705. [CrossRef]
71. Wang, Z.; Srinivasa, A.; Reddy, J.; Dubrowski, A. Topology Optimization of Lightweight Structures With Application to Bone Scaffolds and 3D Printed Shoes for Diabetics. *J. Appl. Mech.* **2022**, *89*, 041009. [CrossRef]
72. Wang, C.; Qian, X. Simultaneous optimization of build orientation and topology for additive manufacturing. *Addit. Manuf.* **2020**, *34*, 101246. [CrossRef]
73. Kazemi, H.; Norato, J.A. Topology optimization of lattices with anisotropic struts. *Struct. Multidiscip. Optim.* **2021**, *63*, 1653–1668. [CrossRef]
74. Ravari, M.K.; Kadkhodaei, M.; Badrossamay, M.; Rezaei, R. Numerical investigation on mechanical properties of cellular lattice structures fabricated by fused deposition modeling. *Int. J. Mech. Sci.* **2014**, *88*, 154–161. [CrossRef]
75. Okarma, K.; Fastowicz, J.; Lech, P.; Lukin, V. Quality Assessment of 3D Printed Surfaces Using Combined Metrics Based on Mutual Structural Similarity Approach Correlated with Subjective Aesthetic Evaluation. *Appl. Sci.* **2020**, *10*, 6248. [CrossRef]
76. Nar, K.; Majewski, C.; Lewis, R. A comprehensive characterisation of Laser Sintered Polyamide-12 surfaces. *Polym. Test.* **2022**, *106*, 107450. [CrossRef]
77. Ribeiro, A.; Blokzijl, M.M.; Levato, R.; Visser, C.W.; Castilho, M.; Hennink, W.E.; Vermonden, T.; Malda, J. Assessing bioink shape fidelity to aid material development in 3D bioprinting. *Biofabrication* **2017**, *10*, 014102. [CrossRef] [PubMed]
78. Montgomery, S.M.; Demoly, F.; Zhou, K.; Qi, H.J. Pixel-Level Grayscale Manipulation to Improve Accuracy in Digital Light Processing 3D Printing. *Adv. Funct. Mater.* **2023**, *33*, 2213252. [CrossRef]
79. Woodward, I.R.; Fromen, C.A. Scalable, process-oriented beam lattices: Generation, characterization, and compensation for open cellular structures. *Addit. Manuf.* **2021**, *48*, 102386. [CrossRef]
80. Arabnejad, S.; Johnston, R.B.; Pura, J.A.; Singh, B.; Tanzer, M.; Pasini, D. High-strength porous biomaterials for bone replacement: A strategy to assess the interplay between cell morphology, mechanical properties, bone ingrowth and manufacturing constraints. *Acta Biomater.* **2016**, *30*, 345–356. [CrossRef]
81. Alghamdi, A.; Maconachie, T.; Downing, D.; Brandt, M.; Qian, M.; Leary, M. Effect of additive manufactured lattice defects on mechanical properties: An automated method for the enhancement of lattice geometry. *Int. J. Adv. Manuf. Technol.* **2020**, *108*, 957–971. [CrossRef]
82. Wang, Z.; Liu, P.; Ji, Y.; Mahadevan, S.; Horstemeyer, M.F.; Hu, Z.; Chen, L.; Chen, L.-Q. Uncertainty quantification in metallic additive manufacturing through physics-informed data-driven modeling. *JOM* **2019**, *71*, 2625–2634. [CrossRef]
83. Moniruzzaman, M.; O'Neal, C.; Bhuiyan, A.; Egan, P.F. Design and Mechanical Testing of 3D Printed Hierarchical Lattices Using Biocompatible Stereolithography. *Designs* **2020**, *4*, 22. [CrossRef]
84. Kulkarni, N.N.; Ekwaro-Osire, S.; Egan, P.F. Fabrication, Mechanics, and Reliability Analysis for Three-Dimensional Printed Lattice Designs. *ASCE-ASME J. Risk Uncertain Eng. Syst. Part B Mech. Eng.* **2021**, *8*, 011107. [CrossRef]
85. Khatri, N.R.; Egan, P.F. Energy Absorption of 3D Printed ABS and TPU Multimaterial Honeycomb Structures. *3D Print. Addit. Manuf.* **2023**. [CrossRef]
86. Mazur, M.; Leary, M.; Sun, S.; Vcelka, M.; Shidid, D.; Brandt, M. Deformation and failure behaviour of Ti-6Al-4V lattice structures manufactured by selective laser melting (SLM). *Int. J. Adv. Manuf. Technol.* **2016**, *84*, 1391–1411. [CrossRef]
87. Altamimi, S.; Lee, D.-W.; Barsoum, I.; Rowshan, R.; Jasiuk, I.M.; Abu Al-Rub, R.K. On Stiffness, Strength, Anisotropy, and Buckling of 30 Strut-Based Lattices with Cubic Crystal Structures. *Adv. Eng. Mater.* **2022**, *24*, 2101379. [CrossRef]
88. Leary, M.; Mazur, M.; Williams, H.; Yang, E.; Alghamdi, A.; Lozanovski, B.; Zhang, X.; Shidid, D.; Farahbod-Sternahl, L.; Witt, G.; et al. Inconel 625 lattice structures manufactured by selective laser melting (SLM): Mechanical properties, deformation and failure modes. *Mater. Des.* **2018**, *157*, 179–199. [CrossRef]
89. Chua, C.; Sing, S.L.; Chua, C. Characterisation of in-situ alloyed titanium-tantalum lattice structures by laser powder bed fusion using finite element analysis. *Virtual Phys. Prototyp.* **2023**, *18*, e2138463. [CrossRef]
90. Abdulhadi, H.S.; Fadeel, A.; Alwattar, T.A.; Mian, A. Developing scaling laws to predict compressive mechanical properties and determine geometrical parameters of modified BCC lattice structures. *Eng. Rep.* **2023**, *5*, e12566. [CrossRef]
91. Alghamdi, A.; Downing, D.; Tino, R.; Almalki, A.; Maconachie, T.; Lozanovski, B.; Brandt, M.; Qian, M.; Leary, M. Buckling phenomena in AM lattice strut elements: A design tool applied to Ti-6Al-4V LB-PBF. *Mater. Des.* **2021**, *208*, 109892. [CrossRef]
92. Liu, Y.; Yang, H.; Zhao, Y.F.; Zheng, G. A heterogeneous lattice structure modeling technique supported by multiquadric radial basis function networks. *J. Comput. Des. Eng.* **2022**, *9*, 68–81. [CrossRef]
93. Wang, R.; Wu, Z. Recent advancement in finite element analysis of spinal interbody cages: A review. *Front. Bioeng. Biotechnol.* **2023**, *11*, 1041973. [CrossRef] [PubMed]
94. Arefin, A.M.; Lahowetz, M.; Egan, P.F. Simulated tissue growth in tetragonal lattices with mechanical stiffness tuned for bone tissue engineering. *Comput. Biol. Med.* **2021**, *138*, 104913. [CrossRef]

95. Zagidullin, R.; Zezin, N.; Rodionov, N. Improving the quality of FDM 3D printing of UAV and aircraft parts and assemblies by parametric software changes. In Proceedings of the IOP Conference Series: Materials Science and Engineering; IOP Publishing: Bristol, UK, 2021; Volume 1027, p. 012031.
96. Schmitt, M.; Mehta, R.M.; Kim, I.Y. Additive manufacturing infill optimization for automotive 3D-printed ABS components. *Rapid Prototyp. J.* **2020**, *26*, 89–99. [CrossRef]
97. Gardner, L.; Kyvelou, P.; Herbert, G.; Buchanan, C. Testing and initial verification of the world's first metal 3D printed bridge. *J. Constr. Steel Res.* **2020**, *172*, 106233. [CrossRef]
98. Scheele, S.C.; Hartmann, C.; Siegrist, M.; Binks, M.; Egan, P.F. Consumer Assessment of 3D Printed Food Shape, Taste, and Fidelity using Chocolate and Marzipan Materials. *3D Print. Addit. Manuf.* **2021**, *9*, 473–782. [CrossRef]
99. Lee, H.; Jang, Y.; Choe, J.K.; Lee, S.; Song, H.; Lee, J.P.; Lone, N.; Kim, J. 3D-printed programmable tensegrity for soft robotics. *Sci. Robot.* **2020**, *5*, eaay9024. [CrossRef]
100. Skawiński, I.; Goetzendorf-Grabowski, T. FDM 3D printing method utility assessment in small RC aircraft design. *Aircr. Eng. Aerosp. Technol.* **2019**, *91*, 865–872. [CrossRef]
101. Nicholson, G.; Roberts, C. University of Sheffield/AMRC, FDM-Printed Fixed Wing UAV. Available online: https://www.amrc.co.uk/case-studies/fdm-printed-fixed-wing-uav (accessed on 19 May 2023).
102. Ravindrababu, S.; Govdeli, Y.; Wong, Z.W.; Kayacan, E. Evaluation of the influence of build and print orientations of unmanned aerial vehicle parts fabricated using fused deposition modeling process. *J. Manuf. Process.* **2018**, *34*, 659–666. [CrossRef]
103. Wang, Y.-C.; Chen, T.; Yeh, Y.-L. Advanced 3D printing technologies for the aircraft industry: A fuzzy systematic approach for assessing the critical factors. *Int. J. Adv. Manuf. Technol.* **2019**, *105*, 4059–4069. [CrossRef]
104. Blachowicz, T.; Pająk, K.; Recha, P.; Ehrmann, A. 3D printing for microsatellites-material requirements and recent developments. *AIMS Mater. Sci.* **2020**, *7*, 926–938. [CrossRef]
105. Li, H.; Meng, H.; Lan, M.; Zhou, J.; Xu, M.; Zhao, X.; Xiang, B. Development of a novel material and casting method for in situ construction on Mars. *Powder Technol.* **2021**, *390*, 219–229. [CrossRef]
106. Korniejenko, K.; Pławecka, K.; Kozub, B. An Overview for Modern Energy-Efficient Solutions for Lunar and Martian Habitats Made Based on Geopolymers Composites and 3D Printing Technology. *Energies* **2022**, *15*, 9322. [CrossRef]
107. Mohanavel, V.; Ali, K.A.; Ranganathan, K.; Jeffrey, J.A.; Ravikumar, M.; Rajkumar, S. The roles and applications of additive manufacturing in the aerospace and automobile sector. *Mater. Today Proc.* **2021**, *47*, 405–409. [CrossRef]
108. Szalai, S.; Herold, B.; Kurhan, D.; Németh, A.; Sysyn, M.; Fischer, S. Optimization of 3D printed rapid prototype deep drawing tools for automotive and railway sheet material testing. *Infrastructures* **2023**, *8*, 43. [CrossRef]
109. Nascimento, D.L.d.M.; Nepomuceno, R.M.; Caiado, R.G.G.; Maqueira, J.M.; Moyano-Fuentes, J.; Garza-Reyes, J.A. A sustainable circular 3D printing model for recycling metal scrap in the automotive industry. *J. Manuf. Technol. Manag.* **2022**, *33*, 876–892. [CrossRef]
110. Komara, A.I.; Budiwantoro, B.; Setiawan, R. Cellular Structure Design and Manufacturability for Electric Vehicle: A Review. *Int. J. Sustain. Transp. Technol.* **2022**, *5*, 70–79. [CrossRef]
111. Aslan, B.; Yıldız, A.R. Optimum design of automobile components using lattice structures for additive manufacturing. *Mater. Test.* **2020**, *62*, 633–639. [CrossRef]
112. Furet, B.; Poullain, P.; Garnier, S. 3D printing for construction based on a complex wall of polymer-foam and concrete. *Addit. Manuf.* **2019**, *28*, 58–64. [CrossRef]
113. Ji, G.; Ding, T.; Xiao, J.; Du, S.; Li, J.; Duan, Z. A 3D printed ready-mixed concrete power distribution substation: Materials and construction technology. *Materials* **2019**, *12*, 1540. [CrossRef]
114. Mechtcherine, V.; Nerella, V.N.; Will, F.; Näther, M.; Otto, J.; Krause, M. Large-scale digital concrete construction–CONPrint3D concept for on-site, monolithic 3D-printing. *Autom. Constr.* **2019**, *107*, 102933. [CrossRef]
115. Salet, T.A.; Ahmed, Z.Y.; Bos, F.P.; Laagland, H.L. Design of a 3D printed concrete bridge by testing. *Virtual Phys. Prototyp.* **2018**, *13*, 222–236. [CrossRef]
116. Vantyghem, G.; De Corte, W.; Shakour, E.; Amir, O. 3D printing of a post-tensioned concrete girder designed by topology optimization. *Autom. Constr.* **2020**, *112*, 103084. [CrossRef]
117. Kanyilmaz, A.; Demir, A.G.; Chierici, M.; Berto, F.; Gardner, L.; Kandukuri, S.Y.; Kassabian, P.; Kinoshita, T.; Laurenti, A.; Paoletti, I.; et al. Role of metal 3D printing to increase quality and resource-efficiency in the construction sector. *Addit. Manuf.* **2022**, *50*, 102541. [CrossRef]
118. Alhumayani, H.; Gomaa, M.; Soebarto, V.; Jabi, W. Environmental assessment of large-scale 3D printing in construction: A comparative study between cob and concrete. *J. Clean. Prod.* **2020**, *270*, 122463. [CrossRef]
119. Mahmoud, R.; Nguyen, Q.; Christopher, G.; Egan, P.F. 3D printed food design and fabrication approach for manufacturability, rheology, and nutrition trade-offs. In Proceedings of the ASME IDETC Design Automation Conference, Virtual Conference, 17–19 August 2021.
120. Scheele, S.C.; Binks, M.; Christopher, G.; Maleky, F.; Egan, P.F. Printability, texture, and sensory trade-offs for 3D printed potato with added proteins and lipids. *J. Food Eng.* **2023**, *351*, 111517. [CrossRef]
121. Derossi, A.; Paolillo, M.; Verboven, P.; Nicolai, B.; Severini, C. Extending 3D food printing application: Apple tissue microstructure as a digital model to create innovative cereal-based snacks. *J. Food Eng.* **2022**, *316*, 110845. [CrossRef]

122. Yang, G.; Tao, Y.; Wang, P.; Xu, X.; Zhu, X. Optimizing 3D printing of chicken meat by response surface methodology and genetic algorithm: Feasibility study of 3D printed chicken product. *Lwt* **2022**, *154*, 112693. [CrossRef]
123. Lin, Y.-J.; Punpongsanon, P.; Wen, X.; Iwai, D.; Sato, K.; Obrist, M.; Mueller, S. FoodFab: Creating food perception illusions using food 3D printing. In Proceedings of the 2020 CHI Conference on Human Factors in Computing Systems, Honolulu, HI, USA, 25–30 April 2020; 2021; pp. 1–13.
124. Motoki, K.; Park, J.; Spence, C.; Velasco, C. Contextual acceptance of novel and unfamiliar foods: Insects, cultured meat, plant-based meat alternatives, and 3D printed foods. *Food Qual. Prefer.* **2022**, *96*, 104368. [CrossRef]
125. Alkhatib, F.; Mahdi, E.; Cabibihan, J.-J. Design and analysis of flexible joints for a robust 3D printed prosthetic hand. In Proceedings of the 2019 IEEE 16th International Conference on Rehabilitation Robotics (ICORR), Toronto, ON, Canada, 24–28 June 2019; pp. 784–789.
126. Ahrendt, D.; Karam, A.R. Development of a computer-aided engineering–supported process for the manufacturing of customized orthopaedic devices by three-dimensional printing onto textile surfaces. *J. Eng. Fibers Fabr.* **2020**, *15*, 1558925020917627. [CrossRef]
127. Keukenkamp, R.; van Netten, J.J.; Busch-Westbroek, T.E.; Nollet, F.; Bus, S.A. Users' needs and expectations and the design of a new custom-made indoor footwear solution for people with diabetes at risk of foot ulceration. *Disabil. Rehabil.* **2022**, *44*, 8493–8500. [CrossRef]
128. Alaña, M.; Lopez-Arancibia, A.; Ghouse, S.; Rodriguez-Florez, N.; de Galarreta, S.R. Additively manufactured lattice structures with controlled transverse isotropy for orthopedic porous implants. *Comput. Biol. Med.* **2022**, *150*, 105761. [CrossRef]
129. Egan, P.; Ferguson, S.; Shea, K. Design of hierarchical 3D printed scaffolds considering mechanical and biological factors for bone tissue engineering. *J. Mech. Des.* **2017**, *139*, 061401. [CrossRef]
130. Egan, P.F.; Shea, K.A.; Ferguson, S.J. Simulated tissue growth for 3D printed scaffolds. *Biomech. Model. Mechanobiol.* **2018**, *17*, 1481–1495. [CrossRef]
131. Cendrero, A.M.; Martínez, F.F.; Requejo, W.G.S.; Lantada, A.D. Open-source library of tissue engineering scaffolds. *Mater. Des.* **2022**, *223*, 111154. [CrossRef]
132. Van Belleghem, S.; Mahadik, B.; Snodderly, K.; Mote, Z.; Jiang, B.; Yu, J.R.; McLoughlin, S.; He, X.; Nam, A.J.; Fisher, J.P. Dual Extrusion Patterning Drives Tissue Development Aesthetics and Shape Retention in 3D Printed Nipple-Areola Constructs. *Adv. Healthc. Mater.* **2021**, *10*, 2101249. [CrossRef]
133. Gul, J.Z.; Sajid, M.; Rehman, M.M.; Siddiqui, G.U.; Shah, I.; Kim, K.-H.; Lee, J.-W.; Choi, K.H. 3D printing for soft robotics–a review. *Sci. Technol. Adv. Mater.* **2018**, *19*, 243–262. [CrossRef]
134. Takishima, Y.; Yoshida, K.; Khosla, A.; Kawakami, M.; Furukawa, H. Fully 3D-printed hydrogel actuator for jellyfish soft robots. *ECS J. Solid State Sci. Technol.* **2021**, *10*, 037002. [CrossRef]
135. Ntagios, M.; Nassar, H.; Pullanchiyodan, A.; Navaraj, W.T.; Dahiya, R. Robotic hands with intrinsic tactile sensing via 3D printed soft pressure sensors. *Adv. Intell. Syst.* **2020**, *2*, 1900080. [CrossRef]
136. Desai, J.P.; Sheng, J.; Cheng, S.S.; Wang, X.; Deaton, N.J.; Rahman, N. Toward patient-specific 3D-printed robotic systems for surgical interventions. *IEEE Trans. Med. Robot. Bionics* **2019**, *1*, 77–87. [CrossRef]
137. Joyee, E.B.; Pan, Y. Additive manufacturing of multi-material soft robot for on-demand drug delivery applications. *J. Manuf. Process.* **2020**, *56*, 1178–1184. [CrossRef]
138. Kantaros, A.; Diegel, O.; Piromalis, D.; Tsaramirsis, G.; Khadidos, A.O.; Khadidos, A.O.; Khan, F.Q.; Jan, S. 3D printing: Making an innovative technology widely accessible through makerspaces and outsourced services. *Mater. Today Proc.* **2022**, *49*, 2712–2723. [CrossRef]
139. Tran, P.; Ngo, T.D.; Ghazlan, A.; Hui, D. Bimaterial 3D printing and numerical analysis of bio-inspired composite structures under in-plane and transverse loadings. *Compos. Part B Eng.* **2017**, *108*, 210–223. [CrossRef]
140. Zhang, X.; Li, M.; Lim, J.H.; Weng, Y.; Tay, Y.W.D.; Pham, H.; Pham, Q.-C. Large-scale 3D printing by a team of mobile robots. *Autom. Constr.* **2018**, *95*, 98–106. [CrossRef]
141. Lee, S.; Zhang, Z.; Gu, G.X. Deep Learning Accelerated Design of Mechanically Efficient Architected Materials. *ACS Appl. Mater. Interfaces* **2023**, *15*, 22543–22552. [CrossRef]
142. Beltagui, A.; Sesis, A.; Stylos, N. A bricolage perspective on democratising innovation: The case of 3D printing in makerspaces. *Technol. Forecast. Soc. Chang.* **2021**, *163*, 120453. [CrossRef]
143. Kim, N.P.; Kim, J.; Han, M.S. The convergence of three-dimensional printing and nail-art technology. *J. Cosmet. Med.* **2019**, *3*, 94–101. [CrossRef]
144. Rouf, S.; Raina, A.; Haq, M.I.U.; Naveed, N.; Jeganmohan, S.; Kichloo, A.F. 3D printed parts and mechanical properties: Influencing parameters, sustainability aspects, global market scenario, challenges and applications. *Adv. Ind. Eng. Polym. Res.* **2022**, *5*, 143–158. [CrossRef]
145. Beltagui, A.; Kunz, N.; Gold, S. The role of 3D printing and open design on adoption of socially sustainable supply chain innovation. *Int. J. Prod. Econ.* **2020**, *221*, 107462. [CrossRef]
146. Butt, J. Exploring the interrelationship between additive manufacturing and Industry 4.0. *Designs* **2020**, *4*, 13. [CrossRef]
147. Ashima, R.; Haleem, A.; Bahl, S.; Javaid, M.; Mahla, S.K.; Singh, S. Automation and manufacturing of smart materials in Additive Manufacturing technologies using Internet of Things towards the adoption of Industry 4.0. *Mater. Today: Proc.* **2021**, *45*, 5081–5088. [CrossRef]

148. Nicoloso, L.G.D.V.; Pelz, J.; Barrack, H.; Kuester, F. Towards 3D printing of a monocoque transtibial prosthesis using a bio-inspired design workflow. *Rapid Prototyp. J.* **2021**, *27*, 67–80. [CrossRef]
149. Jaradat, M.; Soliman, E.; Taha, M.R. 3D-printed bio-inspired mechanically interlocked viscoelastic dampers for energy dissipation. *Mater. Des.* **2023**, *228*, 111826. [CrossRef]
150. Verma, S.; Kumar, A.; Lin, S.-C.; Jeng, J.-Y. A bio-inspired design strategy for easy powder removal in powder-bed based additive manufactured lattice structure. *Virtual Phys. Prototyp.* **2022**, *17*, 468–488. [CrossRef]
151. Mehboob, H.; Tarlochan, F.; Mehboob, A.; Chang, S.-H.; Ramesh, S.; Harun, W.S.W.; Kadirgama, K. A novel design, analysis and 3D printing of Ti-6Al-4V alloy bio-inspired porous femoral stem. *J. Mater. Sci. Mater. Med.* **2020**, *31*, 78. [CrossRef] [PubMed]
152. Song, K.; Li, D.; Zhang, C.; Liu, T.; Tang, Y.; Xie, Y.M.; Liao, W. Bio-inspired hierarchical honeycomb metastructures with superior mechanical properties. *Compos. Struct.* **2023**, *304*, 116452. [CrossRef]
153. Hamzehei, R.; Zolfagharian, A.; Dariushi, S.; Bodaghi, M. 3D-printed bio-inspired zero Poisson's ratio graded metamaterials with high energy absorption performance. *Smart Mater. Struct.* **2022**, *31*, 035001. [CrossRef]
154. Namvar, N.; Moloukzadeh, I.; Zolfagharian, A.; Demoly, F.; Bodaghi, M. Bio-inspired design, modeling, and 3D printing of lattice-based scale model scooter decks. *Int. J. Adv. Manuf. Technol.* **2023**, *126*, 2887–2903. [CrossRef]
155. Soltani, A.; Noroozi, R.; Bodaghi, M.; Zolfagharian, A.; Hedayati, R. 3D printing on-water sports boards with bio-inspired core designs. *Polymers* **2020**, *12*, 250. [CrossRef] [PubMed]
156. De Marzi, A.; Vibrante, M.; Bottin, M.; Franchin, G. Development of robot assisted hybrid additive manufacturing technology for the freeform fabrication of lattice structures. *Addit. Manuf.* **2023**, *66*, 103456. [CrossRef]
157. Feucht, T.; Lange, J.; Erven, M.; Costanzi, C.B.; Knaack, U.; Waldschmitt, B. Additive manufacturing by means of parametric robot programming. *Constr. Robot.* **2020**, *4*, 31–48. [CrossRef]
158. Bhatt, P.M.; Kulkarni, A.; Malhan, R.K.; Gupta, S.K. Optimizing part placement for improving accuracy of robot-based additive manufacturing. In Proceedings of the 2021 IEEE International Conference on Robotics and Automation (ICRA), Xi'an, China, 30 May–5 June 2021; IEEE: Piscataway, NJ, USA; pp. 859–865.
159. Tiryaki, M.E.; Zhang, X.; Pham, Q.-C. Printing-while-moving: A new paradigm for large-scale robotic 3D Printing. In Proceedings of the 2019 IEEE/RSJ International Conference on Intelligent Robots and Systems (IROS), Macau, China, 3–8 November 2019; pp. 2286–2291.
160. Shen, H.; Pan, L.; Qian, J. Research on large-scale additive manufacturing based on multi-robot collaboration technology. *Addit. Manuf.* **2019**, *30*, 100906. [CrossRef]
161. Dörfler, K.; Dielemans, G.; Lachmayer, L.; Recker, T.; Raatz, A.; Lowke, D.; Gerke, M. Additive Manufacturing using mobile robots: Opportunities and challenges for building construction. *Cem. Concr. Res.* **2022**, *158*, 106772. [CrossRef]
162. Babu, S.S.; Mourad, A.-H.I.; Harib, K.H.; Vijayavenkataraman, S. Recent developments in the application of machine-learning towards accelerated predictive multiscale design and additive manufacturing. *Virtual Phys. Prototyp.* **2023**, *18*, e2141653. [CrossRef]
163. Jiang, J.; Xiong, Y.; Zhang, Z.; Rosen, D.W. Machine learning integrated design for additive manufacturing. *J. Intell. Manuf.* **2022**, *33*, 1073–1086. [CrossRef]
164. Yadav, D.; Chhabra, D.; Garg, R.K.; Ahlawat, A.; Phogat, A. Optimization of FDM 3D printing process parameters for multi-material using artificial neural network. *Mater. Today Proc.* **2020**, *21*, 1583–1591. [CrossRef]
165. Khan, M.F.; Alam, A.; Siddiqui, M.A.; Alam, M.S.; Rafat, Y.; Salik, N.; Al-Saidan, I. Real-time defect detection in 3D printing using machine learning. *Mater. Today Proc.* **2021**, *42*, 521–528. [CrossRef]
166. Yu, Z.; Thakolkaran, P.; Shea, K.; Stanković, T. Artificial neural network supported design of a lattice-based artificial spinal disc for restoring patient-specific anisotropic behaviors. *Comput. Biol. Med.* **2023**, *152*, 106475. [CrossRef]
167. Goh, G.D.; Sing, S.L.; Lim, Y.F.; Thong, J.L.J.; Peh, Z.K.; Mogali, S.R.; Yeong, W.Y. Machine learning for 3D printed multi-materials tissue-mimicking anatomical models. *Mater. Des.* **2021**, *211*, 110125. [CrossRef]
168. Ong, J.J.; Castro, B.M.; Gaisford, S.; Cabalar, P.; Basit, A.W.; Pérez, G.; Goyanes, A. Accelerating 3D printing of pharmaceutical products using machine learning. *Int. J. Pharm. X* **2022**, *4*, 100120. [CrossRef]

designs

MDPI

Review

A Review of the Current State of the Art of Polyether Ether Ketone (PEEK) Composite Based 3D-Printed Biomedical Scaffolds

Rajesh Surendran [1], Sithara Sreenilayam Pavithran [2,3,*], Anugop Balachandran [4], Sony Vijayan [3], Kailasnath Madanan [4] and Dermot Brabazon [2,3]

1 Department of Physics, Mar Ivanios College (Autonomous), Thiruvananthapuram 695015, India; rajesh.s@mic.ac.in
2 I-Form, Advanced Manufacturing Research Centre, Dublin City University, Glasnevin, D09 Dublin, Ireland; dermot.brabazon@dcu.ie
3 Advanced Processing Technology Research Centre, School of Mechanical and Manufacturing Engineering, Dublin City University, Glasnevin, D09 Dublin, Ireland; sony.vijayan@dcu.ie
4 International School of Photonics, Cochin University of Science and Technology, Kochi 682022, India; anugopb@cusat.ac.in (A.B.); kailas@cusat.ac.in (K.M.)
* Correspondence: sithara.sreenilayam@dcu.ie

Abstract: Three-dimensional printing or additive manufacturing (AM) has enabled innovative advancements in tissue engineering through scaffold development. The use of scaffolds, developed by using AM technology for tissue repair (like cartilage and bone), could enable the growth of several cell types on the same implant. Scaffolds are 3D-printed using polymer-based composites. polyether ether ketone (PEEK)-based composites are ideal for scaffold 3D printing due to their excellent biocompatibility and mechanical properties resembling human bone. It is therefore considered to be the next-generation bioactive material for tissue engineering. Despite several reviews on the application of PEEK in biomedical fields, a detailed review of the recent progress made in the development of PEEK composites and the 3D printing of scaffolds has not been published. Therefore, this review focuses on the current status of technological developments in the 3D printing of bone scaffolds using PEEK-based composites. Furthermore, this review summarizes the challenges associated with the 3D printing of high-performance scaffolds based on PEEK composites.

Keywords: 3D printing; biomaterial; implants; PEEK composites; additive manufacturing

Citation: Surendran, R.; Pavithran, S.S.; Balachandran, A.; Vijayan, S.; Madanan, K.; Brabazon, D. A Review of the Current State of the Art of Polyether Ether Ketone (PEEK) Composite Based 3D-Printed Biomedical Scaffolds. *Designs* **2023**, *7*, 128. https://doi.org/10.3390/designs7060128

Academic Editor: Mahdi Bodaghi

Received: 26 August 2023
Revised: 23 October 2023
Accepted: 31 October 2023
Published: 10 November 2023

1. Introduction

Additive manufacturing (AM) technologies have advanced greatly over the past few decades. In these technologies, products or three-dimensional (3D) parts are developed from their computer-aided-design (CAD) models [1–3]. The parts to be printed are built up in a layer-by-layer scheme on a plane designated as the X–Y plane, and the layers are added in the orthogonal Z-direction. 3D printing opens up new possibilities for fabricating complex structures using multiple materials [4,5] and reduces the design–manufacturing cycle, which in turn reduces the production costs and the material wastage. According to the nature of materials and parts development, there are seven different classes of AM techniques (Figure 1). Many parameters affect the quality of the final parts in the AM processes. To manufacture high-quality AM products, a good understanding of the process and material properties is essential. In recent years, the quality control/checking of AM structures has received much attention from manufacturing industries to make sure that the parts developed have the required specific functional properties [6–8]. For example, unwanted porosity in AM parts adversely affects their mechanical performance [9,10]. Recent studies have demonstrated the production of high-density parts (>99.8%) with the utilization of improved control systems [11]. The development of new AM materials for manufacturing critical AM structural parts is becoming increasingly important. AM technologies are used in the aerospace [12], automotive [13], medical science [14], and printed electronics [15,16] industries. The application of AM technologies in the biomedical field includes the production of customized prostheses, surgical and assistive tools, implants, instruments, pre-operative surgical planning and disease diagnosis and treatments [17]. Using 3D printing methods, functional tissue can be assembled from cells and scaffold materials. In tissue engineering, a 3D-printed construct would be ideal since it would be able to direct cells to migrate and proliferate, resulting in functional tissue.

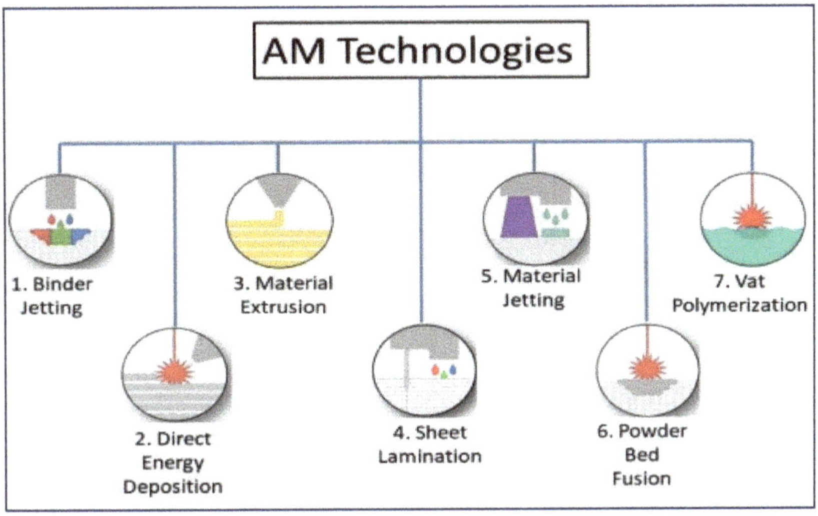

Figure 1. Additive manufacturing (AM) classification [18].

A combination of material engineering and tissue engineering can be used to treat human body parts losses [19]. In tissue engineering technology, scaffolds, or 3D porous biomaterials, play an important role [20]. In scaffold fabrication, biomaterials made from synthetic polymers, natural polymers, ceramics, composites, and metals are engineered in such a way as to facilitate the cellular interactions required for the generation of new functional tissues [21–23]. Polycarbonate, poly(D,L-lactic-co-glycolic acid), and poly(e-caprolactone) are examples of synthetic polymer-based biomaterials that can be applied to scaffold development. There are several natural polymers that can be used for this application, including silk, keratin, chitosan, and alginate. In terms of ceramics, calcium sulphate and calcium silicate are suitable for scaffold fabrication. Titanium (Ti–6Al–4V) and cobalt–chromium (Co–Cr) are two examples of metal materials used for the above-mentioned application. Each scaffold material has its own specific properties (e.g., mechanical, chemical, or physical), and it is selected according to the properties required for the specific biomedical implant and fabrication process [24–26]. To maintain mechanical characteristics similar to those of nearby tissue, it is necessary for the scaffold to temporarily resist external loads or stresses during the regeneration process of new tissue.

Tissue engineering scaffolds can be manufactured using a variety of techniques. Traditional manufacturing processes, such as phase separation [27], freeze-drying [28,29], salt leaching [30,31], or gas forming [32,33], do not offer an optimal solution to maintain its porous structure effectively. Pore size and porosity in scaffolds affect the delivery of oxygen and nutrients and enhance the growth of cells [34]. Through its porous structure, the implant gains mechanical stability by enabling mechanical interlocking between scaffolds and neighboring tissues [35]. Additionally, the pores' network structure aids in new tissue growth. Despite being good for nutrient exchange, high porosity negatively affects a scaffold's mechanical properties [36]. The mechanical and nutrient mass transport functions of a scaffold system must be balanced to achieve optimal performance. The final porosity and size of pores should therefore be considered at the design stage of a scaffold based on its intended application. Recently, there has been much interest in bone scaffold generation with customized architecture, strength, macro/micro-structure, wettability, cellular responses, etc. The AM process can be used to produce patient-specific scaffolds [37]. Recent studies by Ali Bahri et al. have reported porous scaffold 3D printing with controlled pore sizes and well-defined external and internal structures [38,39]. Metals like cobalt–chromium, titanium, and stainless steel alloys are typically used in orthopaedic scaffold 3D printing for bone repair or replacement since they are relatively strong and biocompatible [40,41]. The selection of a metal for use as an implant scaffold material is restricted due to its biodegradation inside the body, as well as the lack of the 3D printing production technique having been developed for printing particular metals [42,43]. Even though scaffolds are mechanically strong with small amounts of metal, biodegradability and compatibility are important issues to consider [44–46]. This problem can be solved by developing biodegradable metal materials. Metal matrix composites based on zinc, iron, magnesium, and calcium are an example of such materials. [47,48]. For scaffold development, iron- and magnesium (Fe-Mn)-based metal matrix composites have already been used. Fe-Mn scaffolds developed by inkjet 3D printing show high tensile properties like bone, and are biodegradable. Such new scaffolds should allow related implant site-specific biological cells to proliferate and differentiate. Due to their good biocompatibility and mechanical properties, ceramic materials containing both non-metallic and metallic components are used for scaffold 3D printing [49]. The ability to generate apatite mineralization makes ceramics attractive for scaffold development [50]. The ceramic material form that is normally found in human bone and teeth is hydroxyapatite (HA) [51]. In scaffold 3D printing and regenerative medicine, this material or other materials with similar properties have received much attention due to their mechanical properties. The positive biocompatibility and cell growth assistance properties of HA have been demonstrated in many studies [52,53]. The ceramics used in 3D printing scaffolds include calcium phosphate, calcium sulphate, calcium silicate, and tricalcium phosphate (Figure 2a) [54–56]. By using calcium phosphate ceramic in combination with HA and tricalcium phosphate, studies have reported scaffold manufacturing with a geometric accuracy of better than 97.5% when compared to computer-aided design (CAD) files. To fabricate models using 3D printing techniques, CAD designs are required. As this offers a high-resolution design and pore sizes as low as 300 μm, such ceramics are considered as suitable materials for scaffold fabrication and cell growth. However, for the load-bearing capabilities of ceramics in the 3D printing of scaffolds, further research and development studies are required.

Polymer-based composites are of key importance for scaffold 3D printing. Both thermosets and thermoplastics have been used extensively for this application [37,57]. Examples of such polymers are gelatin methacrylate (GelMA), poly(ethylene glycol) diacrylate (PEGDA), poly(D,L-lactic-co-glycolic acid) (PLGA), polyglycolic acid (PGA), poly(hydroxy butyrate), polycaprolactone (PCL) and poly(propylene fumarate) [58,59]. Among these, the synthetic polymers PLGA and PCL show low toxicity during material degradation [60]. To avoid materials that cause inflammation in tissue regeneration, it is essential to understand the inflammatory response [61,62], as a number of these polymers exhibit fast biodegradation [63]. PCL- and PLGA-based scaffolds have been 3D-printed with pore sizes of ~300 μm (Figure 2b,c) [64].

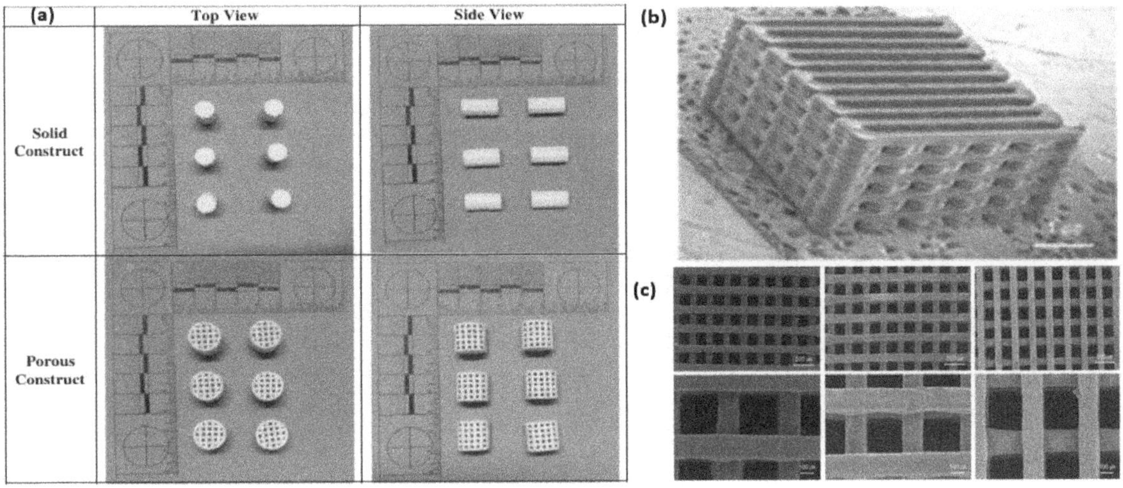

Figure 2. (**a**) Calcium sulphate and HA-based porous structures developed using thermal inkjet 3D printer [55]. Copyright © 2014 with permission from Elsevier. PCL/PLGA scaffold developed using fused deposition modeling (FDM) printer showing (**b**) macroscopic and (**c**) higher magnification SEM images [64]. Copyright © 2012 with permission from Elsevier.

Collagen, a type of natural polymer, is the most common candidate used for tissue engineering, along with corn starch, dextran, proteins, and polysaccharides [65]. Collagen (an insoluble fibrin found in human body tissues)-based 3D-printed scaffolds have been reported in many studies [66,67]. Various methods can be used to develop collagen scaffolds [68]. In 2022, Zhengwei Li et al. reported a technique to produce an in situ mineralized scaffold based on collagen for guided bone regeneration [39]. The mineralization process in this work involved electrospinning mineral ions in a collagen solution. In one of the works, Xiao-Hong Li et al. reported a collagen–silk fibroin-based scaffold implant for nerve-repairing applications [69]. In their study, they focused on two techniques, such as freeze-drying technology and 3D printing, and it was found that rats' neurological function, based on locomotor performance and electrophysiological analysis, was significantly enhanced by the implant made via the 3D printing method compared to the other implant. One study by Xiaoyi Lan et al. showed a bioprinting technique to develop collagen-based tissue-engineered nasal cartilage (Figure 3b) [70]. For this work, they used human nasoseptal chondrocytes and bovine type I collagen hydrogel for FRESH-inspired bioprinting, and suggested that this is a promising strategy for exploring further for providing autologous nasal cartilages for nasal cartilage reconstructive surgeries. The use of composite materials provides higher mechanical properties to the 3D-printed scaffold. The 3D printing of scaffolds made of bioactive calcium phosphate glass and polylactic acid (PLA) composite was reported in one study. In this case, an extrusion-based 3D printer was used to develop scaffolds with high porosity [71]. The utilization of composites is important in the development of the 3D-printed extracellular matrix of scaffolds. Figure 3a is a schematic of the scaffold 3D printing process [72] and Figure 3b is the bioprinting process based on collagen ink [70]. Table 1 presents some examples of biomedical scaffold materials, methods, and applications.

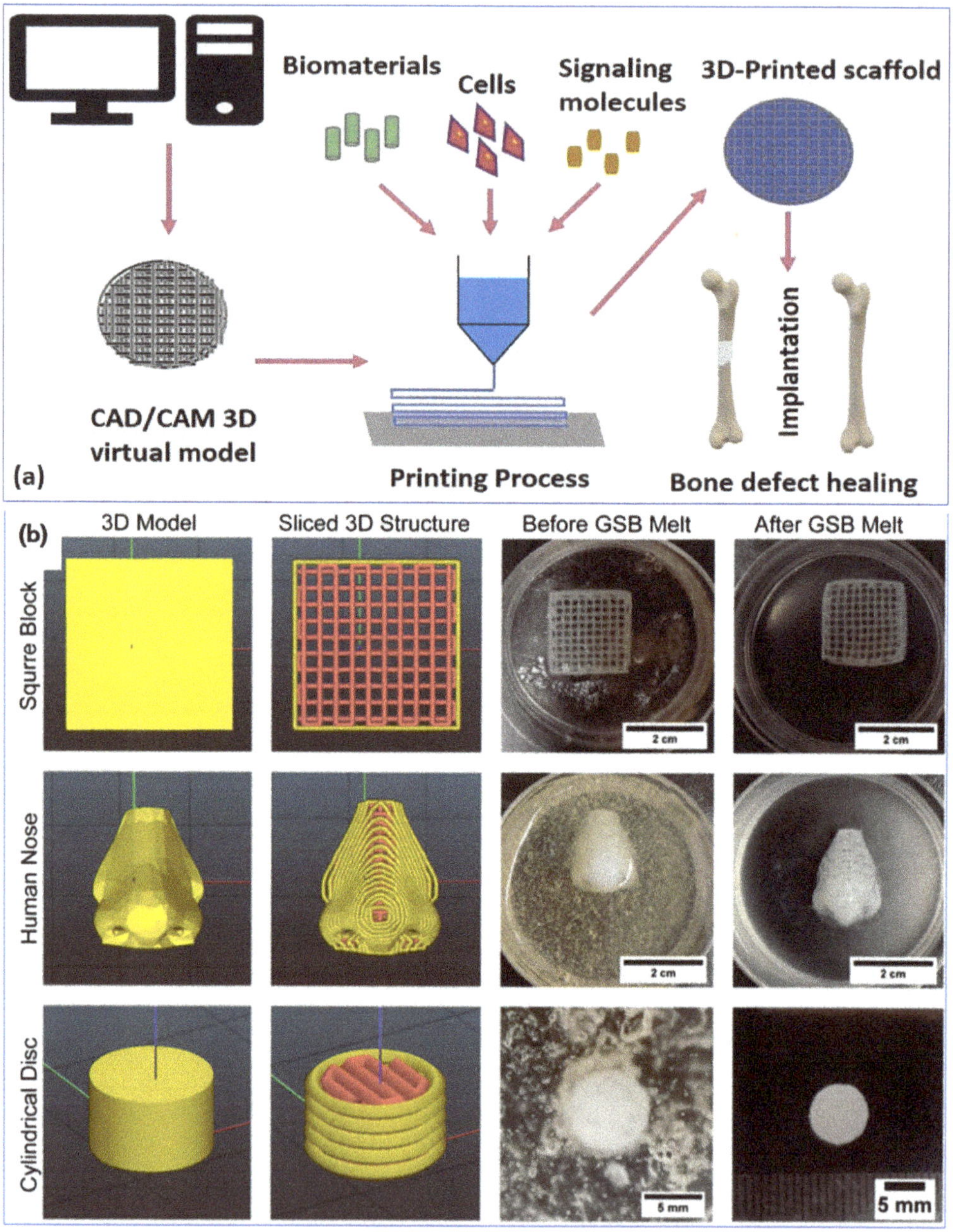

Figure 3. (**a**) Schematic of the scaffold 3D printing process [72]. Copyright © 2021 with permission from Elsevier. (**b**) Bioinspired tissues made up of collagen bio-ink in a gelatin bath using a FRESH bioprinting technique [70]. Copyright © 2021 the authors.

Table 1. Selected examples of materials, methods, and applications of biomedical scaffolds.

	Material	Application	Fabrication Methods	Ref.
Metal/metal alloys	Ti-6Al-4V	Bone implant	electron beam melting and selective laser melting	[73]
	Ti-6Al-4V	Dental implants	Laser Beam Melting	[74]
	Ti-6Al-4V	Segmental bone reconstruction	Electron beam melting	[75]
	Stainless steel	Bone implant	Selective Laser Melting	[76]
Ceramics	Tricalcium phosphate	Bone implant fabrication	Inkjet printing	[77]
	α-tricalcium phosphate	Maxillofacial bone defect reconstruction	3D printing	[78]
	Calcium carbonate	Adhesion, growth, and proliferation of osteoblast MC_3T_3 cells	Supercritical CO_2-based process	[79]
	β-tricalcium phosphate	Hard tissue repair	3D gel-printing	[80]
	Hydroxy apatite	Bone repair	DLP	[81]
	Calcium phosphate	Skull bone tissue reconstruction	Inkjet printing	[82]
	$CaO\text{-}SiO_2\text{-}P_2O_5\text{-}B_2O_3$ glass-ceramic	Maxillofacial bone defect reconstruction	3D printing	[83]
	Alumina	bone implant	Lithography-based Ceramics Manufacturing (LCM) technology	[84]
Composites	b-tricalcium phosphate and poly (D,L)-lactide	Fabrication of biodegradable bone implants	Selective Laser Melting	[85]
	Ti-6Al-4V, magnesium-calcium silicate (Mg-CS), and chitosan (CH)	Orthopedic	Laser melting deposition	[86]
	Polylactic acid/ biphasic calcium phosphate	Bone substitutes	Fused Deposition Modeling	[87]
	mPCL and TCP	Long bone reconstruction	3D printing	[88]
	Barium titanate and hydroxyapatite	Bone implant	3D printing	[89]
	Poly-lactic acid and nano-hydroxyapatite	bone scaffold	Fused Deposition Modelling	[90]
	30%HA-70% b-TCP BCP	Dental bone defect augmentation	3D printing	[91]
Polymers	Collagen and fibrinogen	Cartilage	Inkjet printing	[92]
	Methacrylated hyaluronan and methacrylated gelatin	Cardiac	Extrusion	[93]
	Gelatin and fibrinogen	Vascular	Extrusion	[94]
	Cell-laden collagen core and alginate sheet	Liver	Extrusion	[95]
	Gelatin, alginate, EGF, and dermal homogenates	Sweat gland	Extrusion	[96]

Many companies have started using the polyether ether ketone (PEEK) material for 3D printing as it exhibits unique qualities. Where traditional production methods or metallic materials are difficult to use, PEEK is a good choice for the low-volume production of required designs. Due to its low moisture absorption, easy sterilization, and biocompatibility, the use of PEEK materials in medical devices has increased in recent years. 3D-printed PEEK based implants and scaffolds have been developed and used without any complications. Recent review articles on PEEK discuss its use in soft and hard tissue engineering [97], as well as orthopedics [98]. As a detailed review of the recent developments in PEEK composites and the 3D printing of scaffolds has not been published, this review provides an overview of the recent advances in PEEK and PEEK composite materials for bone scaffold production as well as the 3D printing techniques used for processing these materials.

2. Poly Ether Ether Ketone (PEEK) for Scaffolds

Polyether ether ketone or PEEK is a biocompatible polymer extensively used for various applications, especially in the biomedical field. PEEK, chemically recognized as a linear poly (arylether ketone), is a high-performance, melt-processable aromatic polymer [99,100]. PEEK molecules are oriented in a planar zig-zag conformation with an orthorhombic crystal structure [101]. Because of its semicrystalline nature, PEEK finds a wide range of biomedical supporting applications such as lumbar cervical, thoracic, spinal, trauma and orthopedic implants. PEEK is chemically inert, and it is insoluble at room temperature (water solubility: 0.5 w/w %.) in all conventional solvents except for 98% sulfuric acid. It is very important to consider the chemical reaction between bone tissue and implant, as this leads to degradation and cytotoxicity (meaning how toxic the implant is to cells). Cells are damaged or even killed by cytotoxic materials. With the addition of various fillers, the modulus of elasticity or Young's modulus of PEEK, i.e., 3–4 GPa, can be increased to be like that of human cortical bone (7–30 GPa). The porosity of cortical bone, or dense or compact bone, is only 5–15% [102]. The fundamental elastic modulus property measures a material's resistance to elastic deformation under load or stress, and that determines a material's capability to hold its shape [102]. As the mechanical environment influences bone growth, the scaffold's Young's modulus, or modulus of elasticity, plays a critical role in enhancing bone formation. A mismatching of the Young's modulus values between the adjacent bone and the implant can lead to fracturing, osteopenia, or stress shielding [103,104]. PEEK is widely used in high-temperature engineering applications, because of its comparatively high glass transition temperature (~145 °C) and higher melting point (~334 °C) [105]. This material shows excellent temperature stability, which could be due to its relatively stiff backbone [106]. A biomaterial's thermal stability determines how long it will last inside a body. This property is therefore an important consideration when it comes to tissue engineering. Because of its high melting and glass transition temperatures [105], this material is stable in the human body. Magnetic resonance imaging (MRI) compatibility, high resistance to gamma and electron beam radiation, and natural radiolucency are all features of this PEEK material and its composites. PEEK's properties, including high mechanical strength, thermal and radiation stability, radiolucency, chemical and wear resistance, and biocompatibility, make it a viable candidate for bone restoration and tissue engineering applications. Biocompatibility and its suitability for the biomedical field were reported in the early 1990's [107–109]. More recently, the use of PEEK for the replacement of metal implants has been demonstrated [110,111]. The early PEEK applications were for intervertebral cages. The PEEK-based cages overcame the problems that arose due to the traditional metal intervertebral cages, which were stress-shielding due to the differential elastic modulus between the metal cage and human bone. This implementation provided the foundation for PEEK's current usage in spinal implants [112,113]. Composites based on PEEK and HA have been studied by several researchers [114–121]. The formation of apatite in simulated body fluid has a direct relationship with its volume content. The mechanical properties, cell attachment, proliferation and spreading, and activity of alkaline phosphatase have been improved due to the amount of filler content in the PEEK matrix [121].

Milling, injection molding, compression molding, forming and sintering are the conventional techniques used to fabricate PEEK composites-based parts; however, precision, the complexity and control of internal geometry, and high processing costs are the limitations of these conventional processing techniques [122–131]. To overcome the above limitations, 3D printing has been used as a forming process for fabricating PEEK-based components to produce complex shapes, which are beyond the scope of conventional technologies. The AM technique is specifically suited for the manufacturing of biomedical implants based on PEEK and its composites as patient-specific scaffolds that overcome the constraints of traditional manufacturing approaches. Selective Laser Sintering (SLS) and Fused Deposition Modeling (FDM) are widely used 3D printing methods to process PEEK and its composites [132–135]. The FDM technique, based on filament extrusion, usually uses low-melting-temperature polymers such as acrylonitrile butadiene styrene, thermoplastic polyurethane (TPU) and polylactic acid (PLA). In the work of Berretta et al., the high-temperature printing of a PEEK composite filament made up of 5% and 1% carbon nanotubes (CNTs) with a MendleMax v2.0 FDM system was examined (Figure 4a–d) [132]. According to their findings, the quality of the FDM parts depends on CNT mixing within the PEEK, printing conditions, and filaments used. In another work, the authors laser-sintered the PEEK cranial implant and analyzed its weight, dimensional accuracy, and mechanical properties (Figure 4e) [136]. They investigated the effects of the build direction (e.g., oblique, horizontal, inverted horizontal or vertical orientations) on the properties of the PEEK cranial implant. The studies show that a cranial implant developed in the horizontal direction exhibits higher geometric accuracy and compressive strength than those produced in the vertical direction.

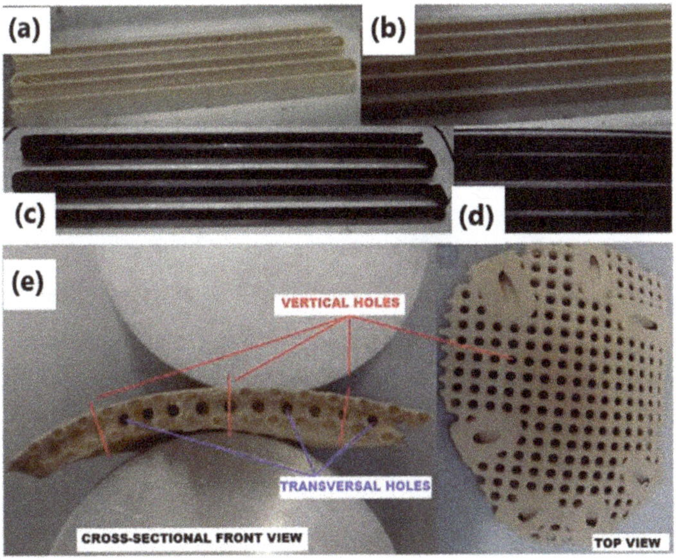

Figure 4. FDM single layer part based on (**a**,**b**) neat and (**c**,**d**) 1% CNT PEEK [57]. (**b**,**d**) Higher magnification pictures of (**a**,**c**). Copyright © 2017 with permission from Elsevier. (**e**) A 3D-printed cranial implant using the HT-LS system EOSINT P 800 [136]. Copyright © 2017 with permission from Elsevier.

Using PEEK, carbon fiber (CF)/PEEK, and glass fiber (GF)/PEEK composites, Wang et al. studied the FDM printing parameters' effects on the mechanical characteristics of the PEEK-composited FDM-produced parts [133]. This work focused mainly on the effects of nozzle temperature, platform temperature, layer thickness and printing speed on the mechanical performance of the 3D-printed parts, such as impact strength, flexural strength and tensile strength. As compared to CF/PEEK and GF/PEEK-based 3D-printed parts, pure PEEK parts showed lower tensile and flexural strengths. According to their study, the thickness of layers and printing speed affect the printed layers' strength. They noted that printing stability could be improved by lowering the print speed. Strontium- (Sr) or calcium (Ca)-doped HA, and

CNT/graphene-doped HA, are the major filler materials used to make PEEK composites for bone scaffolds. The addition of biomaterials and the incorporation of porosity into PEEK is very effective in improving osseointegration and bone implant interfaces [137–142].

In 2020, Oladapo et al. developed a cHAp/PEEK composite for the FDM-printing of scaffolds (Figure 5a) and studied the effects of cHAp on the PEEK surface [137]. The bioactivity of the developed implants was investigated by immersing them in a simulated body fluid for several days (Figure 5b–g) [137]. They found reduced biocompatibility and osseointegration around PEEK when compared to the PEEK/cHAp composite. The mechanical properties showed an improvement, with 15 wt. % cHAp as the optimum filler loading. The composite shows a better spread, adhesion, proliferation, and greater alkaline phosphatase activity. Furthermore, the osseointegration activity around the composite was higher than that around PEEK. In their study, the PEEK/cHAp composite was found to be more biocompatible and osseointegrable than PEEK. In another work, Faizal Manzoor et al. prepared PEEK-based filaments containing 10 wt. % of pure nanosized HA and PEEK/nano HA doped with Sr as well as Zn through hot-melt extrusion, which were subsequently 3D-printed via FDM [143]. Despite the reduction in crystallization temperature and increase in melting point with the addition of filler into pure PEEK, no noticeable changes in crystallinity were observed. Apart from the small drop in the tensile strength (~14%) and Young's modulus of approximately 5% in the PEEK/HA in comparison to the pure polymeric phase, the composites showed only slight differences in mechanical properties with the addition of the inorganic phases into the PEEK matrix. Moreover, the formation of apatite was observed on the surfaces of samples containing all the tested fillers, such as nano HA, SrHA and Zn HA. The in vitro bioactivity of 3D-printed samples was evaluated via a simulated body fluid immersion test for up to 28 days (Figure 5j) [143]. Zheng et al. reported the 3D printing of a PEEK/HA composite using a composite filament [144]. The composite filament was prepared through mechanical mixing followed by extrusion with filler loading of up to 30%. The HA phase was uniformly distributed on the surface of the PEEK matrix. A high amount of particle agglomeration was observed at this maximum filler loading. The strength and failure strain decreased with increasing HA content, whereas the modulus increased. The tensile modulus of the composite increased by around 68% when the HA content was increased to 30 wt. %. The tensile strength and elongation were highest for parts printed in the horizontal direction.

Recently, in 2020, Sikder et al. developed a melt blended amorphous magnesium phosphate (AMP) PEEK composite for dental AM manufacturing as well as orthopedic implants [145]. Osteogenic gene expression, cell viability, and proliferation were studied using mouse pre-osteoblasts (MC3T3-E1). Pure PEEK was used for in vivo analyses. The results show a homogeneous mixing of AMP particles within the matrix based on PEEK, with no degradation of the phase. AMP-PEEK composites exhibit low shear viscosities, which are suitable for 3D printing. The bioactivity was found to be improved with an increase in pre-osteoblast adhesion and proliferation through the incorporation of AMP in the PEEK matrix. Important gene expression markers like type I collagen (Col1), osteocalcin (OCN) and osteopontin (OPN) were increased by the presence of magnesium ions. In addition, AMP-filled PEEK composites exhibited enhanced osseointegration with a significant increase in new bone formation surrounding the composite implants. Even though PEEK filled with HA are the most studied and utilized materials for the development of scaffolds, several other materials were also studied to assess their ability to support tissue engineering. Even though FDM is a rapidly growing 3D printing technology, the mechanical properties and biocompatibility of FDM-printed PEEK and its composites need further investigation. Han et al. examined the printing of carbon fiber-reinforced PEEK (CFR-PEEK) composite using FDM and evaluated the resulting sample's mechanical properties [146]. In general, the printed CFR-PEEK composites were found to have higher mechanical tensile and bending strengths than the pure PEEK, without compromising much in terms of compressive strength. The cytotoxicity studies indicated that the printing process generated non-toxic effects even after 24 h of incubation. The carbon fiber reinforcement could improve mechanical properties, while having little influence on cytotoxicity and cell adhesion. Fahad et al. fabricated PEEK composites filled with both CNTs and graphene nanoplatelets (GNPs) [147]. It was reported that the 3D-printed PEEK nanocomposites maintained the desired degree of crystallinity with enhanced mechanical properties. An increase in the bioactivity and appetite growth in SBF was noticed for the reinforcements with CNT and GNP, which could be described by the ion's electrostatic interaction in SBF with SO_3H, a functional group generated by sulfonation.

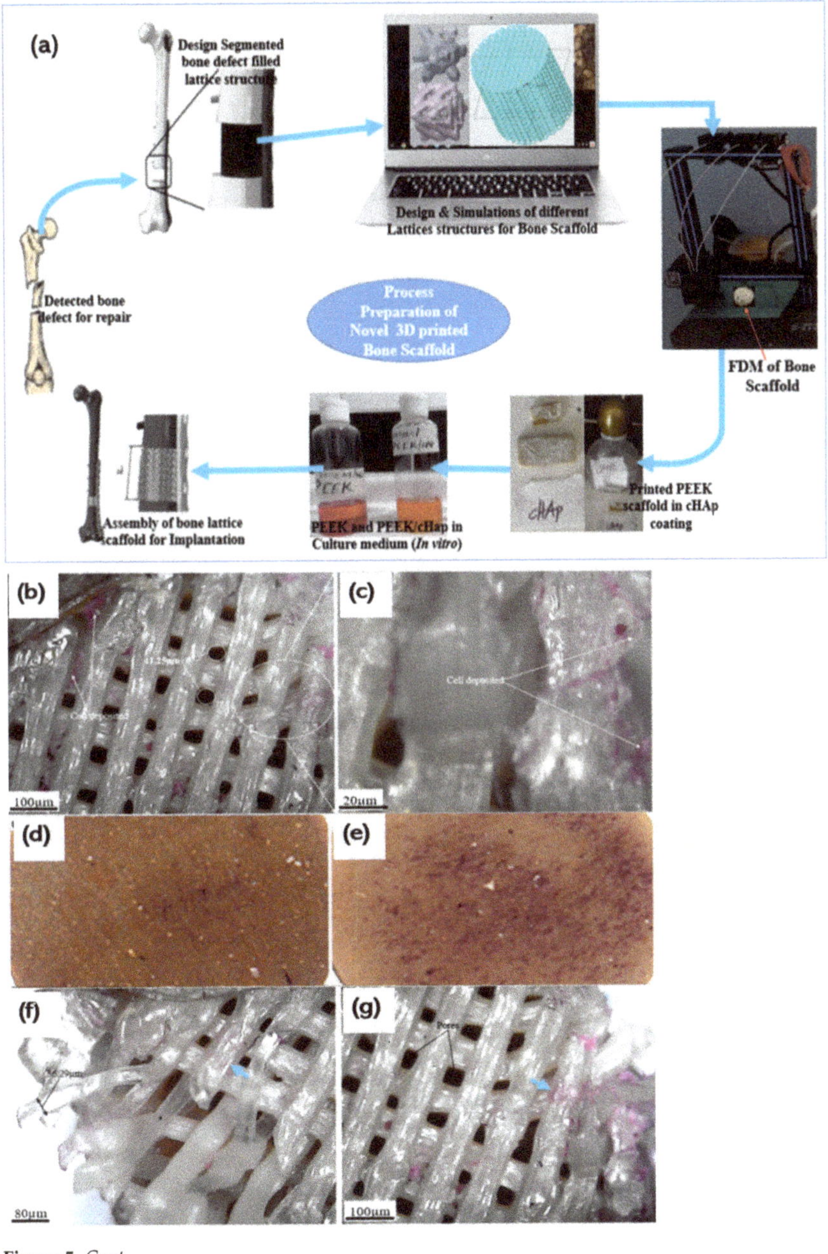

Figure 5. *Cont.*

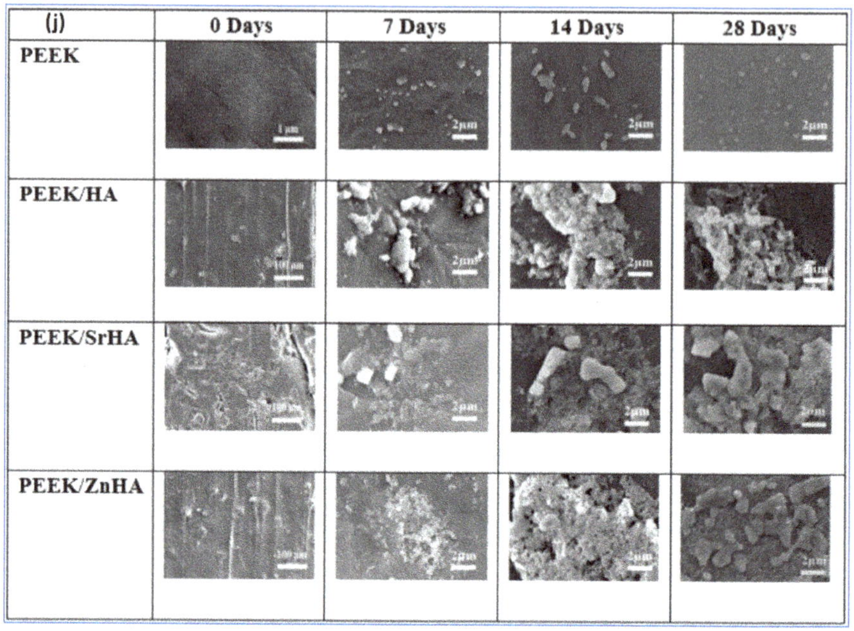

Figure 5. (**a**) Schematic of PEEK-based bone implant printed using FDM. (**b**–**g**) Cell attachment to the surface of the PEEK and PEEK and cHAp scaffold [137]. Copyright © 2020 with permission from Elsevier. (**j**) SEM micrographs of apatite layer development on the PEEK and PEEK composite surface after keeping them in a simulated body fluid for different days [143]. Copyright © 2021 with permission from Elsevier.

To ensure mechanical suitability for medical implants, Petersmann et al. studied various 3D-printable polymers, such as PLA, PEEK, polypropylene (PP), poly(methyl methacrylate) (PMMA), poly(vinylidene fluoride) (PVDF) and glycol-modified poly(ethylene terephthalate) (PETG), and in the temperature range desired for specific applications [148]. The high strength of PMMA and PEEK makes them suitable for load-bearing components; for example, cranial implants [149–151]. However, PP and PVDF show high flexibility, which could make them suitable for the sutural material application [152]. Pierantozzi et al. fabricated scaffolds based on PCL, PCL-HA and PCL-Sr HA using the FDM technique (Figure 6) [153]. Based on micro-CT analysis (Figure 6a), it became evident that the different designs produced showed high fidelity to the CAD model in terms of the porosity values (which were in the range of cancellous bone). Micro-CT and scanning electron microscopy (SEM) together confirm the uniform dispersion of ceramic particles within the PCL matrix, both on the surface and in the interior. These scaffolds also show good biocompatibility and support for cell growth, attachment, and proliferation. Among the studied composites, the SrHA-based scaffold showed high mineralization when compared to the PCL-HA and PCL-based scaffolds. In terms of mechanical properties, mineralization plays a key role. The use of mineralized micro- or nanomaterials in composites will enhance the mechanical properties of biomaterials and mimic the functions of natural bone tissue. Therefore, the authors of this study suggested that SrHA-based composites would be an attractive candidate for the development of bone scaffolds in tissue engineering.

Several other biocompatible polymers have been examined to make composites using different fillers, such as HA carbonated HA and nano HA. These composites show good wettability from biological fluids. It is pointed out by several authors that the degree of crystallinity and glass transition temperature were almost independent of the filler content [154–156]. Modal et al. reported the development of nanocomposite-based biomaterials containing acrylated epoxidized soyabean oil, PEGDA, and nano HA through the masked stereolithography (mSLA) technique for tissue engineering applications. The developed composites showed chemical stability, mechanical stability, good viability, and the proliferation of osteoblasts that differentiated from the mouse pre-osteoblast, MC3T3-E1 [157]. PCL in combination with PLGA and tricalcium phosphate in a particular ratio (i.e., 4:4:2) showed superior bone restoration capability [158]. However, PEEK-based composites have a combination of advantages over alternatives, which make them an excellent choice as a bone scaffold material, including their excellent biocompatibility, low density (1.32 g/cm^3), chemical resistance, and mechanical properties that better match human bone.

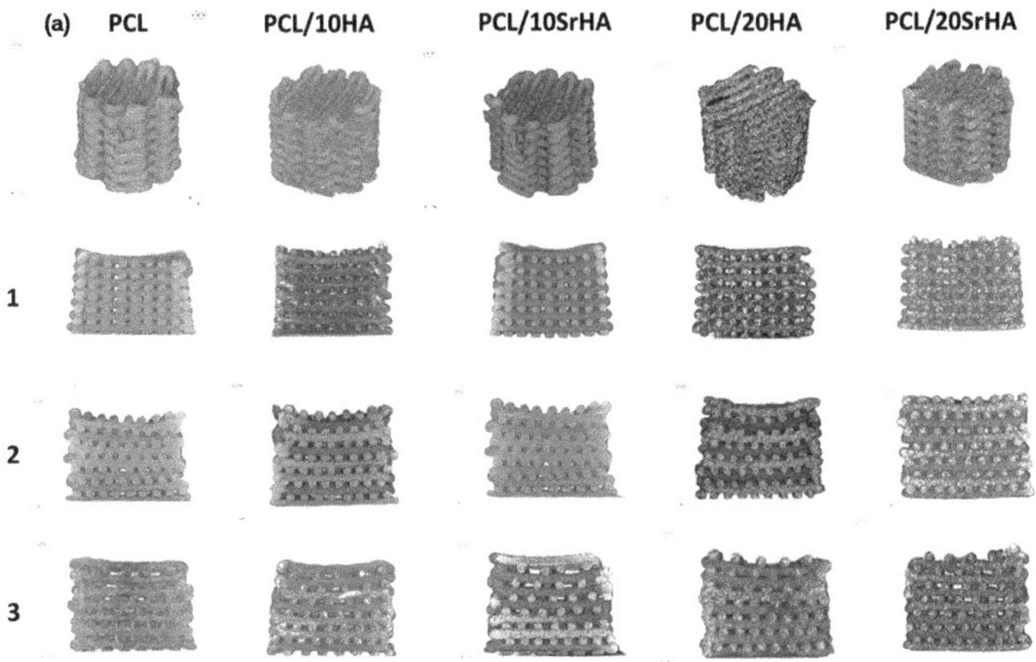

Figure 6. *Cont.*

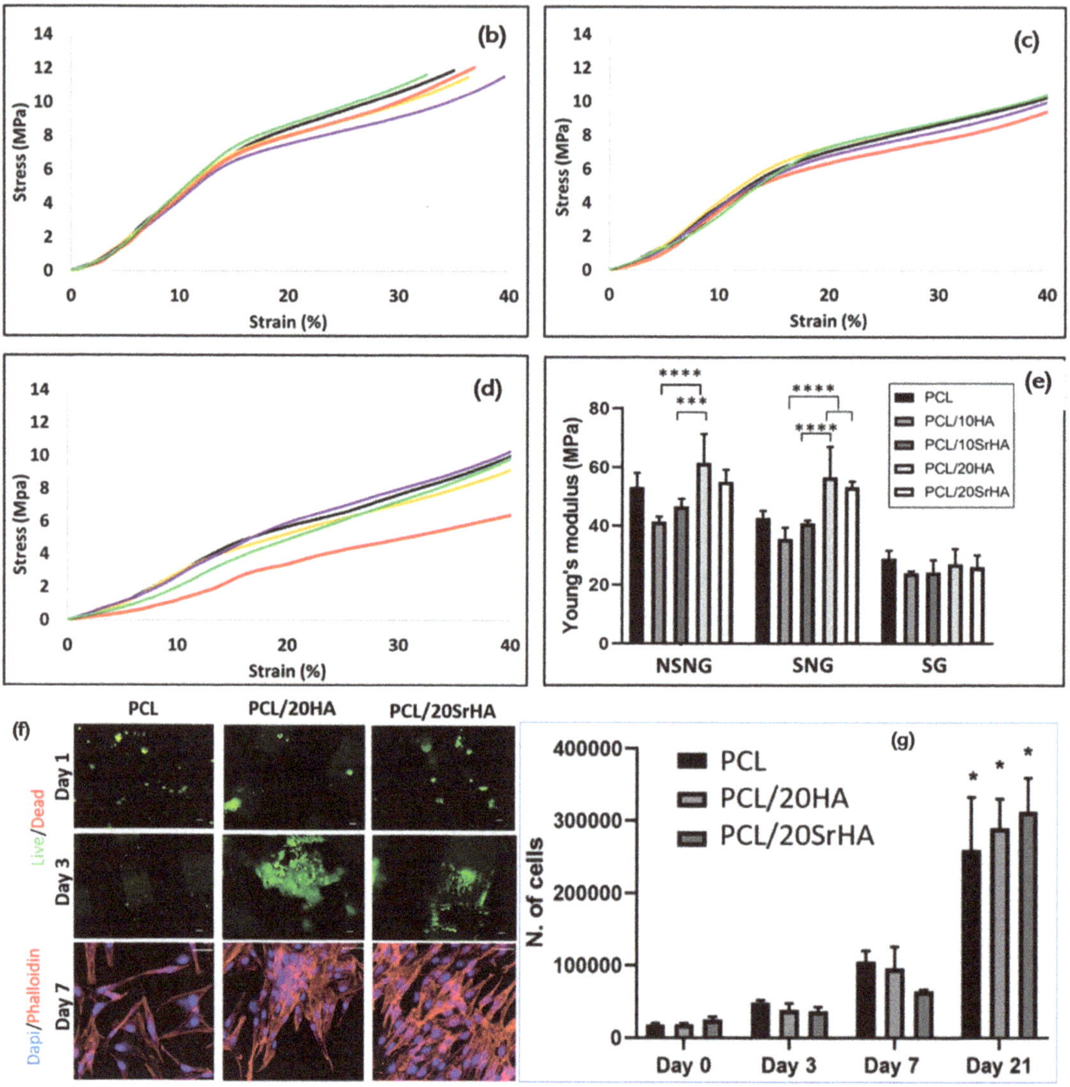

Figure 6. (**a**) Micro-CT pictures of 3D-printed scaffolds. (**b–d**) Compressive stress–strain curves of 3D-printed PCL scaffolds (black line), PCL and 10HA (purple line), PCL and 10SrHA (yellow line), PCL and 20HA (green line), and PCL and 20SrHA (red line) and (**e**) Young's modulus values (statistics: $p < 0.0001$ (****), p $\frac{1}{4}$ 0.0005 (***)). (**f**) Evaluation of cytocompatibility of 3D-printed scaffolds based on PCL, PCL/20HA and PCL and 20SrHA. (**g**) Cell number count and standard calibration curve (statistics: $p < 0.0001$ (*)) [153]. Copyright © 2021 with permission from Elsevier.

3. Conclusions

This article reviewed the recent developments in the PEEK composites-based scaffolds developed using 3D printing methods and summarized the mechanical, biological, and biophysical properties of scaffolds for tissue engineering using pure PEEK as well as PEEK composites. It is shown how PEEK osteointegration is accelerated by surface coatings that incorporate bioactive HA with different particle morphologies. Recent studies show that PEEK-based composites are the ideal choice to scaffold 3D printing due to their biocompatibility, low density, and good mechanical characterization, which match well with natural bone. The cytotoxicity of materials is an important consideration when selecting materials for tissue engineering applications. The cytotoxicity of PEEK can be tailored using appropriate filler materials, which is an advantage. The material has potential for tissue engineering applications, but there are issues like weak bonding between PEEK implants and surrounding tissues. The problem is expected to be solved by using innovative PEEK-based composites and porous structures. Thus, more research is needed to resolve this issue and to allow this material to be applied more in tissue engineering. The use of 3D printing technology in scaffold fabrication when compared to traditional techniques provides advantages such as the ability to create versatile and customized scaffolds with highly complex architectures, and to mimic the extracellular matrix. However, 3D-printed implants based on PEEK have yet to be commercialized. Prior to the clinical translation of 3D-printed customized PEEK implants, some obstacles need to be overcome in terms of regulation and technical standards. Clinical translation is currently hindered by a lack of technical data.

Author Contributions: Conceptualization, R.S.; writing—original draft preparation, R.S., A.B., S.V. and S.S.P.; writing—review and editing, D.B. and K.M. All authors have read and agreed to the published version of the manuscript.

Funding: Science Foundation Ireland: 16/1571 RC/3872, 19/US-C2C/3579; Erasmus + KA107 ICM: IRLDUBLIN04.

Data Availability Statement: Data are contained within the article.

Conflicts of Interest: The authors declare no conflict of interest.

References

1. Ngo, T.D.; Kashani, A.; Imbalzano, G.; Nguyen, K.T.Q.; Hui, D. Additive manufacturing (3D printing): A review of materials, methods, applications and challenges. *Compos. Part B Eng.* **2018**, *143*, 172–196. [CrossRef]
2. Gibson, I.; Rosen, D.W.; Stucker, B. *Additive Manufacturing Technologies*; Springer: New York, NY, USA, 2015.
3. Rafiee, M.; Farahani, R.D.; Therriault, D. Multi-Material 3D and 4D Printing: A Survey. *Adv. Sci.* **2020**, *7*, 1902307. [CrossRef] [PubMed]
4. Goh, G.L.; Zhang, H.; Chong, T.H.; Yeong, W.Y. 3D Printing of Multilayered and Multimaterial Electronics: A Review. *Adv. Electron. Mater.* **2021**, *7*, 2100445. [CrossRef]
5. Anketa, J.; Ikshita, C.; Ishika, W.; Ankush, R.; Mir, I.U.-H. 3D printing—A review of processes, materials and applications in industry 4.0. *Sustain. Oper. Comput.* **2022**, *3*, 33–42.
6. Qi, X.B.; Chen, G.F.; Li, Y.; Cheng, X.; Li, C.P. Applying Neural-Network-Based Machine Learning to Additive Manufacturing: Current Applications, Challenges, and Future Perspectives. *Engineering* **2019**, *5*, 721–729. [CrossRef]
7. Wang, C.; Tan, X.P.; Tor, S.B.; Lim, C.S. Machine learning in additive manufacturing: State-of-the-art and perspectives. *Addit. Manuf.* **2020**, *36*, 101538. [CrossRef]
8. Qin, J.; Hu, F.; Liu, Y.; Witherell, P.; Wang, C.C.L.; Rosen, D.W.; Simpson, T.W.; Lu, Y.; Tang, Q. Research and application of machine learning for additive manufacturing. *Addit. Manuf.* **2022**, *52*, 102691. [CrossRef]
9. Aboulkhair, N.T.; Everitt, N.M.; Ashcroft, I.; Tuck, C. Reducing porosity in AlSi10Mg parts processed by selective laser melting. *Addit. Manuf.* **2014**, *1*, 77–86. [CrossRef]
10. Liu, T.; Guessasma, S.; Zhu, J.; Zhang, W.; Nouri, H.; Belhabib, S. Microstructural defects induced by stereolithography and related compressive behaviour of polymers. *J. Mater. Process. Technol.* **2018**, *251*, 37–46. [CrossRef]
11. Sing, S.L.; Wiria, F.E.; Yeong, W.Y. Selective laser melting of titanium alloy with 50 wt% tantalum: Effect of laser process parameters on part quality. *Int. J. Refract. Metal Hard Mater.* **2018**, *77*, 120–127. [CrossRef]
12. Froes, F.; Boger, R. *Additive Manufacture for Aerospace Industry*; Elsevier: Alpharetta, GA, USA, 2019; pp. 1–6.
13. Charles, A.; Hofer, A.; Elkaseer, A.; Scholz, S.G. Additive Manufacturing in the Automotive Industry and the Potential for Driving the Green and Electric Transition. *Smart Innov. Syst. Technol.* **2022**, *262*, 339–346.
14. Li, C.; Pisignano, D.; Zhao, Y.; Xue, J. Advances in Medical Applications of Additive Manufacturing. *Engineering* **2020**, *6*, 1222–1231. [CrossRef]
15. Sreenilayam, S.P.; Ul Ahad, I.; Nicolosi, V.; Garzon, V.A.; Brabazon, D. Advanced materials of printed wearables for physiological parameter monitoring. *Mater. Today* **2020**, *32*, 147. [CrossRef]
16. Sreenilayam, S.P.; Ul Ahad, I.; Nicolosi, V.; Brabazon, D. MXene materials based printed flexible devices for healthcare, biomedical and energy storage applications. *Mater. Today* **2021**, *43*, 99–131. [CrossRef]

17. Tom, T.; Sreenilayam, S.P.; Brabazon, D.; Jose, J.P.; Joseph, B.; Madanan, K.; Thomas, S. Additive manufacturing in the biomedical field-recent research developments. *Results Eng.* **2022**, *16*, 100661. [CrossRef]
18. Obeidi, M.A. Metal additive manufacturing by laser-powder bed fusion: Guidelines for process optimisation. *Results Eng.* **2022**, *15*, 100473. [CrossRef]
19. Langer, R.; Tirrell, D.A. Designing materials for biology and medicine. *Nature* **2004**, *428*, 487–492. [CrossRef]
20. Chan, B.P.; Leong, K.W. Scaffolding in tissue engineering: General approaches and tissue-specific considerations. *Eur. Spine J.* **2008**, *17*, 467–479. [CrossRef]
21. Nikolova, M.P.; Chavali, M.S. Recent advances in biomaterials for 3D scaffolds: A review. *Bioact. Mater.* **2019**, *4*, 271–292. [CrossRef]
22. Chen, M.; Le, D.Q.; Baatrup, A.; Nygaard, J.V.; Hein, S.; Bjerre, L.; Kassem, M.; Zou, X.; Bunger, C. Self-assembled composite matrix in a hierarchical 3-D scaffold for bone tissue engineering. *Acta Biomater.* **2011**, *7*, 2244. [CrossRef]
23. Peter, S.J.; Miller, M.J.; Yasko, A.W.; Yaszemski, M.J.; Mikos, A.G. Polymer concepts in tissue engineering. *J. Biomed. Mater. Res.* **1998**, *43*, 422. [CrossRef]
24. Bose, S.; Vahabzadeh, S.; Bandyopadhyay, A. Bone tissue engineering using 3D printing. *Mater. Today* **2013**, *16*, 496. [CrossRef]
25. Liu, W.Y.; Li, Y.; Liu, J.Y.; Niu, X.F.; Wang, Y.; Li, D.Y. Application and Performance of 3D Printing in Nanobiomaterials. *J. Nanomater.* **2013**, *2013*, 681050. [CrossRef]
26. Taboas, J.M.; Maddox, R.D.; Krebsbach, P.H.; Hollister, S. Indirect solid free form fabrication of local and global porous, biomimetic and composite 3D polymer-ceramic scaffolds. *J. Biomater.* **2003**, *24*, 181. [CrossRef]
27. Mikos, A.; Temenoff, J. Formation of highly porous biodegradable scaffolds for tissue engineering. *Electron. J. Biotechnol.* **2000**, *3*, 23. [CrossRef]
28. Whang, K.; Thomas, C.; Healy, K.; Number, G. A novel method to fabricate bioabsorbable scaffolds. *Polymer* **1995**, *36*, 837. [CrossRef]
29. O'Brien, F.J.; Harley, B.A.; Yannas, I.V.; Gibson, L. Influence of freezing rate on pore structure in freeze-dried collagen-GAG scaffolds. *Biomaterials* **2004**, *25*, 1077. [CrossRef] [PubMed]
30. Ma, P.X.; Langer, R. Fabrication of biodegradable polymer foams for cell transplantation and tissue engineering. *Methods Mol. Med.* **1999**, *18*, 47.
31. Ma, P.X. Scaffolds for tissue fabrication. *Mater. Today* **2004**, *7*, 30. [CrossRef]
32. Nam, Y.S.; Yoon, J.J.; Park, T.G. A novel fabrication method of macroporous biodegradable polymer scaffolds using gas foaming salt as a porogen additive. *J. Biomed. Mater. Res.* **2000**, *53*, 1. [CrossRef]
33. Keskar, V.; Marion, N.W.; Mao, J.J.; Gemeinhart, R.A. In vitro evaluation of macroporous hydrogels to facilitate stem cell infiltration, growth, and mineralization. *Tissue Eng. Part A* **2009**, *15*, 1695. [CrossRef] [PubMed]
34. Salerno, A.; Di Maio, E.; Iannace, S.; Netti, P. Tailoring the pore structure of PCL scaffolds for tissue engineering prepared via gas foaming of multi-phase blends. *J. Porous Mater.* **2012**, *19*, 181. [CrossRef]
35. Karageorgiou, V.; Kaplan, D. Porosity of 3D biomaterial scaffolds and osteogenesis. *Biomaterials* **2005**, *26*, 5474. [CrossRef] [PubMed]
36. Hollister, S. Porous scaffold design for tissue engineering. *Nat. Mater.* **2005**, *4*, 518. [CrossRef]
37. Cubo-Mateo, N.; Rodríguez-Lorenzo, L.M. Design of Thermoplastic 3D-Printed Scaffolds for Bone Tissue Engineering: Influence of Parameters of "Hidden" Importance in the Physical Properties of Scaffolds. *Polymers* **2020**, *12*, 1546. [CrossRef]
38. Mitra, A.-E.; Trevor, C.B.; Ali, B. RAFT-Mediated 3D Printing of "Living" Materials with Tailored Hierarchical Porosity. *ACS Appl. Polym. Mater.* **2022**, *4*, 4940.
39. Li, Z.; Du, T.; Gao, C.; Tang, L.; Chen, K.; Liu, J.; Yang, J.; Zhao, X.; Niu, X.; Ruan, C. In-situmineralized homogeneous collagen-based scaffolds for potential guided bone regeneration. *Biofabrication* **2022**, *14*, 045016. [CrossRef]
40. Vasconcellos, L.M.R.D.; Oliveira, M.V.D.; Graça, M.L.D.A.; Vasconcellos, L.G.O.D.; Carvalho, Y.R.; Cairo, C.A.A. Porous Titanium Scaffolds Produced by Powder Metallurgy for Biomedical Applications. *Mater. Res.* **2008**, *11*, 275. [CrossRef]
41. Chou, D.T.; Wells, D.; Hong, D.; Lee, B.; Kuhn, H.; Kumta, P.N. Novel processing of iron–manganese alloy-based biomaterials by inkjet 3-D printing. *Acta Biomater.* **2013**, *9*, 8593. [CrossRef]
42. Wataha, J.C.; Hobbs, D.T.; Wong, D.J.; Dogan, S.; Zhang, H.; Chung, K.H. Elvington, Titanates deliver metal ions to human monocytes. *J. Mater. Sci. Mater. Med.* **2010**, *21*, 1289. [CrossRef]
43. Davis, R.R.; Hobbs, D.T.; Khashaba, R.; Sehkar, P.; Seta, F.N.; Messer, R.L.; Lewis, J.B.; Wataha, J.C. Titanate particles as agents to deliver gold compounds to fibroblasts and monocytes. *J. Biomed. Mater. Res. Part A* **2010**, *93*, 864.
44. Jayakumar, R.; Ramachandran, R.; Divyarani, V.V.; Chennazhi, K.P.; Tamura, H.; Nair, S.V. Fabrication of chitin-chitosan/nano TiO2-composite scaffolds for tissue engineering applications. *Int. J. Biol. Macromol.* **2011**, *48*, 336. [CrossRef] [PubMed]
45. Balla, V.K.; Bodhak, S.; Bose, S.; Bandyopadhyay, A. Porous tantalum structures for bone implants: Fabrication, mechanical and in vitro biological properties. *Acta Biomater.* **2010**, *6*, 3349. [CrossRef] [PubMed]
46. Dabrowski, B.; Swieszkowski, W.; Godlinski, D.; Kurzydlowski, K.J. Highly porous titanium scaffolds for orthopaedic applications. *J. Biomed. Mater. Res. Part B Appl. Biomater.* **2010**, *95*, 53. [CrossRef] [PubMed]
47. Zhuang, H.; Han, Y.; Feng, A. Preparation, mechanical properties and *in vitro* biodegradation of porous magnesium scaffolds. *Mater. Sci. Eng. C* **2008**, *28*, 1462. [CrossRef]

48. Vorndran, E.; Wunder, K.; Moseke, C.; Biermann, I.; Muller, F.A.; Zorn, K.; Gbureck, U. Hydraulic setting $Mg_3(PO_4)_2$ powders for 3D printing technology. *Adv. Appl. Ceram.* **2011**, *110*, 476. [CrossRef]
49. Seitz, H.; Rieder, W.; Irsen, S.; Leukers, B.; Tille, C. Three-dimensional printing of porous ceramic scaffolds for bone tissue engineering. *J. Biomed. Mater. Res. Part B Appl. Biomater.* **2005**, *74*, 782. [CrossRef]
50. Wu, C.; Fan, W.; Zhou, Y.; Luo, Y.; Gelinsky, M.; Chang, J.; Xiao, Y. 3D-printing of highly uniform $CaSiO_3$ ceramic scaffolds: Preparation, characterization, and *in vivo* osteogenesis. *J. Mater. Chem.* **2012**, *22*, 12288. [CrossRef]
51. Roy, D.M.; Kurtossy, L.S. Hydroxyapatite formed from Coral Skeletal Carbonate by Hydrothermal Exchange. *Nature* **1974**, *247*, 220. [CrossRef]
52. Leukers, B.; Gulkan, H.; Irsen, S.H.; Milz, S.; Tille, C.; Schieker, M.; Seitz, H. Hydroxyapatite scaffolds for bone tissue engineering made by 3D printing. *J. Mater. Sci. Mater. Med.* **2005**, *16*, 1121. [CrossRef]
53. Warnke, P.H.; Seitz, H.; Warnke, F.; Becker, S.T.; Sivananthan, S.; Sherry, E.; Liu, Q.; Wiltfang, J.; Douglas, T. Ceramic scaffolds produced by computer-assisted 3D printing and sintering: Characterization and biocompatibility investigations. *J. Biomed. Mater. Res. Part B Appl. Biomater.* **2010**, *93*, 212. [CrossRef] [PubMed]
54. Detsch, R.; Schaefer, S.; Deisinger, U.; Ziegler, G.; Seitz, H.; Leukers, B. In Vitro-Osteoclastic Activity Studies on Surfaces of 3D Printed Calcium Phosphate Scaffolds. *J. Biomater. Appl.* **2011**, *26*, 359. [CrossRef] [PubMed]
55. Zhou, Z.; Buchanan, F.; Mitchell, C.; Dunne, N. Printability of calcium phosphate: Calcium sulphate powders for the application of tissue engineered bone scaffolds using the 3D printing technique. *Mater. Sci. Eng. C* **2014**, *38*, 1. [CrossRef] [PubMed]
56. Castilho, M.; Moseke, C.; Ewald, A.; Gbureck, U.; Groll, J.; Pires, I.; Teßmar, J.; Vorndran, E. Direct 3D powder printing of biphasic calcium phosphate scaffolds for substitution of complex bone defects. *Biofabrication* **2014**, *6*, 015006. [CrossRef] [PubMed]
57. Pugliese, R.; Beltrami, B.; Regondi, S.; Lunetta, C. Polymeric biomaterials for 3D printing in medicine: An overview. *Ann. 3D Print. Med.* **2021**, *2*, 100011. [CrossRef]
58. Billiet, T.; Gevaert, E.; De Schryver, T.; Cornelissen, M.; Dubruel, P. The 3D printing of gelatin methacrylamide cell-laden tissue-engineered constructs with high cell viability. *Biomaterials.* **2014**, *35*, 49. [CrossRef]
59. Hribar, K.C.; Soman, P.; Warner, J.; Chung, P.; Chen, S. Light-assisted direct-write of 3D functional biomaterials. *Lab Chip* **2014**, *14*, 268. [CrossRef]
60. Leong, K.F.; Cheah, C.M.; Chua, C.K. Solid freeform fabrication of three-dimensional scaffolds for engineering replacement tissues and organs. *Biomaterials* **2003**, *24*, 2363. [CrossRef]
61. Mountziaris, P.M.; Spicer, P.P.; Kasper, F.K.; Mikos, A.G. Harnessing and modulating inflammation in strategies for bone regeneration. *Tissue Eng. Part B Rev.* **2011**, *17*, 393. [CrossRef]
62. Rajan, V.; Murray, R. The duplicitous nature of inflammation in wound repair. *Wound Pract. Res.* **2008**, *16*, 122.
63. Sung, H.J.; Meredith, C.; Johnson, C.; Galis, Z.S. The effect of scaffold degradation rate on three-dimensional cell growth and angiogenesis. *Biomaterials* **2004**, *25*, 5735. [CrossRef] [PubMed]
64. Hong, J.M.; Kim, B.J.; Shim, J.H.; Kang, K.S.; Kim, K.J.; Rhie, J.W.; Cha, H.J.; Cho, D.W. Enhancement of bone regeneration through facile surface functionalization of solid freeform fabrication-based three-dimensional scaffolds using mussel adhesive proteins. *Acta Biomater.* **2012**, *8*, 2578. [CrossRef] [PubMed]
65. Inzana, J.A.; Olvera, D.; Fuller, S.M.; Kelly, J.P.; Graeve, O.A.; Schwarz, E.M.; Kates, S.L.; Awad, H.A. 3D printing of composite calcium phosphate and collagen scaffolds for bone regeneration. *Biomaterials* **2014**, *35*, 4026. [CrossRef] [PubMed]
66. Elangovan, S.; D'Mello, S.R.; Hong, L.; Ross, R.D.; Allamargot, C.; Dawson, D.V.; Stanford, C.M.; Johnson, G.K.; Sumner, D.R.; Salem, A.K. The enhancement of bone regeneration by gene activated matrix encoding for platelet derived growth factor. *Biomaterials* **2014**, *35*, 737. [CrossRef]
67. Bielajew, B.J.; Hu, J.C.; Athanasiou, K.A. Collagen: Quantification, biomechanics, and role of minor subtypes in cartilage. *Nat. Rev. Mater.* **2020**, *5*, 730. [CrossRef]
68. Lee, A.; Hudson, A.R.; Shiwarski, D.J.; Tashman, J.W.; Hinton, T.J.; Yerneni, S.; Bliley, J.M.; Campbell, B.G.; Feinberg, A.W. 3D bioprinting of collagen to rebuild components of the human heart. *Science* **2019**, *365*, 482–487. [CrossRef]
69. Li, X.H.; Zhu, X.; Liu, X.Y.; Xu, H.H.; Jiang, W.; Wang, J.J.; Chen, F.; Zhang, S.; Li, R.-X.; Chen, X.-Y.; et al. The corticospinal tract structure of collagen/silk fibroin scaffold implants using 3d printing promotes functional recovery after complete spinal cord transection in rats. *J. Mater. Sci. Mater. Med.* **2021**, *32*, 31. [CrossRef]
70. Lan, X.; Liang, Y.; Erkut, E.J.N.; Kunze, M.; Mulet-Sierra, A.; Gong, T.; Osswald, M.; Ansari, K.; Seikaly, H.; Boluk, Y.; et al. Bioprinting of human nasoseptal chondrocytes-laden collagen hydrogel for cartilage tissue engineering. *Faseb J.* **2021**, *35*, e21191. [CrossRef]
71. Serra, T.; Planell, J.A.; Navarro, M. High-resolution PLA-based composite scaffolds via 3-D printing technology. *Acta Biomater.* **2013**, *9*, 5521. [CrossRef]
72. Yadav, L.R.; Chandran, S.V.; Lavanya, K.; Selvamurugan, N. Chitosan-based 3D-printed scaffolds for bone tissue engineering. *Int. J. Biol. Macromol.* **2021**, *183*, 1925. [CrossRef]
73. Zhao, B.; Wang, H.; Qiao, N.; Wang, C.; Hu, M. Corrosion resistance characteristics of a Ti–6Al–4V alloy scaffold that is fabricated by electron beam melting and selective laser melting for implantation in vivo. *Mater. Sci. Eng. C* **2016**, *70*, 832. [CrossRef] [PubMed]

74. Yang, F.; Chen, C.; Zhou, Q.; Gong, Y.; Li, R.; Li, C.; Klämpfl, F.; Freund, S.; Wu, X.; Sun, Y.; et al. Laser beam melting 3D printing of Ti6Al4V based porous structured dental implants: Fabrication, biocompatibility analysis and photoelastic study. *Sci. Rep.* **2017**, *7*, 45360. [CrossRef] [PubMed]
75. Surmeneva, M.A.; Surmenev, R.A.; Chudinova, E.A.; Koptioug, A.; Tkachev, M.S.; Gorodzha, S.N.; Rännar, L.E. Fabrication of multiple-layered gradient cellular metal scaffold via electron beam melting for segmental bone reconstruction. *Mater. Des.* **2017**, *133*, 195. [CrossRef]
76. Fousová, M.; Kubásek, J.; Vojtěch, D.; Fort, J.; Čapek, J. 3D printed porous stainless steel for potential use in medicine. *IOP Conf. Ser. Mater. Sci. Eng.* **2017**, *179*, 012025. [CrossRef]
77. Igawa, K.; Mochizuki, M.; Sugimori, O.; Shimizu, K.; Yamazawa, K.; Kawaguchi, H.; Nakamura, K.; Takato, T.; Nishimura, R.; Suzuki, S.; et al. Tailor-made tricalcium phosphate bone implant directly fabricated by a three-dimensional ink-jet printer. *J. Artif. Organs.* **2006**, *9*, 234. [CrossRef]
78. Saijo, H.; Fujihara, Y.; Kanno, Y.; Hoshi, K.; Hikita, A.; Chung, U.; Takato, T. Clinical experience of full custom-made artificial bones for the maxillofacial region. *Regener. Ther.* **2016**, *5*, 72–78. [CrossRef]
79. Woldetsadik, A.D.; Sharma, S.K.; Khapli, S.; Jagannathan, R.; Magzoub, M. Hierarchically porous calcium carbonate scaffolds for bone tissue engineering. *ACS Biomater. Sci. Eng.* **2017**, *3*, 2457. [CrossRef]
80. Zhang, Y.; Shao, H.; Lin, T.; Peng, J.; Wang, A.; Zhang, Z.; Wang, L.; Liu, S.; Yu, X. Effect of Ca/P ratios on porous calcium phosphate salt bioceramic scaffolds for bone engineering by 3D gel-printing method. *Ceram. Int.* **2019**, *45*, 20493. [CrossRef]
81. Liu, Z.; Liang, H.; Shi, T.; Xie, D.; Chen, R.; Han, X.; Shen, L.; Wang, C.; Tian, Z. Additive manufacturing of hydroxyapatite bone scaffolds via digital light processing and in vitro compatibility. *Ceram. Int.* **2019**, *45*, 11079. [CrossRef]
82. Zhang, B.; Sun, H.; Wu, L.; Ma, L.; Xing, F.; Kong, Q.; Fan, Y.; Zhou, C.; Zhang, X. 3D printing of calcium phosphate bioceramic with tailored biodegradation rate for skull bone tissue reconstruction. *Bio-Des. Manuf.* **2019**, *2*, 161. [CrossRef]
83. Lee, U.L.L.; Lim, J.Y.Y.; Park, S.N.N.; Choi, B.H.H.; Kang, H.; Choi, W.C.C. A clinical trial to evaluate the efficacy and safety of 3D printed bioceramic implants for the reconstruction of zygomatic bone defects. *Materials* **2020**, *13*, 4515. [CrossRef] [PubMed]
84. Vijayavenkataraman, S.; Kuan, L.Y.; Lu, W.F. 3D-printed ceramic triply periodic minimal surface structures for design of functionally graded bone implants. *Mater. Des.* **2020**, *191*, 108602. [CrossRef]
85. Lindner, M.; Hoeges, S.; Meiners, W.; Wissenbach, K.; Smeets, R.; Telle, R.; Poprawe, R.; Fischer, H. Manufacturing of individual biodegradable bone substitute implants using selective laser melting technique. *J. Biomed. Mater. Res. A* **2011**, *97A*, 466. [CrossRef] [PubMed]
86. Tsai, C.H.; Hung, C.H.; Kuo, C.N.; Chen, C.Y.; Peng, Y.N.; Shie, M.Y. Improved bioactivity of 3D printed porous titanium alloy scaffold with chitosan/magnesium-calcium silicate composite for orthopaedic applications. *Materials* **2019**, *12*, 203. [CrossRef]
87. Nevado, P.; Lopera, A.; Bezzon, V.; Fulla, M.R.; Palacio, J.; Zaghete, M.A.; Biasotto, G.; Montoya, A.; Rivera, J.; Robledo, S.M.; et al. Preparation and in vitro evaluation of PLA/biphasic calcium phosphate filaments used for fused deposition modelling of scaffolds. *Mater. Sci. Eng. C* **2020**, *114*, 111013. [CrossRef]
88. Kobbe, P.; Laubach, M.; Hutmacher, D.W.; Alabdulrahman, H.; Sellei, R.M.; Hildebrand, F. Convergence of scaffold-guided bone regeneration and RIA bone grafting for the treatment of a critical-sized bone defect of the femoral shaft. *Eur. J. Med. Res.* **2020**, *25*, 70. [CrossRef]
89. Polley, C.; Distler, T.; Detsch, R.; Lund, H.; Springer, A.; Boccaccini, A.R.; Seitz, H. 3D printing of piezoelectric barium titanate-hydroxyapatite scaffolds with interconnected porosity for bone tissue engineering. *Materials* **2020**, *13*, 1773. [CrossRef]
90. Wang, W.; Zhang, B.; Li, M.; Li, J.; Zhang, C.; Han, Y.; Wang, L.; Wang, K.; Zhou, C.; Liu, L.; et al. 3D printing of PLA/n-HA composite scaffolds with customized mechanical properties and biological functions for bone tissue engineering. *Compos. B Eng.* **2021**, *224*, 109192. [CrossRef]
91. Mangano, C.; Giuliani, A.; De Tullio, I.; Raspanti, M.; Piattelli, A.; Iezzi, G. Case report: Histological and histomorphometrical results of a 3-D printed biphasic calcium phosphate ceramic 7 years after insertion in a human maxillary alveolar ridge. *Front. Bioeng. Biotechnol.* **2021**, *9*, 614325. [CrossRef]
92. Xu, T.; Binder, K.W.; Albanna, M.Z.; Dice, D.; Zhao, W.; Yoo, J.J.; Atala, A. Hybrid printing of mechanically and biologically improved constructs for cartilage tissue engineering applications. *Biofabrication* **2013**, *5*, 015001. [CrossRef]
93. Duan, B.; Kapetanovic, E.; Hockaday, L.A.; Butcher, J.T. Three-dimensional printed trileaflet valve conduits using biological hydrogels and human valve interstitial cells. *Acta Biomater.* **2014**, *10*, 1836. [CrossRef] [PubMed]
94. Kolesky, D.B.; Homan, K.A.; Skylar-Scott, M.A.; Lewis, J.A. Three-dimensional bioprinting of thick vascularized tissues. *Proc. Natl. Acad. Sci. USA* **2016**, *113*, 3179. [CrossRef] [PubMed]
95. Yeo, M.; Lee, J.S.; Chun, W.; Kim, G.H. An innovative collagen-based cell-printing method for obtaining human adipose stem cell-laden structures consisting of coresheath structures for tissue engineering. *Biomacromolecules* **2016**, *17*, 1365. [CrossRef] [PubMed]
96. Huang, S.; Yao, B.; Xie, J.; Fu, X. 3D bioprinted extracellular matrix mimics facilitate directed differentiation of epithelial progenitors for sweat gland regeneration. *Acta Biomater.* **2016**, *32*, 170. [CrossRef] [PubMed]
97. Chengzhu, L.; Yuchao, L.; Sie, C.T. Polyetheretherketone and Its Composites for Bone Replacement and Regeneration. *Polymers* **2020**, *12*, 2858.
98. Sun, C.; Kang, J.; Yang, C.; Zheng, J.; Su, Y.; Dong, E.; Liu, Y.; Yao, S.; Shi, C.; Pang, H.; et al. Additive manufactured polyether-ether-ketone implants for orthopaedic applications: A narrative review. *Biomater Transl.* **2022**, *3*, 116–133.

99. Ma, R.; Guo, D. Evaluating the bioactivity of a hydroxyapatite-incorporated polyetheretherketone biocomposite. *J. Orthop. Surg. Res.* **2019**, *14*, 1. [CrossRef]
100. Velisaris, C.N.; Seferis, J.C. Crystallization kinetics of polyetheretherketone (peek) matrices. *Polym. Eng. Sci.* **1986**, *26*, 1574. [CrossRef]
101. Lee, Y.; Porter, R.S.; Lin, J.S. On the double-melting behavior of poly (ether ether ketone). *Macromolecules* **1989**, *22*, 1756. [CrossRef]
102. Elise, F.M.; Ginu, U.U.; Amira, I.H. Bone Mechanical Properties in Healthy and Diseased States. *Annu. Rev. Biomed. Eng.* **2018**, *20*, 119–143.
103. Limmahakhun, S.; Oloyede, A.; Sitthiseripratip, K.; Xiao, Y.; Yan, C. Stiffness and strength tailoring of cobalt chromium graded cellular structures for stress-shielding reduction. *Mater. Des.* **2017**, *114*, 633–641. [CrossRef]
104. Liverani, E.; Rogati, G.; Pagani, S.; Brogini, S.; Fortunato, A.; Caravaggi, P. Mechanical interaction between additive-manufactured metal lattice structures and bone in compression: Implications for stress shielding of orthopaedic implants. *J. Mech. Behav. Biomed. Mater.* **2021**, *121*, 104608. [CrossRef] [PubMed]
105. Cebe, P.; Chung, S.Y.; Hong, S.D. Effect of thermal history on mechanical properties of polyetheretherketone below the glass transition temperature. *Appl. Polym. Sci.* **1987**, *33*, 487. [CrossRef]
106. Rajani, K.V.; Rajesh, S.; Murali, K.P.; Mohanan, P.; Ratheesh, R. Preparation and microwave characterization of PTFE/PEEK Blends. *Polym. Compos.* **2009**, *30*, 296. [CrossRef]
107. Burg, K.J.L.; Shalaby, S.W. PES and PEEK. In *Encyclopedia of Materials: Science and Technology*, 2nd ed.; Elsevier: Amsterdam, The Netherlands, 2001; pp. 6837–6839.
108. Jin, W.; Chu, P.K. Orthopedic Implants, Reference Module. In *Biomedical Sciences—Encyclopedia of Biomedical Engineering*; Elsevier: Amsterdam, The Netherlands, 2019; pp. 425–439.
109. Qin, L.; Yao, S.; Zhao, J.; Zhou, C.; Oates, T.W.; Weir, M.D.; Wu, J.; Xu, H.H.K. Review on Development and Dental Applications of Polyetheretherketone-Based Biomaterials and Restorations. *Materials* **2021**, *14*, 408. [CrossRef] [PubMed]
110. Schwitalla, A.D.; Zimmermann, T.; Spintig, T.; Kallage, I.; Müller, W.D. Fatigue limits of different PEEK materials for dental implants. *J. Mech. Behav. Biomed. Mater.* **2017**, *69*, 163. [CrossRef]
111. Mishra, S.; Chowdhary, R. PEEK materials as an alternative to titanium in dental implants: A systematic review. *Clin. Implant. Dent. Relat. Res.* **2019**, *21*, 208. [CrossRef]
112. Maldonado-Naranjo, A.L.; Healy, A.T.; Kalfas, I.H. Polyetheretherketone (PEEK) intervertebral cage as a cause of chronic systemic allergy: A case report. *Spine J.* **2015**, *5*, e1–e3. [CrossRef]
113. Basgul, C.; Yu, T.; MacDonald, D.; Siskey, R.; Marcolongo, M.; Kurtz, S. Structure-property relationships for 3D-printed PEEK intervertebral Lumbar cages produced using fused filament fabrication. *J. Mater. Res.* **2018**, *33*, 2040. [CrossRef]
114. Bakar, M.A.; Cheang, P.; Khor, K. Mechanical properties of injection molded hydroxyapatite-polyetheretherketone biocomposites. *Compos. Sci. Technol.* **2003**, *63*, 421. [CrossRef]
115. Bakar, M.A.; Cheng, M.; Tang, S.; Yu, S.; Liao, K.; Tan, C.; Khor, K.; Cheang, P. Tensile properties, tension–tension fatigue and biological response of polyetheretherketone–hydroxyapatite composites for load-bearing orthopedic implants. *Biomaterials* **2003**, *24*, 2245.
116. Converse, G.L.; Yue, W.; Roeder, R.K. Processing, and tensile properties of hydroxyapatite-whisker-reinforced polyetheretherketone. *Biomaterials* **2007**, *28*, 927. [CrossRef] [PubMed]
117. de Araújo Nobre, M.; Ferro, A.; Maló, P. Adult patient risk stratification using a risk score for periodontitis. *J. Clin. Med.* **2019**, *8*, 307. [CrossRef] [PubMed]
118. Garcia-Gonzalez, D.; Jayamohan, J.; Sotiropoulos, S.; Yoon, S.H.; Cook, J.; Siviour, C.; Arias, A.; Jerusalem, A. On the mechanical behaviour of PEEK and HA cranial implants under impact loading. *J. Mech. Behav. Biomed. Mater.* **2017**, *69*, 342. [CrossRef] [PubMed]
119. Wang, L.; Weng, L.; Song, S.; Sun, Q. Mechanical properties and microstructure of polyetheretherketone–hydroxyapatite nanocomposite materials. *Mater. Lett.* **2010**, *64*, 2201. [CrossRef]
120. Wang, L.; Weng, L.; Song, S.; Zhang, Z.; Tian, S.; Ma, R. Characterization of polyetheretherketone-hydroxyapatite nanocomposite materials. *Mater. Sci. Eng. A* **2011**, *528*, 3689. [CrossRef]
121. Yu, S.; Hariram, K.P.; Kumar, R.; Cheang, P.; Aik, K.K. In vitro apatite formation and its growth kinetics on hydroxyapatite/ polyetheretherketone biocomposites. *Biomaterials* **2005**, *26*, 2343. [CrossRef]
122. Ma, R.; Tang, T. Current Strategies to Improve the Bioactivity of PEEK. *Int. J. Mol. Sci.* **2014**, *15*, 5426. [CrossRef]
123. Rinaldi, M.; Ghidini, T.; Cecchini, F.; Brandao, A.; Nanni, F. Additive layer manufacturing of poly (ether ether ketone) via FDM. *Compos. Part B Eng.* **2018**, *145*, 162. [CrossRef]
124. Jaafar, J.; Siregar, J.P.; Tezara, C.; Hazim, M.; Hamdan, M.; Rihayat, T. A review of important considerations in the compression molding process of short natural fiber composites. *Int. J. Adv. Manuf. Technol.* **2019**, *105*, 3437. [CrossRef]
125. Bastan, F.E. Fabrication and characterization of an electrostatically bonded PEEK-hydroxyapatite composites for biomedical applications. *J. Bio. Med. Mater. Res. B Appl. Bio Meter.* **2020**, *108*, 2513.
126. Conrad, T.L.; Roeder, R.K. Effects of porogen morphology on the architecture, permeability, and mechanical properties of hydroxyapatite whisker reinforced polyetheretherketone scaffolds. *J. Mech. Behav. Biomed. Mater.* **2020**, *106*, 103730. [CrossRef] [PubMed]

127. Deng, Y.; Liu, X.; Xu, A.; Wang, L.; Luo, Z.; Zheng, Y.; Deng, F.; Wei, J.; Tang, Z.; Wei, S. Effect of surface roughness on osteogenesis in vitro and osseointegration in vivo of carbon fiber-reinforced polyetheretherketone–nanohydroxyapatite composite. *Int. J. Nanomed.* **2015**, *10*, 1425.
128. Nisa, V.S.; Rajesh, S.; Murali, K.P.; Priyadersini, V.; Potty, S.N.; Ratheesh, R. Preparation, Characterization and Dielectric properties of temperature stable $SrTiO_3$/PEEK, composites for microwave substrate applications. *Compos. Sci. Technol.* **2008**, *68*, 106. [CrossRef]
129. Rego, B.T.; Neto, W.A.R.; de Paula, A.C.C.; Goes, A.M.; Bretas, R.E.S. Mechanical properties and stem cell adhesion of injection-molded poly (ether ether ketone) and hydroxyapatite nanocomposites. *J. Appl. Polym. Sci.* **2015**, *132*, 41748. [CrossRef]
130. Ventola, C.L. Medical applications for 3D printing: Current and projected uses. *Pharm. Ther.* **2014**, *39*, 704.
131. Saeed, K.; McIlhagger, A.; Harkin-Jones, E.; Kelly, J.; Archer, E. Predication of the in-plane mechanical properties of continuous carbon fibre reinforced 3D printed polymer composites using classical laminated-plate theory. *Compos. Struct.* **2021**, *259*, 113226. [CrossRef]
132. Berretta, S.; Davies, R.; Shyng, Y.; Wang, Y.; Ghita, O. Fused Deposition Modelling of high temperature polymers: Exploring CNT PEEK composites. *Polym. Test.* **2017**, *63*, 251. [CrossRef]
133. Wang, P.; Zou, B.; Ding, S.; Li, L.; Huang, C. Effects of FDM-3D printing parameters on mechanical properties and microstructure of CF/ PEEK and GF/PEEK. *Chinese J. Aeronotics* **2021**, *34*, 236. [CrossRef]
134. Schmidt, M.; Pohle, D.; Rechtenwald, T. Selective laser sintering of PEEK. *CIRP Ann.* **2007**, *56*, 205. [CrossRef]
135. Wang, P.; Zou, B.; Xiao, H.; Ding, S.; Huang, C. Effects of printing parameters of fused deposition modeling on mechanical properties, surface quality, and microstructure of PEEK. *J. Mater. Process. Technol.* **2019**, *271*, 62. [CrossRef]
136. Berretta, S.; Evans, K.; Ghita, O. Additive manufacture of PEEK cranial implants: Manufacturing considerations versus accuracy and mechanical performance. *Mater. Des.* **2018**, *139*, 141. [CrossRef]
137. Oladapo, B.I.; Ismail, S.O.; Bowoto, O.K.; Omigbodun, F.T.; Olawumi, M.A.; Muhammad, M.A. Lattice design and 3D-printing of PEEK with Ca10(OH)(PO4)3 and in-vitro bio-composite for bone implant. *Int. J. Biol. Macromol.* **2020**, *165*, 50. [CrossRef]
138. Rasheva, Z.; Zhang, G.; Burkhart, T.A. Correlation between the tribological and mechanical properties of short carbon fibers reinforced PEEK materials with different fiber orientations. *Tribol. Int.* **2010**, *43*, 1430. [CrossRef]
139. Evans, N.T.; Torstrick, F.B.; Lee, C.S.D.; Dupont, K.M.; Safranski, D.L.; Chang, W.A.; Macedo, A.E.; Lin, A.S.P.; Boothby, J.M.; Whittingslow, D.C.; et al. High-strength, surface-porous polyether-ether-ketone for load-bearing orthopedic implants. *Acta Biomater.* **2015**, *13*, 159. [CrossRef]
140. Hassanajili, S.; Karami-Pour, A.; Oryan, A.; Talaei-Khozani, T. Preparation and characterization of PLA/PCL/HA composite scaffolds using indirect 3D printing for bone tissue engineering. *Mater. Sci. Eng. C* **2019**, *104*, 109960. [CrossRef]
141. Chen, Q.; Liu, Y.; Yao, Q.Q.; Yu, S.S.; Zheng, K.; Pischetsrieder, M.; Boccaccini, A.R. Multi layered bioactive composite coating drug delivery capability by electrophoretic deposition combined with layer-by-layer deposition. *Adv. Biomater. Devices Med.* **2014**, *1*, 18.
142. Torres, Y.; Romero, C.; Chen, Q.; Pérez, G.; Rodríguez-Ortiz, J.A.; Pavón, J.J.; Álvarez, L.; Arévalo, C.; Boccaccini, A.R. Electrophoretic deposition of PEEK/45S5 bioactive glass coating on porous titanium substrate: Influence of processing conditions and porosity parameters. *Key Eng. Mater.* **2016**, *704*, 343. [CrossRef]
143. Manzoor, F.; Golbang, A.; Jindal, S.; Dixon, D.; McIlhagger, A.; Harkin-Jones, E.; Crawford, D.; Mancuso, E. 3D printed PEEK/HA composites for bone tissue engineering applications: Effect of material formulation on mechanical performance and bioactive potential. *J. Mech. Behav. Biomed. Mater.* **2021**, *121*, 104601. [CrossRef]
144. Zheng, J.; Kang, J.; Sun, C.; Yang, C.; Wang, L.; Li, D. Effects of printing path and material components on mechanical properties of 3D-printed polyether-ether-ketone/hydroxyapatite composites. *J. Mech. Behav. Biomed. Mater.* **2021**, *118*, 104475. [CrossRef]
145. Sikder, P.; Ferreira, J.A.; Ehsan, A.; Fakhrabadic Kantorski, K.Z.; Liberatore, M.W.; Bottino, M.C.; Bhaduri, S.B. Bioactive amorphous magnesium phosphate-polyetheretherketone composite filaments for 3D printing. *Dent. Mater.* **2020**, *36*, 865. [CrossRef] [PubMed]
146. Han, X.; Yang, D.; Yang, C.; Spintzyk, S.; Scheideler, L.; Li, P.; Li, D.; Geis-Gerstorfer, J.; Rupp, F. Carbon Fiber Reinforced PEEK Composites Based on 3D-Printing Technology for Orthopedic and Dental Applications. *J. Clin. Med.* **2019**, *8*, 240. [CrossRef] [PubMed]
147. Alam, F.; Varadarajan, K.M.; Koo, J.H.; Wardle, B.L.; Kumar, S. Additively Manufactured Polyetheretherketone (PEEK) with Carbon Nanostructure Reinforcement for Biomedical Structural Applications. *Adv. Eng. Mater.* **2020**, *22*, 2000483. [CrossRef]
148. Petersmann, S.; Spoerk, M.; De Steene, W.V.; Üçal, M.; Wiener, J.; Pinter, G.; Arbeiter, F. Mechanical properties of polymeric implant materials produced by extrusion-based additive manufacturing. *J. Mech. Behav. Biomed. Mater.* **2020**, *104*, 103611. [CrossRef]
149. Navarro, M.; Michiardi, A.; Castano, O.; Planell, J.A. Biomaterials in orthopaedics. *J. R. Soc. Interface* **2008**, *5*, 1137. [CrossRef]
150. Petersmann, S.; Spoerk, M.; Huber, P.; Lang, M.; Pinter, G.; Arbeiter, F. Impact optimization of 3D-printed poly(methyl metharylate) for cranial implants. *Macromol. Mater. Eng.* **2019**, *304*, 1900263. [CrossRef]
151. Panayotov, I.V.; Orti, V.; Cuisinier, F.; Yachouh, J. Polyetheretherketone (PEEK) for medical applications. *J. Mater. Sci. Mater. Med.* **2016**, *27*, 118. [CrossRef]
152. Ratner, B.D.; Hoffman, A.S.; Schoen, F.J.; Lemons, J.E. *Biomaterials Science: An Introduction to Materials in Medicin*, 2nd ed.; Elsevier: Amsterdam, The Netherlands, 2004.

153. Pierantozzi, D.; Scalzone, A.; Jindal, S.; Stıpniece, L.; Salma-Ancane, K.; Dalgarno, K.; Gentile, P.; Mancuso, E. 3D printed Sr-containing composite scaffolds: Effect of structural design and material formulation towards new strategies for bone tissue engineering. *Compos. Sci. Technol.* **2020**, *191*, 108069. [CrossRef]
154. Oladapoa, B.I.; Ismail, S.O.; Zahedi, M.; Khan, A.; Usman, H. 3D printing and morphological characterisation of polymeric composite scaffolds. *Eng. Struct.* **2020**, *216*, 110752. [CrossRef]
155. Petrovskaya, T.S.; Toropkov, N.E.; Mironov, E.G.; Azarmi, F. 3D printed biocompatible polylactidehydroxyapatite based material for bone implants. *Mater. Manuf. Process.* **2018**, *33*, 1899. [CrossRef]
156. Corcione, C.E.; Gervaso, F.; Scalera, F.; Montagna, F.; Sannino, A.; Maffezzol, A. The feasibility of printing polylactic acid–nanohydroxyapatite composites using a low-cost fused deposition modelling 3D printer. *J. Appl. Polym. Sci.* **2017**, *134*, 44656. [CrossRef]
157. Mondal, D.; Haghpanah, Z.; Huxman, C.J.; Tanter, S.; Sun, D.; Gorbet, M.; Willett, T.L. mSLA-based 3D printing of acrylated epoxidized soybean oil—nano-hydroxyapatite composites for bone repair. *Mater. Sci. Eng. C* **2021**, *130*, 112456. [CrossRef] [PubMed]
158. Hwang, K.S.; Choi, J.W.; Kim, J.H.; Chung, H.Y.; Jin, S.; Shim, J.H.; Yun, W.S.; Jeong, C.M.; Huh, J.B. Comparative Efficacies of Collagen-Based 3D Printed PCL/PLGA/β-TCP Composite Block Bone Grafts and Biphasic Calcium Phosphate Bone Substitute for Bone Regeneration. *Materials* **2017**, *10*, 421. [CrossRef] [PubMed]

designs

MDPI

Article

An Assembly-Oriented Design Framework for Additive Manufacturing

Germain Sossou, Frédéric Demoly *, Samuel Gomes and Ghislain Montavon

ICB UMR 6303 CNRS, University Bourgogne Franche-Comté, UTBM, 90010 Belfort, France; germain.sossou@outlook.fr (G.S.); samuel.gomes@utbm.fr (S.G.); ghislain.montavon@utbm.fr (G.M.)
* Correspondence: frederic.demoly@utbm.fr

Abstract: The shape complexity capability of additive manufacturing (AM) is currently the main thrust of the design for AM (DFAM) research. In order to aid designers embracing that complexity-for-free characteristics of AM, many design approaches have been put forth. However, AM does not only benefit parts' designs: its capability can be harnessed at assembly level to design performant and innovative products. Most of the few contributions on the topic are concerned with part consolidation of existing assemblies, but other advantages such as assembly-free mechanisms, multi-material components, or even component embedding can also improve product design complexity. This paper aims to put forth a thorough DFAM framework for new product development (made of multiple parts) and which consider all the assembly-related characteristics of AM. It considers what can be called AM-based architecture minimization, which includes, among others, part consolidation and assembly-free mechanisms as well. Within context of an 'AM-factory', in which the most appropriate machine(s) is/are selected for easing a whole assembly manufacturing before the detailed geometric definition is committed. For the sake of completeness, a methodology based on functional flows has also been investigated for the parts' design. A gripper as case study has been introduced to illustrate the framework.

Keywords: design for additive manufacturing; additive manufacturing; assembly; part consolidation; assembly-free mechanisms

Citation: Sossou, G.; Demoly, F.; Gomes, S.; Montavon, G. An Assembly-Oriented Design Framework for Additive Manufacturing. *Designs* **2022**, *6*, 20. https://doi.org/10.3390/designs6010020

Academic Editor: Obeidi Muhannad

Received: 21 January 2022
Accepted: 17 February 2022
Published: 18 February 2022

1. Introduction

Originally considered as a physical prototyping process, additive manufacturing (AM) is now being investigated and used for end-of-use products. Such endeavor can be traced to the expansion of both the techniques following the principle of AM and the materials processed by them (as shown in Table 1). As such, a new paradigm shift is taking place in the way we manufacture goods. For instance, while conventional (i.e., subtractive or formative) processes would require a product to be manufactured sequentially in many simple shape parts and then to be assembled, the layer-by-layer manufacturing peculiar to AM allows single step manufacturing of products featuring complex shapes (difficult or impossible to manufacture with conventional processes).

However, for this paradigm shift to be complete, it is needed to undertake significant efforts from the early design stages and do over current models, methods, and tools accordingly, and/or promote new ones. In a way similar to how research efforts have been targeted at design for X [1]—design for manufacture (DFM) or design for assembly (DFA) [2]—to aid designers in delivering solutions that are easily manufacturable and ready for a seamless assembly, research has to be carried out to design in embracing the full potentials of AM. As such, a new trend in the design theory and methodology realm has emerged under the umbrella term of design for additive manufacturing (DFAM) [3,4].

Table 1. Additive manufacturing processes classification.

Family Process	Description	Typical Processed Material	Typical Technique
Material extrusion	A material is semi-solid state is extruded through a nozzle/needle and is cured.	Polymers, ceramics, metals, wood	Fused Filament Fabrication (FFF), Robocasting (Direct Ink Writing),
Powder bed fusion	A thermal source selectively fuses layers of powder.	Polymers, ceramics, metals	Selective Laser Sintering (SLS), Selective Laser Melting (SLM)
Photopolymerization	Layers of photopolymers are selectively cured upon exposure to a radiation.	Photocurable polymers	Sterelithography (SLA)
Directed energy deposition	A focused high power laser beam melts a material powder as it is being deposited.	Metals	Laser Engineered Net Shaping (LENS), Direct Metal Deposition (DMD), 3D laser cladding
Sheet lamination	Material sheets are bonded; each sheet (representing a cross section of the CAD model) is selectively cut with an energy source.	Papers, metals, polymers	Laminated Object Manufacturing (LOM), Ultrasonic Consolidation (UC)
Material jetting	Droplets of a material (or a mix of two materials) are selectively deposited in thin layers from a print head, and cured either by a source of energy or by environmental conditions.	Polymers, wax	Multi-Jet Modeling (Drop-On-Demand), PolyJet™
Binder jetting	A binder is selectively deposited, from a printhead, onto a powder bed, forming a section of the CAD model.	Plastics, metals, composites, ceramics, polymers	3D printing

AM is mainly praised for its capability to manufacture intricate shapes or interlocking features that are unfeasible with conventional processes. This is reflected in the vast majority of work in the DFAM-related literature. Most of the current contributions to the field seem to be mainly dedicated to parts' design where the goals are basically twofold: (i) providing tools and methodologies for embracing the shape complexity capability of AM in order to improve performance, and (ii) constraining the design process by considering AM specific manufacturing constraints. As such, these contributions are essential for a better adoption of AM by designers and engineers [5,6]. However, as components rarely work by themselves and are usually part of a whole assembly, the design freedom allowed by AM should be extended to the assembly level as well. The AM capabilities are not limited to shape complexity, nor are they only beneficial to a part's performance. DFAM contributions—aimed at seizing these benefits—are rather scarce, but the issue is being given more and more research interest. As a matter of fact, when DFAM is considered from an assembly point of view, it is mainly for redesign cases [7]. Indeed almost none of the current works is led to the consideration of the AM benefits for assembly design in a new product development (NPD) context. Nevertheless, a few contributions [8–12] are targeted at conceptual design for AM. Furthermore, the few assembly-oriented DFAM (A-DFAM) contributions do restrict the scientific issues to part consolidation (PC) [13]. These two aforementioned situations are fully encapsulated in Yang et al.'s [11] statements: '[A-DFAM] mainly focuses on the redesign of assemblies by taking advantage of AM-enabled capabilities ... ' and 'The general motivation of assembly-level DFAM research is to investigate the possibility of part consolidation which brings benefits of reduced part count, enhanced performance (e.g., frictionless), and decreased cost ... '. While PC is indeed an AM-benefit that can greatly simplify an assembly, other benefits are still poorly explored. Fully functional assemblies of components can seamlessly be printed [14], with the advantages of consolidated parts, multi-material parts, moving mechanisms, electronic components embedding, and so on. The purpose of this article is to address the aforementioned gaps in the literature, namely considering assembly-level DFAM beyond PC and redesign cases. More specifically the paper aims at putting forth a thorough DFAM framework for NPD (made of multiple parts) and which considers all the assembly-related characteristics of AM. As these characteristics do vary across AM techniques and machines, another novelty of our proposal is that AM techniques and machines are selected to ease the product manufacturing, PC decisions are made based on these techniques' actual capabilities.

The content of the paper is organized as follows. In Section 2, the literature is reviewed, the rationale behind the proposed framework is explained, and a summary of the characteristics of AM processes is provided; the proposed methodology is presented in Section 3; demonstration on a case study is presented within Section 4; finally, conclusions and future work are drawn.

2. Literature Review

2.1. Design for Additive Manufacturing

The main capability of AM is praised for is its ability to seamlessly build virtually any shape at no extra (tooling) cost. This can explain why the vast majority of current DFAM works is targeted at part design. These contributions consist in optimizing part geometry in order to enhance performance through topology optimization or latticed design [13,15,16]. Other contributions are dedicated at making designers aware of the AM specific constraints in order to reduce iterations and tailor what can be designed to what is truly manufacturable.

DFAM is barely being considered from an assembly design perspective. This branch of DFAM can be split in two subbranches according to Yang et al. [11], namely consistent assembly (CA) and reduced assembly (RA). CA deals with all the design alterations that can be done to make an existing assembly printable without decreasing its number of components. RA—which is the scope of this paper—is concerned with how to take advantage of AM in order to reduce assembly complexity (through PC for instance).

The few works related to RA tend to gravitate around PC. An earliest DFAM contribution is the well-known Airbus aircraft duct [17], whose part count has been drastically reduced from 16 components (and fasteners) to one component. Another AM-based PC case study is the one made by Schmelzle et al. [18]. In order to aid designers seizing such an opportunity, some methodologies have been developed. Yang et al. [19] have presented a PC method that aims at redesigning assemblies of non-moving parts. The method requires a computer-aided design (CAD) file of the original design definition along with functional performance requirements. It is implicitly assumed that the parts of the assembly are originally secured together with conventional assembly methods. Worth mentioning in their methodology is that function integration is made by merging the product functional surfaces into a single design space. Moreover, in order to leverage AM capability beyond PC, the so-generated design space is used to define an optimized shape of the consolidated assembly. In a similar study, Rodrigue and Rivette [20] proposed a design methodology targeted at assemblies design for AM. The goal of their proposal is to assist the designer in the embodiment design stage, so that all the advantages offered by AM are considered; these include mainly PC and multi-material part. Using the Boothroyd and Dewhurst (B&D) rules for design for manufacture and assembly (DFMA) [21], they proposed simple guidelines for PC in an AM context. In addition, once the assembly's parts that need to stand alone were identified, their geometries were optimized for functionality improvement using topology optimization. Yang and Zhao [22] have taken another look at the B&D DFMA rules for part count reduction in the light of AM capabilities, where a new set of design rules have been put forth. These new rules were then used to propose a design framework following a screening-and-refinement strategy. In the screening step, the derived rules are used to filter a given existing assembly in order to find potential parts candidates for consolidation. The (second) refinement step consists in consolidating and optimizing the selected candidates leveraging part-related AM capabilities. The overall goal of their framework is to synthesize optimal solutions with the minimum cost-to-performance ratio by optimizing the part-material-architecture triplet. In order to automate these new rules' implementation, they developed a numerically assisted approach called part consolidation candidate detection. Given a CAD assembly model, the developed approach aims at first verifying the rules and secondly at finding optimal grouping solutions with the lowest number of groups.

A common characteristic of the aforementioned DFAM contributions is that they require an initial design of an assembly to be consolidated. As such, as shown for the GE fuel nozzle case [23], such work is needed for a better adoption of AM in the industry. Nevertheless, designing a right AM-friendly assembly from scratch could also be part of the road to an AM industrial establishment. A few contributions have also been made in this regard. Yang and Zhao [11] made it clear that conventional design methodologies are not suitable at prompting designers to embrace the design freedom allowed by AM. For instance, generating concepts with conventional design methods (such as axiomatic design) could not lead to products with consolidated parts and thus function integration. In order to somehow fill this gap, the authors proposed a conceptual design framework aimed at aiding design flow management in an AM context. The goal being "to stimulate designers who are novice in AM to think in an AM way". While their proposal may prove helpful at generating innovating concepts, these concepts would still need to be checked against the constraints of the actual AM techniques or machines selected to manufacture them. Similarly, Laverne et al. [10] proposed a conceptual design framework for supporting AM-enabled product innovation. Surveying such DFAM contributions, Rias et al. [24] came up with the conclusion that concepts generated for AM were only partially original with "a maximum of 75% newness". They have then proposed a five-stage creative DFAM method fostering the generation of creative concepts exploiting the unique capabilities of AM; the proposed method is somewhat based on the AM design features database presented in Bin Maidin et al. [9].

All these DFAM contributions are quite similar in that AM techniques/machines' specific capabilities and constraints are not explicitly taken into account. A situation well exemplified by one of the inputs of the PC detection algorithm proposed by Yang et al. [25]: "AM Process General Constraints". As will be shown in the following subsection, these AM characteristics vary from technique to technique. Even for the same technique, two different machines embodying it may have different constraints. For instance with the FFF machine Ultimaker 2, two parts moving relatively to each other or two mating parts made of different materials must be printed separately (and assembled afterwards). However, with a FFF Ultimaker 3 Extended machine—which is capable of multi-material printing and particularly can process sacrificial (water soluble support) material—the two components can be printed together and thus designed differently than if they were to be printed and assembled afterwards. This shows that there is a need to be aware of what techniques or machines are needed to print an assembly, while the latter is being designed.

To sum up this literature review, worth mentioning is the survey of the A-DFAM contributions, led in Yang et al. [26]. They showed that the following issues still to overcome are:

1. The lack of an analysis of what impact AM could have on conceptual design. An issue somehow addressed by the aforementioned conceptual design methods [9–12,24].
2. The lack of an explicit functional analysis (FA) method.
3. Too few decision-making decision support tools for easing PC.
4. The deficiency of functional reasoning approaches to generate AM-enabled features.
5. Too few approaches integrating manufacturing and assembly knowledge into the design stage.

We posit that point 2 cannot really be an issue peculiar to DFAM, even though we acknowledge that as for any design activity FA is paramount. Points 1 and 4 are currently being investigated, but these approaches usually prompt the designer to embrace the design freedom allowed by AM, regardless of what the actual AM techniques or machines are able to do. Therefore, while being efficient at leading to innovative designs, they are highly prone to iterations.

In this paper issues 3 and 5 are addressed. Our contribution to A-DFAM is threefold. First, in contrast to most of the aforementioned contributions, our proposal is dedicated at new product development (even though it may be used for redesign cases as well). Secondly, it encompasses what can be called AM-based architecture minimization, which include PC and assembly-free mechanisms as well. Thirdly, our proposal considers a manufacturing context of an "AM-factory" (i.e., FabLabs, Hubs), where many AM machines embodying different techniques are available. Particularly, the approach aims at selecting the most appropriate machine(s) for easing (possibly) a whole assembly manufacturing before this latter is fully designed.

2.2. *Additive Manufacturing Design-Related Characteristics*

The ultimate goal of DFAM is about harnessing the unique capabilities offered by AM to maximize products' performance while taking into account the constraints related to these processes to ensure a seamless manufacturing. This section is about clarifying these capabilities and constraints, and also it is aimed at providing a comparison between AM process families regarding these characteristics. In addition, this clarification is also intended to specify the levels of a product (e.g., part level), which are likely to be impacted by these characteristics. It will then provide insights into how and when to guide designers, or raise their awareness about specific aspects of AM.

2.2.1. Assembly-Related Additive Manufacturing Characteristics

AM is usually described as having the unique capabilities of shape complexity, hierarchical complexity, material complexity, and functional complexity [27]. All these capabilities are mostly considered at part level, except for the functional complexity, one which may be considered at product level. More specifically, at this level, the unique capabilities of AM can be enumerated as:

- Multi-material manufacturing [28]: the capability to directly manufacture multiple material components either discretely or continuously. Processes fully capable—that is, without any hardware alteration—of this property include DMD [29], LENS [30] for the metals, and FFF [31], 3DP, and PolyJet for the polymeric materials.
- Kinematic pair printing [14]: the capability to directly manufacture assemblies with moving parts. This has been referred to as non-assembly fabrication, in situ fabrication, or assembly-free fabrication. The printing of historical Reuleaux kinematic models [32] is a great illustration of this AM capability. Table 2 shows some kinematic pairs and specific joints, which have been additively manufactured. Critical to this capability is the clearance between the moving parts and access to these clearances for uncured or support material to be removed.
- Around insert building: in some cases it is likely that groups of components are not (or cannot be) manufactured with AM (e.g., engines, batteries, etc.), but are required to be embedded into a part. Some AM processes have this capability to be paused, for a complete part to be laid on the part being manufactured, and to be resumed. This can be viewed as another route for multi-material printing. AM techniques that have been demonstrated to build around inserts include stereolithography [33], ultrasonic consolidation [34], laminated object manufacturing [35], shape deposition manufacturing [36], and PolyJet [37].
- Electronics printing: the capability to deposit electronics components (e.g., conductive inks, sensor, etc.). This ability is somewhat related to the machine capability. In various ways, these capabilities can ease a product architecture. It is the case for the example for the Voxel 8 machine [38].

Table 2. Non-assembly additively manufactured kinematic pairs and joints.

Type	SLA	SLS	FFF	SLM	PolyJet
Revolute	[39–41]	[39,40]	[42,43]	[44–46]	[47,48]
Prismatic	[40]	[40]			
Cylindrical				[45]	
Spherical	[39,40]	[39,40]			
Gear				[44]	[48]
Universal joint	[40]	[40]		[46]	[47]

2.2.2. Part-Related Characteristics

There are number of characteristics that distinguish AM techniques as regards the features and the quality of the output without considering any post-processing step. These are described as follows:

- Material type: as shown in Table 1, materials processed by AM machines include plastics, metals, ceramics, and composites. Whereas some techniques such as SLM or LENS do only process metals, techniques such as PolyJet or SLA are limited to plastics. As such, choosing a type of material is an implicit way of selecting an AM technique.
- Resolution: the features details and minimum wall thickness of a part strongly depend on the machine resolution (XY resolution and layer height or vertical resolution). Resolution varies between techniques, and for the same technique it can also be a matter of hardware.
- Maximum size: any AM technique is limited in buildable size by the machine embodying it, therefore so are the parts (or the number of parts) that can be manufactured at once. This limitation may lead to break down the CAD model to manufacture it in smaller chunks and then reassemble them afterwards [49]. The largest AM machines have built dimensions ranging from 90 × 60 × 30 cm (Merke IV) to 40 × 10 × 6 m (Windsor) and even to (theoretically) infinite dimensions. Knowing which machine will manufacture the part before designing it will definitely restrain the design space, and conversely with a rough idea of the overall dimensions of the part a proper machine may be selected.
- Surface quality: number of factors do affect part's surface finish. Owing to the layer-wise manufacturing method peculiar to AM, stair step effect makes orientation and layer thickness very influential regarding surface finish. On a single part, many different surface finishes may be encountered, varying by a more than nine factors in terms of roughness (Ra). Nevertheless, the technique itself does also influence the surface quality. In techniques processing plastics for instance, SLA machines do provide a better surface quality than FFF ones; in metals, SLM parts have better quality than those manufactured with LENS.

3. Assembly-Oriented Design Framework for AM

When considering AM at product level, the ultimate goals that should be sought are:

- The possibility to consider parts consolidation, in order to minimize the product architecture.
- The consideration of assembly-free mechanisms [39,40], that is, assemblies (of moving parts) that can be printed at once and which after a few minor post-processing tasks (e.g., support removal, surface polishing, etc.) are ready for use.

The aim of this research is to provide a methodology for seizing these opportunities by the conceptual design stage of the product. In other words, the methodology aims at seizing these benefits before a detailed geometry of the product is defined. As shown in the assembly-related AM characteristics section, these goals can be reached for single material mechanisms with some AM machines. Furthermore, multi-material products can be printed by AM machines such as PolyJet machines or the Voxel8 which, during the same print job, can process multiple polymers (including hard, soft, and flexible) and print electronics components as well. However, as the previous analyses have shown, none of the available AM techniques are capable of all the enumerated assembly-related characteristics at the same time. Besides, some components in a product may prove more economically viable if manufactured by conventional processes or simply if outsourced. For these reasons we posit that:

- Owing to the heterogeneous nature of a product, assembly-free AM is not, generally, possible in the current state of the art. The AM of a product may involve many techniques, even in some cases other conventional manufacturing processes as well (hybrid manufacturing [50]), and a few assembly operations.
- The possibility to consider part consolidation cannot be made regardless of the specific characteristics of available AM techniques.

The goal that should, therefore, be sought is to find a product architecture, or an engineering bill of materials (eBOM) in case of redesign, which is minimal (while still functional)—that is, with the least separate components—and which is manufacturable with a single AM technique or with the least AM techniques. Besides, once the product architecture is minimized, a strategy must be retained for the way the parts will be designed to take advantage of the selected AM techniques' characteristics. Our strategy is that, once the part's functional interfaces (i.e., surfaces, lines, etc.) are identified, these are connected both to ensure structural integrity and a proper conveyance of the flows getting through the part, abiding by any design space and manufacturing constraints. Thus, AM-friendly parts' geometries can emerge.

3.1. *Overall Description*

At the core of the proposed assembly-oriented design framework for AM are the aforementioned premises. The framework is dedicated at helping designers mainly through embodiment and detail design stages. It aims for three main goals:

1. Ensuring that the product architecture is kept to its supreme boundary by considering the available AM techniques and machines specific part consolidation capabilities.
2. Determining a manufacturing plan along with any in situ (during manufacturing) or subsequent assembly operations.
3. Providing critical geometric elements building the design space—such as functional interfaces and volume-envelopes—(based on the architecture and the manufacturing plan) to be used for the detail design of the parts.

As shown in Figure 1, these goals are reached through three main steps, which are described in the following sections. Since the framework is adapted for new product development, the way functional analysis is to be conducted is first described for the sake of clarity and consistency.

3.2. *Initial Product Architecture Derivation*

The framework is dedicated to new development of products mainly manufactured and assembled with AM. As such, it is intended to guide the product architect and designer from the overall functionality statement of the product to the detailed design of the product. A first stage is to get an enhanced product architecture from a comprehensive functional analysis (both external and internal). This stage is structured in three steps described as follows.

3.2.1. Functional Analysis

The product environment—that is, everything it is to interact with in order to fulfil its functionality—is analyzed and enumerated (e.g., human, air, raw material, etc.) in terms of external elements (EEs). Such analysis can be summarized using a diagram like the one presented in Figure 2a (for the gripper case study). In this case study, EEs include the user's hand, the object to clamp, and the environment that the product is surrounded by. Two kinds of top level functions are then to be identified: main functions, which are those whose action is related to two EEs, and constraints functions, which act on a single EE (as shown in Figure 2a), these do not actively participate in the overall functionality but are required for the product to be conveniently usable (e.g., maintain gas tightness, stand on a plane platform). The flows (i.e., energy, material, or signal) to be conveyed by the product are enumerated. Using standardized basic functions from the Reconciled Functional Basis [51], each top level function is decomposed to its lowest level and the basic functions derived from it are arranged in an order allowing the fulfillment of the overall function. Finally, the flows are routed through all the standardized functions. The outcome of this step can be summarized with a block diagram (see Figure 2b).

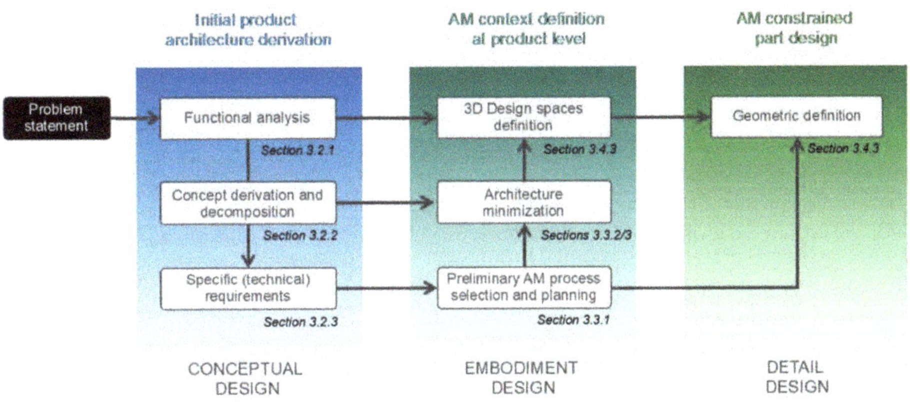

Figure 1. Flowchart of the assembly-oriented design framework for AM (blue boxes relates steps, green ones represents input/output information).

3.2.2. Concept Derivation and Decomposition

A concept is derived for the product with a close attention to the fact that each basic function is fulfilled by one or many components and that many components may fulfil a single function. The derived concept is abstracted with a product architecture that combines three views:

- Part-to-part kinematics relationships view: all the parts are enumerated and the kinematic pair between them are specified.
- Functional flows view: components in the part-to-part relationships graph are first clustered according to the basic functions they fulfil, and then the corresponding flows are routed from the EEs through the components. Some flows may have to be split, while others may be required to merge. Spatial relationships view: using the mereotopological primitives descriptors to describe the physical connections between spatial regions (denoted x and y) such as developed in Demoly et al. [52] as "x is part of y", "x is internal part of y"(IP), "x is tangent of y", and "x overlaps y". Such descriptors provide complimentary information on relationships between non-relatively moving parts specified.

3.2.3. Specific (Technical) Requirements

Specific information is added for the components. Firstly, maintenance/repair requirements and production requirements (i.e., 3D printable, not to be 3D printed, or outsourced component) are specified. Then for each part, candidate to AM, a number of desired characteristics are to be specified. These characteristics (or requirements) are used to generate suitable AM techniques and materials. Figure 3 delineates a data structure for the characteristics of a part to be additively manufactured, along with some values of the characteristics. It is worth highlighting that some of these characteristics (such as minimum wall thickness for instance) are conventionally known after a part's detailed geometry has been made, or these may be constraints set by the available AM machines. However, in our proposal, the designer using the approach is prompted to think about them beforehand, and to somehow commit to them. Nevertheless, not all the characteristics must be specified, one may choose to simply ignore the less crucial ones; the aim of that strategy is to get the proper AM techniques that will manufacture the parts and design the product geometry accordingly (and avoid time consuming design iterations). This choice is consistent with the proactiveness of the proposed approach: the way(s) the product is to be additively manufactured will be known before efforts are put into its detailed design, so that the designer is aware of how the designs can fully leverage the specific capabilities of the selected AM techniques while abiding by their constraints. In addition, to direct detailed design, these specific requirements will also be used to influence decisions made about part consolidation.

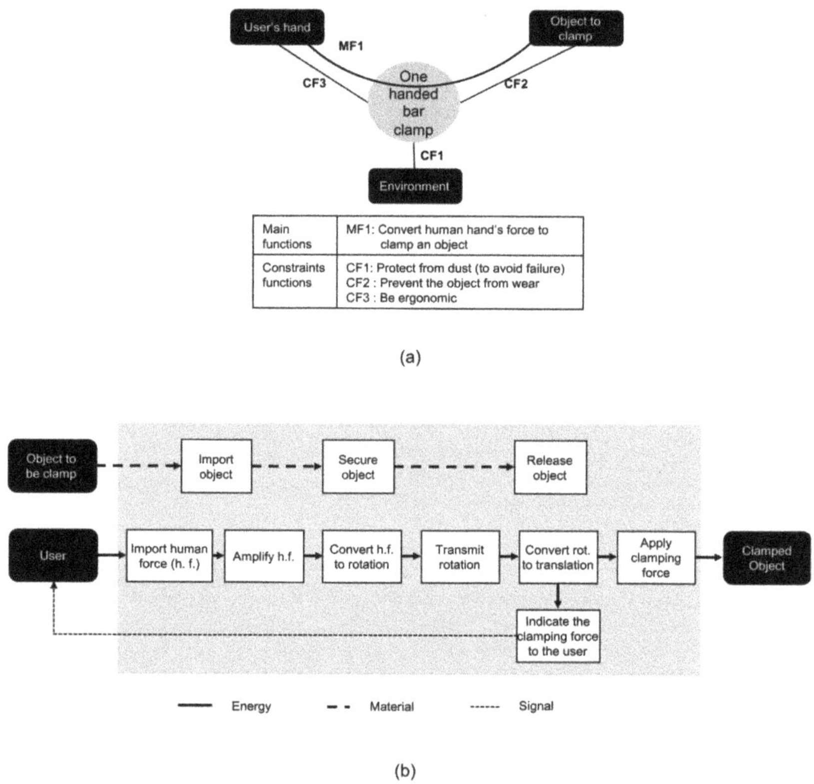

Figure 2. Clamp's functional analysis. (**a**) External functional analysis; (**b**) Functional decomposition.

The pieces of information contained in the derived initial product architecture are:

- For the parts themselves: indication of whether the component will be outsourced or is not to be additively manufactured (e.g., battery, engine, bearings, etc.), indication of whether the component would need to be maintained (in case of a weary component) or often moved, characteristics for AM techniques/machines selection.
- For the part-to-part relationships: kinematic pairs, spatial relationships, and flows.

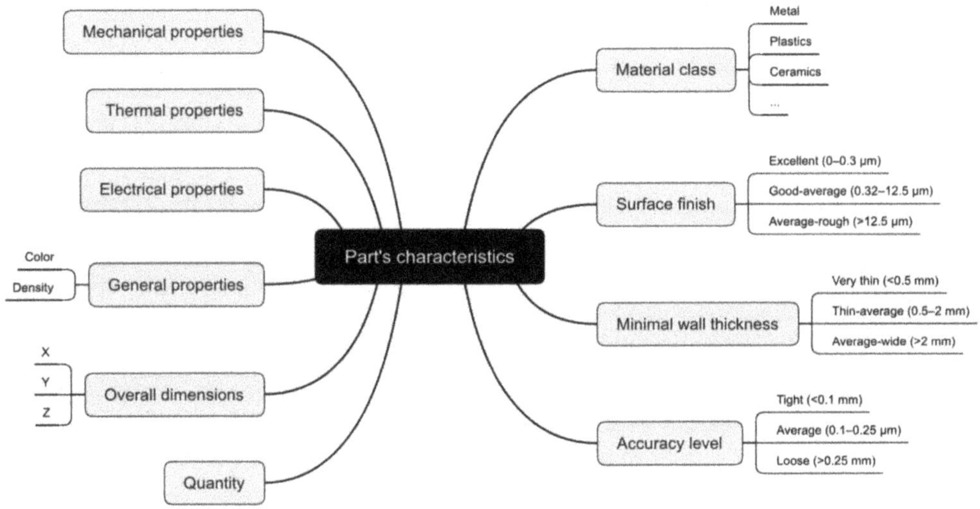

Figure 3. Part characteristics for AM techniques selection.

3.3. AM Context Definition at Product Level

We define a minimal product architecture as an architecture containing the minimum necessary separate parts. This stage of the methodology is intended to make sure that AM capabilities and constraints are taken into account at product level by setting up an AM context. More specifically it aims at:

- Determining an architecture with the least components based on the available AM techniques' capabilities.
- Maintaining functionality regardless of how parts have been combined.
- Determining a manufacturing plan (including assembly operations) so that the detailed design stage is made accordingly, abiding by the selected processes' constraints. Another rationale underpinning this goal is that, those components that are to be manufactured together (be them moving relatively to each other or not) must be designed together (or at least with the same awareness of how manufacturing will occur), since manufacturing direction will be the same for them.

As the product concept is generated regardless of any conventional manufacturing constraint, it is likely that standalone components have to be so for various reasons (i.e., functionality, maintenance, etc.). As such, a product architecture—which is derived within a AM context—may already be a minimal one, that is, without unnecessary standalone components like fasteners. However, to ensure that unnecessary standalone components are eliminated, any architecture should be checked, and the framework can be used for that purpose. It is notable that the method may also be used for redesign cases where the product has been originally designed to be manufactured by conventional processes.

Hence, setting an AM context is based on the initial product architecture (as described in Section 3.2) and it requires knowledge about the specific characteristics of the available AM techniques (as described in Section 2.2). The stage is organized in three steps.

3.3.1. Preliminary Processes Selection and Manufacturing Plan Generation

A number of studies have been carried out about AM process selection for parts [53,54]. As this aspect is not the core of our proposal, we elected to build on previous work addressing this issue. This preliminary technique selection is achieved through two substeps:

1. Materials and techniques selection for each part individually based on the specific requirements stated for each of them. This is where research work from Ghazy [53] on a decision support system has been harnessed to our contribution.

2. Techniques selection for the whole assembly. Each of the part to be additively manufactured is likely to be manufacturable by more than one AM material-technique combinations. This substep is, basically, intended to determine the combinations that will suit most of the parts, as a way to streamline the manufacturing of the whole assembly.

The procedure for techniques and materials selection at part level is made through two main steps as depicted in Figure 4. Requirements on the part that has an influence on the AM technique are first used to generate a set of feasible techniques (among the available AM techniques); these requirements are used to find techniques satisfying them individually, then the so-found sets of techniques are crossed to find techniques meeting all the requirements. In case there is no available techniques meeting all the requirements, those which are conflicting are highlighted and either they are edited or the part is deemed not manufacturable by the available AM techniques. Each technique has a set of required materials that it can process. Techniques generated from the first step, along with parts' requirements related to materials used to select techniques and materials combinations possible for the part. Similarly to the previous step, when no combination is found, either material requirement is edited or the part is deemed not additively manufacturable.

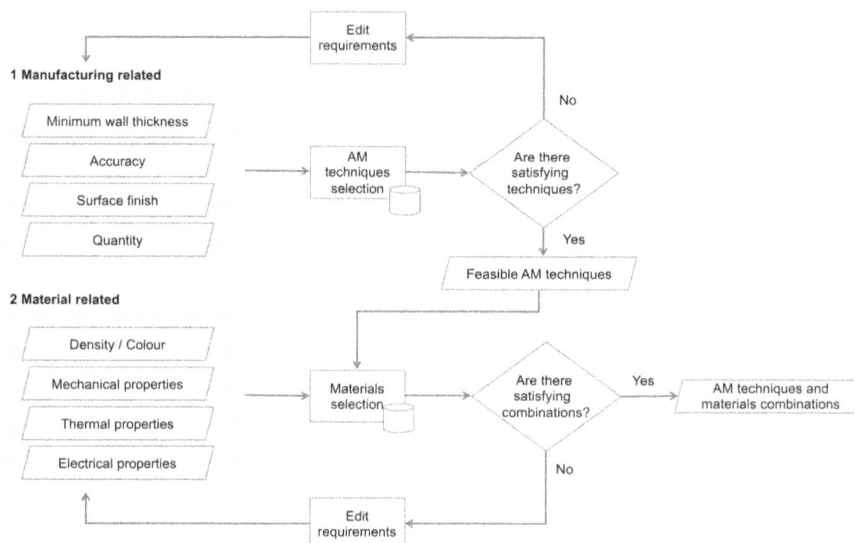

Figure 4. AM preliminary techniques selection (at part level).

For each component C_i, candidate to AM (that excludes outsourced components or those required to be manufactured by a conventional process), a set S_i of available AM techniques that can, possibly, manufacture it, is therefore generated. As an attempt to streamline the product manufacturing, the maximum intersection of the selected AM techniques sets (that is, the largest set where the sets S_i overlap), which is denoted S, is determined. S is then ranked according to the assembly related characteristics described in Section 2.2.1. The best process is then selected and designated as the main AM technique. The same procedure is repeated for the remaining components (those whose S_i elements are not included in S) to find a secondary AM technique. The process may be repeated again in case there are still parts not manufacturable by the selected secondary AM technique. In order to keep the manufacturing scheme simple, a criterion for limiting the number of repetition—that is, the number of secondary AM techniques—is decided. In case where the criterion does not met either the requirements on parts revised, or the whole concept is deemed as not easily additively manufacturable and have to be altered.

At the end of this stage, a new view is then generated for the product architecture: a manufacturing view indicating whether a component is manufacturable by the main AM technique, manufacturable by a secondary AM technique, or not additively manufacturable.

3.3.2. Architecture Minimization between Mating Components

Once the techniques by which components are to be manufactured are determined, decisions are made both about how they can be consolidated to generate a minimal architecture and how the whole product will be manufactured and assembled. In others words, the purpose of this step is to provide an answer to the following question: given two parts non-moving relatively to each other, how can these parts be consolidated in order to simplify the product architecture? As shown in the bottom of Figure 5, three outcomes can result from this analysis: either the parts are simply merged (Consolidated assembly), or one of the part is laid inside the other, while this latter is being manufactured (assembly with part embedding), or they are regularly assembled. The process is as follows:

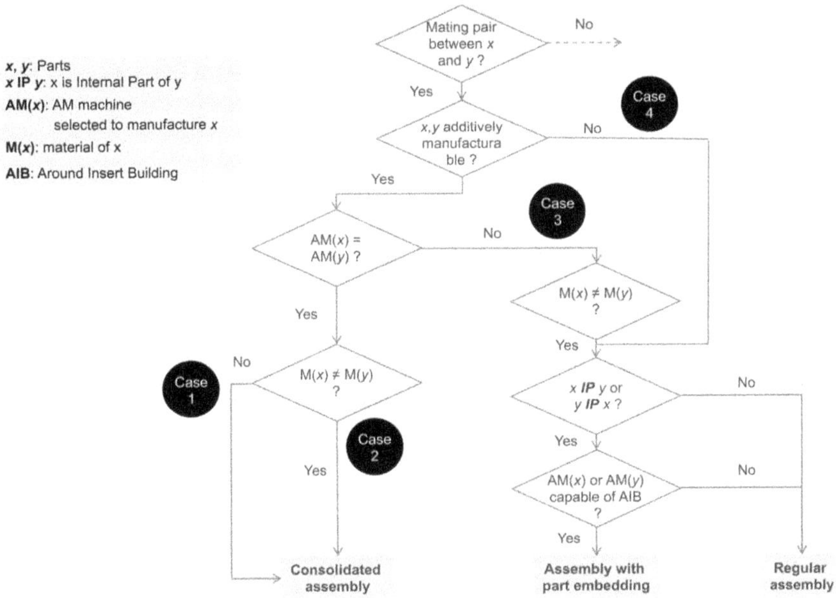

Figure 5. Decision ladder of the product architecture minimization through mating pairs.

1. Components without relative motion are first clustered within sub-assemblies.
2. Consolidation is made by components pairs comparisons in each sub-assembly. The components are subsequently denoted by x and y. Four cases can then occur, as regards manufacturing and materials requirements:
   - Case 1: components are of the same material and are manufactured by the same AM technique (AM(x) = AM(y) and M(x) = M(y)). In such case, they are simply merged as a single component and their respective functional flows are combined.
   - Case 2: components are of different materials processed by the same technique (AM(x) = AM(y) and M(x) $\neq$ M(y)). In that case, they are also merged and flows are combined.
   - Case 3: components are of different materials processed by different techniques (AM(x) $\neq$ AM(y) and M(x) $\neq$ M(y)). In this case, spatial relationships between the components (read from the architecture spatial relationships view) along with the abilities of the two techniques of building around insert are used to make a decision as shown in Figure 5. The two parts are either tangent or overlapping at

some region. In case they are tangent, they are simply manufactured separately and regularly assembled (requiring therefore assembly features for a rigid kinematic pair). If they are overlapping, let us denote 2, the outermost component in the regions where the parts are overlapping. If the technique manufacturing 2 is capable of building around insert, then 1 and 2 are assembled in such a way that 1 is embedded in 2, while this latter is being manufactured, otherwise they must be regularly assembled, that is, manufactured separately and assembled afterwards.

- Case 4: one of the components is outsourced or non-additively manufacturable. We assume that two adjacent outsourced components are considered as a single outsourced component, the case where both compared components are outsourced is thus excluded. In case the parts are overlapping and the outermost component is the outsourced one, then the parts are regularly assembled. Otherwise, the outcome is the same as for Case 3.

3.3.3. Architecture Minimization between Moving Components

Components moving relatively to each other are conventionally manufactured separately and assembled afterwards; however, as shown in Section 2.2, depending on the kinematic pair linking them and the selected AM techniques, the adjacent components can be manufactured together while still be able to move relatively to each other. This step is about identifying all these moving components that can be manufactured together. It will have implications on the final design, in that manufacturing a joint may require a particular orientation and consequently alterations of a conventional joint's design [39]. The step is conducted by comparing pairs of moving components A and B, which are related by joint J: A–J–B. The possible situations are as follows:

- A and B are manufacturable by the same technique. If kinematics pair J is also manufacturable by the technique, then the whole assembly is consolidated during manufacturing. Otherwise—that is, if the kinematics pair is not manufacturable—the components are assembled afterwards and assembly features must be integrated in their designs.
- A and B are not manufacturable by the same technique (or one of the parts is not an AM part). In that case, they are assembled normally.

As depicted in Figure 6, at the end of this stage, the enhanced product architecture derived in the previous stage (Section 3.2) has been used to generate a minimal architecture and a manufacturing plan based on the available AM machines capabilities. The derived minimal architecture shows parts that have been combined, flows that have been merged. The manufacturing plan tells which part is to be manufactured by which AM machine and how the part is to be assembled. The minimized architecture and the manufacturing plan are then the basis of the parts design stage, which is described in the next subsection.

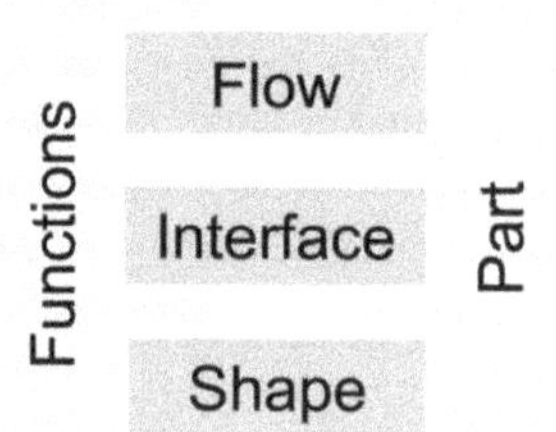

Figure 6. Elements characterizing a part.

3.4. *AM Constrained Part Design*

3.4.1. The Case of Mechanisms to Be Manufactured by the Same Process, in a Single Print Job

It is likely that groups of components, moving relatively to each other or not, are to be manufactured and assembled together (it is actually one of the sought goals of our proposal). This case has been covered in Sossou et al. [55]. The methodology there proposed can be summarized as:

1. Product architecture 3D laying out: the design space of each part is defined and accordingly positioned.
2. Functional interfaces definition: based on the part-to-part kinematics relationships, the design spaces are updated with the proper functional interfaces (FIs).
3. AM contextualization: a step that consists in setting clearances between FIs, choosing a printing configuration, choosing a printing orientation, and allowing access to the clearances.
4. Components' geometries designs.

For more details, the reader is advised to refer to [55].

3.4.2. Foundations of Our Strategy for Harnessing AM Shape Complexity to Part's Performance

Any system, sub-assembly, or component is designed to fulfil a function that represents what the device must do. The device shape, form, or architecture answers the question of how the function is fulfilled. This shows why a device function must be fully understood before time is spent on its embodiment design. Tightly related to a device's function is the notion of flow, which basically is what the device operates on for the function to be fulfilled. Systems, even those made of a single component, rarely fulfilled their function(s) in an isolated manner. In other words, they always fulfil their function by interacting with their surroundings (be it made of a user or others components) by means of one or several flows exchange. The interaction between a component and its surroundings occurs at its interfaces, which can be seen as the flows "gates". It can be summarized that a component is fully defined firstly by its functions (which is what it does), the flows it processes for the functions to be fulfilled, its interfaces through which the flows are exchanged, and its shape holding the interfaces together. This breakdown of component is depicted in Figure 6.

Once flows getting through a component are identified, it is of interest to determine what impact (if any) they have on the shape and interfaces of the component. At their lowest level of specificity, flows are either energy, material, or signal [51,56]. Functions associated with material flows are basically functions that process the materials by somehow changing its state. These are categorized by Ullman [57] in three groups: through flow functions (e.g., rotate, translate, move, lift, channel, etc.), diverge flow functions (disassemble and separate), and converge flow functions (mix and attach). A component fulfilling such function must "clear the way" for the materials and as such its shape is not crucial for the flow conveyance. However, when the processed flow is energy, say in a mechanical form (load for instance), it is obvious that shape, along with interface, play a key role.

Ullman [57] has stated that: "It has been estimated that fewer than 20% of the dimensions on most components in a device are critical to performance. This is because most of the material in a component is there to connect the functional interfaces and therefore is not dimensionally critical. Once the functional interfaces between components have been determined, designing the body of the component is often a sophisticated connect-the-dots problem". The premise governing the methodology proposed to leverage AM shape complexity capability to design effective components is profoundly related to that statement. The extra 80% of components' dimensions that are not critical to the performance may be due to conventional (subtractive) processes' limitations. Furthermore, to better suit our approach, the statement can be extended as " ... a sophisticated connect the dots problem in order to assure a structural integrity and a proper conveyance of the flows related to the component's functionality". A strong statement in Ulman [57], highlighting the importance of interfaces, in agreement with the aforementioned statement and which is worth quoting reads: "Components grow primarily from interfaces".

In a nutshell, a component is bounded by its FIs, these must be connected to ensure structural integrity and to ensure that flows entering or leaving interfaces are conveniently conveyed. The AM shape complexity gives way to ensure that connectedness with the bare minimum matter. This minimalist vision of part design is illustrated in Figure 7 for an aircraft hinge plate; a comparison is made between how this component has been designed for conventional processes and how it could have been designed for AM.

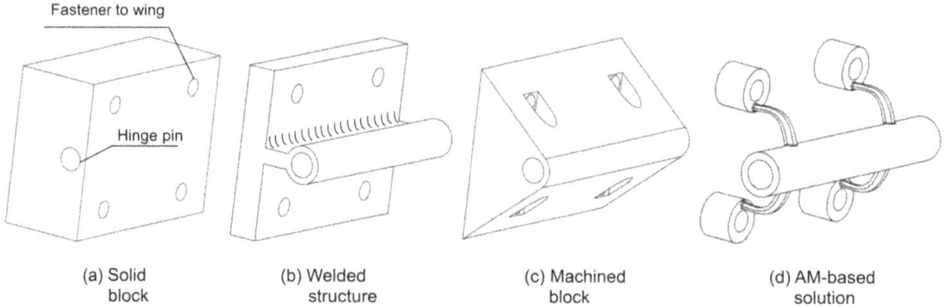

Figure 7. Aircraft hinge plate designs according to various manufacturing processes including AM.

3.4.3. The Proposed Methodology for Part-Oriented DFAM

Consistent with the aforementioned philosophy, a five-steps strategy is proposed to design the geometry in a minimalist way, as outlined in Figure 8. The selected AM techniques constraints are taken into account as well. The strategy is illustrated in Figure 8 with a vise's frame. First, the component's FI shapes are drawn. Second, the FIs' shapes are thickened into functional volumes (FVs); the thicknesses can be governed both by resolution of the considered AM process and tolerances related to the considered FIs. In the third step, paths joining the FVs are defined; this can be done either to ensure the component's connectedness, or to ensure a way for a specific flow (or even a combination of flows), or both. To allow for the subsequent parametric optimization to generate a proper path for the connecting element, this curve can be parameterized. This could be done, for instance, by creating a curve as a spline through three points and by using the four middle point coordinates in a plane as parameters. Fourthly, the connecting elements' shapes along the previously defined paths are designed. To improve the performance of the component, the section can be of different types including solid cross-section, solid cross-section with lattice structure, hollowed cross-section, hollowed cross-section with lattice structure, etc. To avoid sharp corner (and equivalently, stress concentration), corners are smoothed in the fifth step. Each of these steps leads to some geometric parameters. Those are then finally optimized to generate a component that behaves and conveys the flows appropriately. The following scheme has been retained for the parametric optimization:

1. Choice of each parameters bounds.
2. Specification of any constraint on the parameters. These can be geometric constraints such as relationships between parameters, or constraints ensuring a proper behavior such as maximum stress or minimal natural frequency. The functional flows conveyed by the part on one hand, and the chosen material on the other hand can provide indication on what constraints related to an appropriate behavior must be met.
3. Definition of the component's mass as an objective function. Choosing the component's mass as an objective function to minimize is consistent with the endeavor to design components with the bare minimum matter. Other objective functions that are sensitive to the chosen parameters and that are relevant to the sought performance may also be chosen.

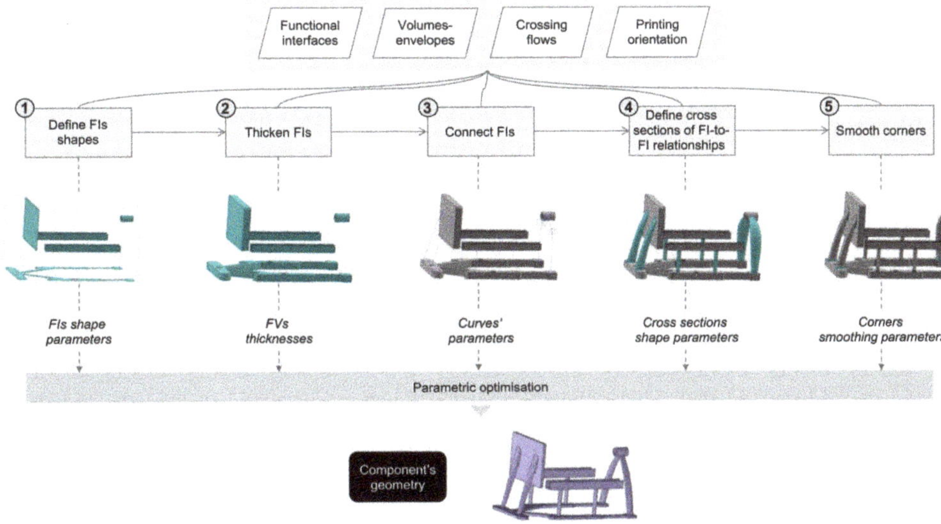

Figure 8. The proposed design methodology for part's design illustrated via a vise's frame based on prior research work [56].

4. Case Study

The proposed framework has been illustrated with a one handed bar clamp, similar to the one that is discussed in Chapters 7 and 9 in Ullmann [57]. The main steps have been shown hereafter.

4.1. Initial Architecture Derivation

After a functional analysis (see Figure 3), as explained in Section 3.2, and generation of a concept, an architecture with 17 components have been retained. The eBOM of the design concept, along with the requirements for each component, can be found in Table 3. Figure 9a shows the part-to-part kinematics relationships view combined with the functional flow view. In Figure 9, edges representing a kinematic relationship are tagged with the name of the kinematic pair involved, the other edges represent the functional flows as indicated in the legend.

Table 3. Parts' requirements.

ID	Part Name	AM	Material Class	Surface Finish	Min Wall Thickness	Accuracy Level	Dimensions (mm)	Mechanical Properties	
								Yield Strength	Durometer
1	Main body	Yes	Plastics	Average-rough	Thin-average	Average	100 × 150 × 30	10–20 MPa	Shore D 60–70
2	Trigger	Yes	Plastics	Average-rough	Thin-average	Average	40 × 90 × 30	10–20 MPa	Shore D 60–70
3	Bar	No							
4	Tail stock	Yes	Plastics	Average-rough	Thin-average	Average	50 × 50 × 30	10–20 MPa	Shore D 60–70
5	Lock	Yes	Plastics	Average-rough	Thin-average	Loose			
6	Pad1	Yes	Plastics	Good-average	Thin-average	Loose			Shore A 40–100
7	Pad2	Yes	Plastics	Good-average	Thin-average	Loose			Shore A 40–100
8	Roll pin	Yes	Plastics, metals	Average-rough	Average				
9	Knurled pin	Yes	Plastics, metals	Average-rough	Average				
10	Release trigger spring1	No							
11	Release trigger spring2	No							
12	Release trigger	Yes	Plastics, metals	Average-rough	Thin-average	Average		10–20 MPa	Shore D 60–70
13	Power spring	No							
14	Jam plates1	Yes	Plastics, metals	Average-rough	Thin-average	Average		10–20 MPa	Shore D 60–70
15	Jam plates2	Yes	Plastics, metals	Average-rough	Thin-average	Average		10–20 MPa	Shore D 60–70
16	Split pins1	No							
17	Split pins2	No							

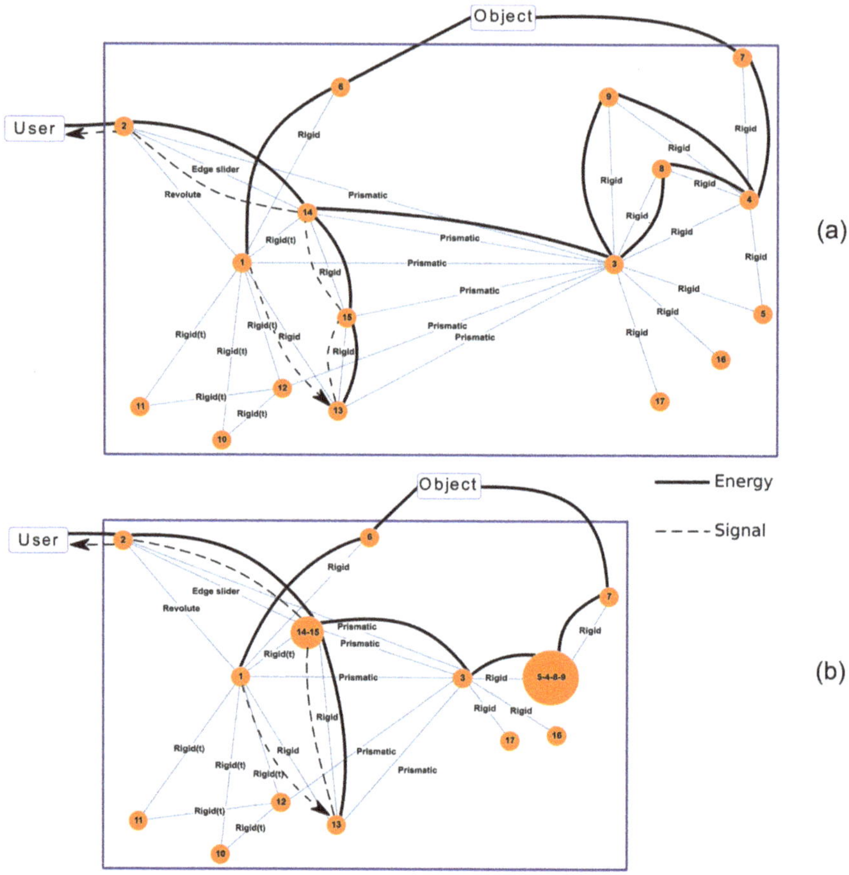

Figure 9. Product architecture showing part-to-part kinematic pairs along with functional flows: (**a**) initial architecture and (**b**) minimized architecture.

4.2. Minimized Architecture and Manufacturing Plan

We used a scenario of an AM factory that has six machines based on the following techniques: FFF, SLA, PolyJet, and SLS. Details of the available machines can be found in Appendix A. All parts candidate to AM were found manufacturable with the available machines. Based on the assembly-related characteristics, and especially the capability to build around inserts, the Ultimaker 3 Extended (U3E) machine has been found to be the best AM machine. The combination U3E-PLA has been selected for the components: 1, 2, 4, 5, 8, 9, 12, 14, and 15 (see Table 3). For components 6 and 7 (pads), a different material—TPU 95A—was selected; it is a semi-flexible and soft (shore A 95) material with high wear and tear resistance. These choices have led to a generated minimized architecture with 13 separate components, as shown in Figure 9b. The manufacturing plan is delineated in Table 4.

Table 4. Manufacturing plan.

Operation	Part or Sub-Assembly	Machine-Material	Comment
1	4-5-8-9 14-15 1-2-12	Ultimaker 3 Extended—PLA	(1, 2, 5) in situ assembly In situ assembly for 1-2 and 1-12
2	6, 7	Ultimaker 3 Extended—TPU-95A	Regular assembly after AM
3	3, 16, 17	(not AM)	Regular assembly to (4-5-8-9)
4	10, 11, 13	(not AM)	Regular assembly to (1-2-12) and 3

4.3. Final Design

The generated minimal architecture has been used to designed the components' geometries. A proprietary tool—the Pegasus CAD Assistant from Demoly et al. [2], was harnessed to semi-automatize the generation of the design spaces and the functional interfaces based on the kinematic relationships as shown in Figure 10a.

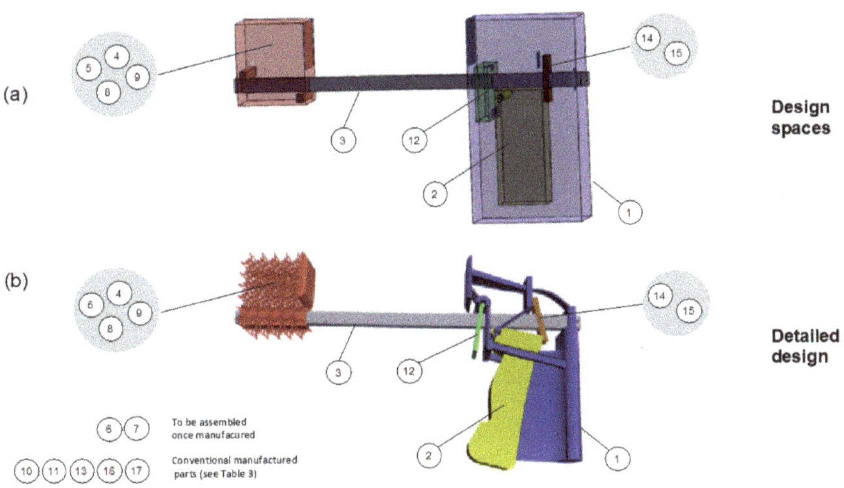

Figure 10. (**a**) Design spaces definition and (**b**) emergence of the detailed geometry of the case study.

Based on these, the parts have been designed by following two approaches including the one proposed in this paper and another one based on lattice structures. As an example, Figure 10b presents a lattice structure developed for the tail stock (i.e., part 4 in Table 3). The design has been made in alignment with the required mechanical behavior of the part as well as the selected technique capabilities. On the other hand, the main body (i.e., part 1 in Table 3) has been defined by using the proposed methodology. The final design results from a flow of integrated design intents, enabling to get an improved solution from a top-down manner.

5. Conclusions and Future Work

In a call to action regarding assembly-level DFAM, Yang et al. [11] stated that "the lack of decision-making support for function integration and part consolidation hinders the search for supreme boundary". In this paper, the highlighted gap has been somewhat filled, in that a support tool for part consolidation—based on the specific characteristics of the AM machines available in a factory and the relationships between the parts—has been proposed.

It is a framework for new product development, with the goal of seizing the assembly-related benefits of AM. These include: PC, in situ assembly of moving components, multi-material printing. It has been shown that seizing these benefits depend on what the available AM machines are actually capable of. A proposal has then been made—based on a set of desired components' characteristics—to select the best AM techniques that allow both for a seamless manufacturing of the product and architecture minimization. Function integration—which has been implicitly covered by part consolidation—could be integrated in the proposed approach by enlarging its scope to conceptual design as suggested in [11]. Considering it has the potential outcome of increasing product performance. Nevertheless, this research has revealed areas of investigation that are worth considering for improving DFAM in general and opportunities for improving our proposal particularly:

- AM materials description. There must be a way to describe materials (be they proprietary or generic) that is consistent with all the existing machines. Some materials may have nearly the same properties, but with quite different names. Some machine manufacturers may call their material with a generic name like ABS, PLA; such names should be made more explicit. Indeed, ABS for an Ultimaker FFF machine may not have the same properties as ABS extruded from a Stratasys FDM (fused deposition modeling) machine.
- Cost as a selection criterion. Criteria such as cost could be included in the preliminary machines selection. Indeed, the approach led to cases where the main selected AM machine and material combination was actually the most expensive one, while there are other available cheaper AM machines that could also seamlessly manufacture the assembly with the prescribed parts' requirements.
- An indicator of viability of AM of assembly. In case many of the components are outsourced, operations such as assembly during AM may dramatically increase the production time. There should therefore be an indicator that quickly checks the architecture and the parts' requirements and tells how valuable it can be to consider the product for AM.

As AM is still in its infancy compared to the conventional manufacturing techniques, it is likely that the full potentials of AM may not be seen in the current generation of designers' artifacts. As such, education is to play a crucial role in allowing the AM paradigm shift completeness [58]. In the proposed approach, machine limitations, especially those related to build space have not been considered in the minimal architecture derivation. However, these can require a product (or even a part) to be digitally cut before AM and reassembled afterwards. A future work could include in the minimum architecture generation of such limitations by considering integration of interlocking features as shown in Song et al. [49].

Author Contributions: Conceptualization, G.S. and F.D.; Methodology, G.S. and F.D.; Writing—original draft preparation, G.S.; Writing—review and editing, F.D.; Supervision, S.G. and G.M. All authors have read and agreed to the published version of the manuscript.

Funding: This research was partially funded by the French 'Investissements d'Avenir' program, project ISITE-BFC (contract ANR-15-IDEX-0003), the EIPHI Graduate School (contract ANR-17-EURE-0002), the S.mart academic network, and the Bourgogne Franche-Comté Region.

Institutional Review Board Statement: Not applicable.

Informed Consent Statement: Not applicable.

Data Availability Statement: Not applicable.

Conflicts of Interest: The authors declare no conflict of interest.

Appendix A

Table A1. Available AM Machines.

Machine	AM Technique	Min Wall Thickness	Accuracy	Surface Finish	Max Dimensions (mm)	Material Type	Processed Materials	Multi-Material	Capability to Print Electronics	Capability to Build around Insert
Ultimaker 3 Extended	FDM	Very thin	Average	Good-average	197 × 215 × 300	Polymers	Nylon, PLA, ABS, CPE, CPE+, PVA, PC, TPU 95A, PP, Breakaway	Yes	No	Yes
Ultimaker 2+	FDM	Very thin	Average	Good-average	223 × 233 × 205	Polymers	PLA, ABS, CPE, CPE+, PC, Nylon, TPU 95A, PP	No	No	Yes
Objet30	PolyJet	Very thin	Tight	Excellent	294 × 192 × 149	Polymers; Wax	VeroWhitePlus™, VeroGray™, VeroBlue™, VeroBlack™, Durus	Yes	No	No
Form 2	SLA	Very thin	Tight	Excellent	145 × 145 × 175	Polymers	CLEAR FLGPCL03, HIGH TEMP FLHTAM01, TOUGH FLTOTL03, DURABLE FLDUCL01, FLEXIBLE FLFLGR02, DENTAL SG FLDGOR01, CASTABLE FLCABL02	No	No	No
EOS P 77	SLS	Very thin	Tight	Good-average	700 × 380 × 580	Polymers	Alumide, PA 1101, PA 1102 black, PA 2200, PA 2201, PA 3200 GF, PrimeCast 101, PrimePart FR (PA 2241 FR),	No	No	No
FORMIGA P 110	SLS	Very thin	Tight	Good-average	200 × 250 × 330	Polymers	PA 2200, PA 2201, PA 3200 GF, PrimeCast 101, PA 2105	No	No	No

References

1. Kuo, T.-C.; Huang, S.H.; Zhang, H.-C. Design for manufacture and design for 'x': Concepts, applications, and perspectives. *Comput. Ind. Eng.* **2001**, *41*, 241–260. [CrossRef]
2. Demoly, F.; Yan, X.-T.; Eynard, B.; Rivest, L.; Gomes, S. An assembly oriented design framework for product structure engineering and assembly sequence planning. *Robot. Comput. Integr. Manuf.* **2011**, *27*, 33–46. [CrossRef]
3. Gao, W.; Zhang, Y.; Ramanujan, D.; Ramani, K.; Chen, Y.; Williams, C.B.; Wang, C.C.L.; Shin, Y.C.; Zhang, S.; Zavattieri, P.D. The status, challenges, and future of additive manufacturing in engineering. *Comput. Aided Des.* **2015**, *69*, 65–89. [CrossRef]
4. Thompson, M.K.; Moroni, G.; Vaneker, T.; Fadel, G.; Campbell, R.I.; Gibson, I.; Bernard, A.; Schulz, J.; Graf, P.; Ahuja, B.; et al. Design for additive manufacturing: Trends, opportunities, considerations, and constraints. *CIRP Ann. Manuf. Technol.* **2016**, *65*, 737–760. [CrossRef]
5. Pradel, P.; Zhu, Z.; Bibb, R.; Moultrie, J. Investigation of design for additive manufacturing in professional design practice. *J. Eng. Des.* **2018**, *29*, 165–200. [CrossRef]
6. Pradel, P.; Zhu, Z.; Bibb, R.; Moultrie, J. A framework for mapping design for additive manufacturing knowledge for industrial and product design. *J. Eng. Des.* **2018**, *29*, 291–326. [CrossRef]
7. Fillingim, K.B.; Nwaeri, R.O.; Paredis, C.J.J.; Rosen, D.; Fu, K. Examining the effect of design for additive manufacturing rule presentation on part redesign quality. *J. Eng. Des.* **2020**, *31*, 427–460. [CrossRef]
8. Becker, R.; Grzesiak, A.; Henning, A. Rethink assembly design. *Assem. Autom.* **2005**, *25*, 262–266. [CrossRef]
9. Bin Maidin, S.; Campbell, I.; Pei, E. Development of a design feature database to support design for additive manufacturing. *Assem. Autom.* **2012**, *32*, 235–244. [CrossRef]
10. Laverne, F.; Segonds, F.; Anwer, N.; Le Coq, M. Assembly based methods to support product innovation in design for additive manufacturing: An exploratory case study. *J. Mech. Des.* **2015**, *137*, 121701. [CrossRef]
11. Yang, S.; Zhao, Y.F. Conceptual design for assembly in the context of additive manufacturing. In Proceedings of the 2016 Annual International Solid Freeform Fabrication Symposium, Austin, TX, USA, 8–10 August 2016.
12. Mokhtarian, H.; Coatanéa, E.; Paris, H.; Mbow, M.M.; Pourroy, F.; Marin, P.R.; Vihinen, J.; Ellman, A. A conceptual design and modeling framework for integrated additive manufacturing. *J. Mech. Des.* **2018**, *140*, 081101. [CrossRef]
13. Orquéra, M.; Campocasso, S.; Millet, D. Some principles to optimise an additively manufactured multi-component product. *J. Eng. Des.* **2019**, *31*, 219–240. [CrossRef]
14. Cuellar, J.S.; Smit, G.; Plettenburg, D.; Zadpoor, A. Additive manufacturing of non-assembly mechanisms. *Addit. Manuf.* **2018**, *21*, 150–158. [CrossRef]
15. Lebaal, N.; Zhang, Y.; Demoly, F.; Roth, S.; Gomes, S.; Bernard, A. Optimised lattice structure configuration for additive manufacturing. *CIRP Ann. Manuf. Technol.* **2019**, *6*, 117–120. [CrossRef]
16. Al Khalil, M.; Lebaal, N.; Demoly, F.; Roth, S. A design and optimization framework of variable-density lattice structures for additive manufacturing. *Mech. Adv. Mater. Struct.* **2021**, 1–15. [CrossRef]
17. Hague, R. *Unlocking the Design Potential of Rapid Manufacturing*; John Wiley & Sons, Ltd.: Hoboken, NJ, USA, 2006.
18. Schmelzle, J.; Kline, E.V.; Dickman, C.J.; Reutzel, E.W.; Jones, G.; Simpson, T.W. (Re)Designing for part consolidation: Understanding the challenges of metal additive manufacturing. *J. Mech. Des.* **2015**, *137*, 111404. [CrossRef]
19. Yang, S.; Tang, Y.; Zhao, Y.F. A new part consolidation method to embrace the design freedom of additive manufacturing. *J. Manuf. Process.* **2015**, *20 Pt 3*, 444–449. [CrossRef]
20. Rodrigue, H.; Rivette, M. An assembly-level design for additive manufacturing methodology. In Proceedings of the IDMME-Virtual Concept, Bordeaux, France, 20–22 October 2010.
21. Boothroyd, G.; Dewhurst, P. *Product Design for Assembly*; Boothroyd Dewhurst, Inc.: Wakefield, RI, USA, 1990.
22. Yang, S.; Zhao, Y.F. Additive manufacturing-enabled part count reduction: A lifecycle perspective. *J. Mech. Des.* **2018**, *140*, 031702. [CrossRef]
23. Kellner, T. *The FAA Cleared the First 3D Printed Part to Fly in a Commercial Jet Engine from GE*; General Electric: Boston, MA, USA, 2015.
24. Rias, A.-L.; Bouchard, C.; Segonds, F.; Abed, S. Design for additive manufacturing: A creative approach. In Proceedings of the 14th International Design Conference, Dubrovnik, Croatia, 16–19 May 2016.
25. Yang, S.; Santoro, F.; Zhao, Y.F. Towards a numerical approach of finding candidates for additive manufacturing-enabled part consolidation. *J. Mech. Des.* **2018**, *140*, 041701. [CrossRef]
26. Yang, S.; Tang, Y.; Zhao, Y.F. Assembly-level design for additive manufacturing: Issues and benchmark. In Proceedings of the ASME 2016 International Design Engineering Technical Conferences and Computers and Information in Engineering Conference, Charlotte, NC, USA, 21–24 August 2016.
27. Gibson, I.; Rosen, D.W.; Stucker, B. *Additive Manufacturing Technologies: Rapid Prototyping to Direct Digital Manufacturing*, 1st ed.; Springer Publishing Company Incorporated: Berlin/Heidelberg, Germany, 2009.
28. Vaezi, M.; Chianrabutra, S.; Mellor, B.; Yang, S. Multiple material additive manufacturing—Part 1: A review. *Virtual Phys. Prototyp.* **2013**, *8*, 19–50. [CrossRef]
29. Wilson, J.M.; Shin, Y.C. Microstructure and wear properties of laser-deposited functionally graded inconel 690 reinforced with tic. *Surf. Coat. Technol.* **2012**, *207*, 517–522. [CrossRef]

30. Bandyopadhyay, A.; Krishna, B.V.; Xue, W.; Bose, S. Application of laser engineered net shaping (lens) to manufacture porous and functionally graded structures for load bearing implants. *J. Mater. Sci. Mater. Med.* **2009**, *20*, 29–34. [CrossRef] [PubMed]
31. Garland, A.; Fadel, G. Design and manufacturing functionally gradient material objects with an off the shelf three-dimensional printer: Challenges and solutions. *J. Mech. Des.* **2015**, *137*, 11140. [CrossRef]
32. Lipson, H.; Moon, F.C.; Hai, J.; Paventi, C. 3-d printing the history of mechanisms. *J. Mech. Des.* **2004**, *127*, 1029–1033. [CrossRef]
33. Kataria, A.; Rosen, D.W. Building around inserts: Methods for fabricating complex devices in stereolithography. *Rapid Prototyp. J.* **2001**, *7*, 253–262. [CrossRef]
34. White, D. Ultrasonic consolidation of aluminum tooling. *Adv. Mater. Processes* **2003**, *161*, 64–65.
35. Liao, Y.S.; Li, H.C.; Chen, M.T. The study of rapid prototyping process with embedded functional inserts. *J. Mater. Process. Technol.* **2007**, *192–193*, 68–74. [CrossRef]
36. Merz, R.; Prinz, F.; Ramaswami, K.; Terk, M.; Weiss, L. Shape deposition manufacturing. In Proceedings of the International Solid Freeform Fabrication Symposium, Austin, TX, USA, 8–10 August 1994.
37. Meisel, N.A.; Elliott, A.M.; Williams, C.B. A procedure for creating actuated joints via embedding shape memory alloys in polyjet 3d printing. *J. Intell. Mater. Syst. Struct.* **2014**, *26*, 1498–1512. [CrossRef]
38. Molitch-Hou, M. The World's First 3D Electronics Printer. Available online: https://3dprintingindustry.com/news/voxel8-unleashes-electronics-3d-printer-ces-world-39060/ (accessed on 15 January 2022).
39. Calì, J.; Calian, D.A.; Amati, C.; Kleinberger, R.; Steed, A.; Kautz, J.; Weyrich, T. 3D-printing of non-assembly, articulated models. *ACM Trans. Graph.* **2012**, *31*, 1–8. [CrossRef]
40. Mavroidis, C.; Delaurentis, K.J.; Won, J.; Alam, M. Fabrication of non-assembly mechanisms and robotic systems using rapid prototyping. *J. Mech. Des.* **2000**, *123*, 516–524. [CrossRef]
41. Song, X.; Chen, Y. Joint design for 3-d printing non-assembly mechanisms. In Proceedings of the ASME International Design Engineering Technical Conferences and Computers and Information in Engineering Conference, Chicago, IL, USA, 12–15 August 2012; pp. 619–631.
42. Stöckli, F.; Modica, F.; Shea, K. Designing passive dynamic walking robots for additive manufacture. *Rapid Prototyp. J.* **2016**, *22*, 842–847. [CrossRef]
43. Wei, X.; Tian, Y.; Joneja, A. A study on revolute joints in 3d-printed non-assembly mechanisms. *Rapid Prototyp. J.* **2016**, *22*, 901–933. [CrossRef]
44. Calignano, F.; Manfredi, D.; Ambrosio, E.P.; Biamino, S.; Pavese, M.; Fino, P. Direct fabrication of joints based on direct metal laser sintering in aluminum and titanium alloys. *Procedia CIRP* **2014**, *21*, 129–132. [CrossRef]
45. Song, C.H.; Yang, Y.Q.; Xiao, Z.F.; Wang, D.; Liu, Y.; Liu, R.C. Design and direct manufacture of non-assembly abacus by selective laser melting. In *International Symposium on Optoelectronic Technology and Application 2014: Laser Materials Processing and Micro/Nano Technologies*; Kaiser, N., Li, B., Ji, Y., Verhaeghe, G., Zhong, M., Lu, B., Wang, H., Leu, M.C., Yan, Y., Xie, H., et al., Eds.; SPIE—International Society for Optical Engineering: Bellingham, WA, USA, 2014.
46. Su, X.; Yang, Y.; Wang, D.; Chen, Y. 2013. Digital assembly and direct fabrication of mechanism based on selective laser melting. *Rapid Prototyp. J.* **2013**, *19*, 166–172. [CrossRef]
47. Chen, Y.; Zhezheng, C. Joint analysis in rapid fabrication of non-assembly mechanisms. *Rapid Prototyp. J.* **2011**, *17*, 408–417. [CrossRef]
48. Wei, Y.; Chen, Y.H.; Yang, Y.; Li, Y.T. Novel design and 3-D printing of nonassembly controllable pneumatic robots. *IEEE/ASME Trans. Mechatron.* **2016**, *21*, 649–659. [CrossRef]
49. Song, P.; Fu, Z.; Liu, L.; Fu, C.-W. Printing 3D objects with interlocking parts. *Comput. Aided Geom. Des.* **2015**, *35–36*, 137–148. [CrossRef]
50. Kerbrat, O.; Mognol, P.; Hascoët, J.Y. A new DFM approach to combine machining and additive manufacturing. *Comput. Ind.* **2011**, *62*, 684–692. [CrossRef]
51. Hirtz, J.; Stone, R.B.; McAdams, D.A.; Szykman, S.; Wood, K.L. A functional basis for engineering design: Reconciling and evolving previous efforts. *Res. Eng. Des.* **2022**, *13*, 65–82. [CrossRef]
52. Demoly, F.; Matsokis, A.; Kiritsis, D. A mereotopological product relationship description approach for assembly oriented design. *Robot. Comput. Integr. Manuf.* **2012**, *28*, 681–693. [CrossRef]
53. Ghazy, M.M.S.A. Development of an Additive Manufacturing Decision Support System (AMDSS). Ph.D. Thesis, University of Newcastle upon Tyne, Newcastle upon Tyne, UK, 2012.
54. Wang, Y.; Blache, R.; Xu, X. Selection of additive manufacturing processes. *Rapid Prototyp. J.* **2017**, *23*, 434–447. [CrossRef]
55. Sossou, G.; Demoly, F.; Montavon, G.; Gomes, S. An additive manufacturing oriented design approach to mechanical assemblies. *J. Comput. Des. Eng.* **2018**, *5*, 3–18. [CrossRef]
56. Pahl, G.; Beitz, W.; Feldhusen, J.; Grote, K.H. *Engineering Design: A Systematic Approach*, 3rd ed.; Springer London Ltd.: London, UK, 2007.
57. Ullman, D. *The Mechanical Design Process*, 4th ed.; McGraw-Hill Education: New York, NY, USA, 2009.
58. Williams, C.; Seepersad, C.C. Design for Additive Manufacturing Curriculum: A Problem- and Project-Based Approach. In Proceedings of the Solid Freeform Fabrication Symposium, Austin, TX, USA, 6–8 August 2012.

designs

MDPI

Article

Caffeine–Acrylic Resin DLP-Manufactured Composite as a Modern Biomaterial

Dorota Tomczak [1], Radosław Wichniarek [2,*] and Wiesław Kuczko [2]

[1] Faculty of Chemical Technology, Poznan University of Technology, Berdychowo 4, 60-965 Poznan, Poland
[2] Faculty of Mechanical Engineering, Poznan University of Technology, Piotrowo 3, 60-138 Poznan, Poland
* Correspondence: radoslaw.wichniarek@put.poznan.pl

Abstract: Materials based on photocurable resins and pharmaceutically active agents (APIs) are gaining interest as a composite drug delivery system. In this study, a composite of caffeine with acrylic resin was obtained using an additive manufacturing method of digital light processing (DLP) as a potential material for transdermal drug delivery. The mechanical properties of the composites and the ability to release caffeine from the resin volume in an aqueous environment were investigated. The amount of caffeine in the resulting samples before and after release was evaluated using a gravimetric method. The global thresholding method was also evaluated for its applicability in examining caffeine release from the composite. It was shown that as the caffeine content increased, the strength properties worsened and the ability to release the drug from the composite increased, which was caused by negligible interfacial interactions between the hydrophilic filler and the hydrophobic matrix. The global thresholding method resulted in similar caffeine release rate values compared to the gravimetric method but only for samples in which the caffeine was mainly located near the sample surface. The distribution of caffeine throughout the sample volume made it impossible to assess the caffeine content of the sample using global thresholding.

Keywords: DLP; caffeine; API; acrylic resin; composite; drug release; hydrophilic; hydrophobic; thresholding

Citation: Tomczak, D.; Wichniarek, R.; Kuczko, W. Caffeine–Acrylic Resin DLP-Manufactured Composite as a Modern Biomaterial. *Designs* **2023**, *7*, 49. https://doi.org/10.3390/designs7020049

Academic Editor: Obeidi Muhannad

Received: 27 February 2023
Revised: 20 March 2023
Accepted: 24 March 2023
Published: 26 March 2023

1. Introduction

Additive manufacturing (AM) machines have been available on the market for many years. AM has already found their application in many industries, and new ideas for implementation and improvement are constantly emerging. The literature has numerous examples of how additive manufacturing is used in the medical field, such as the creation of implants, orthoses, prostheses, and preoperative and intraoperative tools. One very promising and relatively new approach to additive manufacturing is the concept of drug printing. Such medicaments, customized individually for each patient, would be the reverse of the current situation, where patients are fitted to the drug manufacturer dose. El Aita I. et al. [1] indicate that the therapy can be optimized and side effects can be decreased by tailoring the dose to the patient's age, weight, and medical history. According to Herrada-Manchón H. et al. [2], oral dosages produced by AM can come in enticing and delectable forms that are simple to handle and consume. This may enhance medication compliance and help young children cope emotionally with the sickness.

Additive manufacturing applies to a whole range of different manufacturing methods which share the fact that the products made with them are created layer-by-layer. AM is carried out directly on the basis of data on the three-dimensional geometry of the product and without the use of dedicated technological equipment. The most popular methods of AM in pharmaceutical applications include: selective laser sintering (SLS) [3], fused deposition modeling (FDM) [4], three-dimensional printing (3DP) [5], and stereolithography (SLA) [6].

The oldest commercially available additive manufacturing method is SLA. The method is based on the solidification of a resin polymer in process called photopolymerization. SLA uses a point light source which is a laser with a wavelength adapted to the needs of curing the selected resin [7]. A variation of SLA is digital light projection (DLP) which is mainly distinguished by another type of light source, which is the projector [8]. Both methods provide great versatility in pharmaceutical applications. Active pharmaceutical ingredients (APIs) can be mixed in the liquid form of the resin as well as coated with finished SLA/DLP products. Compared to other methods of additive manufacturing, the SLA/DLP process is characterized by high dimensions and shape accuracy of digital geometry. Both methods are suitable for working with thermally stable as well as thermally labile drugs [9].

One of the limitations and at the same time challenges in the use of resin additive manufacturing methods include the problem of toxicity and compatibility of resins with the human body. The photocurable photopolymer should not be harmful to humans but also should not be reactive to APIs [10,11]. What is more, multi-component formulation production is highly constrained, particularly at greater drug loadings [10].

Wang J. et al. [9] have studied the suitability of manufacturing drug loaded tablets by SLA. They claimed that it is possible to use a concentration up to 5.9% (w/w) in the case of mixtures of photocurable monomers such as polyethylene glycol diacrylate (PEGDA) with paracetamol and 4-aminosalicylic acid. They were able to create tablets that were homogeneous in size and displayed the same hue as the original photopolymer solution. Robles-Martinez P. et al. [12] proved that it is possible to use resin AM to produce a tablet containing as many as six different APIs (paracetamol, aspirin, naproxen, prednisolone, chloramphenicol, caffeine). The authors stated that all formulations were suitable for AM, but more research is required to identify the best printer settings for each formulation in order to achieve precise dimensions, and therefore dosing, as well as to create certain drug release patterns as necessary.

Despite the wide range of work performed in the development of biomedical products using additive manufacturing methods such as SLA or DLP, the obtaining of pharmaceutical printed products is still limited mainly due to the incompatibility of the introduced additives to photocurable resins. The active pharmaceutical ingredients should be compatible enough with the polymer matrix material to make it possible to obtain polymer-drug composites, ensure their functional properties, and at the same time allow appropriate drug delivery to the patient. The use of transdermal microneedle systems is one of the methods of drug delivery. The frequently used PEGDA material was used to produce a microneedle system by DLP process and was tested in vitro and ex vivo using human skin showing release characteristics when exposed to external inputs such as temperature and pH [13]. Furthermore, a study by Lim S. H. et al. [14] shows a microneedle patch fabricated from PEGDA and vinyl pyrrolidone with an optimal quantitative ratio of constituent monomers loaded with acetyl-hexapeptide 3 for application to facial skin as an anti-ageing product. The selection of the matrix and manufacturing parameters was dictated by the influence of the materials on the active substance, mechanical properties of the product, water absorption and swelling associated with drug release, and the effect on resin cross-linking. PEGDA, due to its hydrophilic character, allowed the loading of hydrophilic drugs at concentrations greater than 1% [9], which is much more difficult for hydrophobic matrices.

Approximately 60% of developed pharmaceuticals are hydrophilic substances [15], which is a barrier to the manufacture of composite systems with a drug based on hydrophobic materials due to the lack of solubility of the drug in the matrix or negligible interfacial interactions, making it difficult to manufacture the final products without their properties deteriorating. Several studies on the preparation of composites in which the polar character of the constituent substances was important are known. Himawan A. et al. [16] developed a hydrogel film with theophylline based on poly(vinyl acetate), polyvinylpyrrolidone, and citric acid. They showed that the hydrophilic character of the drug limited its release from the film because of its high affinity for the matrix components. A solid lipid nanoparticle formulation of a hydrophilic neuroprotective analogue of cyclic guanosine monophosphate, a hydrophobic polyester-based composite film, obtained by Li h. et al. [17] enabled the system to reach the retina, potentially increasing the efficiency of drug delivery. In special cases, the simultaneous delivery of a hydrophilic and a hydrophobic drug is required to optimize treatment. The preparation of this type of system was described by Arpita R. et al. [18], who obtained a combination of hydrophobic ibuprofen and hydrophilic tetracycline hydrochloride in a hydrogel based on β-cyclodextrin, poly(hydroxypropyl methacrylate), poly(acrylic acid), and PEGDA, which has a matrix of dual polar character due to the appropriate choice of monomers. The resulting composite was tested in vitro and in vivo, confirming its effectiveness in releasing hydrophilic and hydrophobic active ingredients from the matrix.

The advantage of composite systems with different polarity of the API filler and polymeric matrix is the significant reduction of the reactivity between the drug filler and the matrix and the enhanced release of the drug from the system due to the reduction of interfacial interactions. The aim of this work was to obtain advanced composites with a caffeine filler known for its pain-relieving properties, helping to treat obesity or apnea or aiding in the treatment of neurological diseases like Alzheimer's [19]. Caffeine represented hydrophilic drugs loaded in a hydrophobic photopolymer matrix which manufactured by the use of additive manufacturing DLP method constitute a potential material for transdermal systems, i.e., devices for transdermal drug delivery. The hydrophilic filler content in the hydrophobic matrix was expected to reach values comparable to those of hydrophilic matrix-based systems. The use of constituents of different polarity was intended to improve drug release from the system, which is a problem for fully hydrophilic systems.

In order to evaluate the properties relevant to drug delivery through the skin, the strength properties of the samples were investigated along with the drug release rate in an aqueous environment that simulates contact with the human body. Evaluation of the global thresholding method as a method to determine the degree of caffeine release from the crosslinked acrylic resin structure is an innovative aspect of this work. The obtained samples are the first documented example of printed products made of caffeine and photocurable hydrophobic acrylic resin contributing modern knowledge of bio-composite-systems with a hydrophilic filler in the form of API and a hydrophobic matrix made of one of the most commonly used photopolymers in DLP.

2. Materials and Methods

2.1. Materials

Aqua Clear Resin acrylic resin (Phrozen Technology, Hsinchu, Taiwan) was used to make the samples. The resin consists of listed components: acrylate oligomer, 4-(1-oxo-2-propenyl)-morpholine, bis(1,2,2,6,6-pentamethyl-4-piperidyl) sebacate, and diphenyl(2,4,6-trimethyl benzoyl) phosphine oxide. The properties of the Aqua Clear Resin are shown in Table 1.

Table 1. Aqua Clear Resin properties [20].

Parameter	Value
Density	1.11 g/cm^3
Melting point	Below 25 °C—liquid state at operating temperature
Solubility in alcohol	+

Furthermore, caffeine with the sum formula $C_8H_{10}N_4O_2$ and a purity grade of 99% (Pol-Aura, Olsztyn, Poland) was used to make the composite samples. The properties of caffeine relevant to formulation with acrylic resin and its application in DLP manufacturing are given in Table 2. The solubility of caffeine in alcohol was determined by the authors in preliminary studies.

Table 2. Caffeine properties.

Parameter	Value
Density	1.23 g/cm^3 [21]
Melting point	234–236.5 °C [21]
Maximum solubility in alcohol	0.02 g/mL in 75 °C

2.2. Pretreatment Processes

The caffeine was mechanically grinded in a mortar for 5 min. Then, 2.5 g of caffeine was added to 10 mL of alcohol and annealed at 50 °C for 2 h to better disperse the substance in the resin. Due to the annealing temperature and the considerable exceeding of the maximum amount of caffeine soluble in alcohol, a saturated solution of alcohol with caffeine with precipitate was obtained. The mixture prepared in this way was added to 50 g of resin and stirred in an MR Hei-Tec Heidilph magnetic stirrer for 10 min at a rotational speed of 1000 rpm obtaining a 4% (w/w) solution of caffeine in resin and alcohol.

2.3. Composites DLP Manufacturing

Manufacturing of the samples was performed using a 2 × 10 × 100 mm geometrical model. One sample series resulted in 3 samples, which were positioned side by side at a distance of 3 mm adjacent to the table with the largest plane. Each sample had technological chamfers to make it easier to remove the samples from the table and to recognize the position during manufacturing, which made it possible to later compare samples from different series. Nine resin samples designated from 0.1 to 0.9 were made as part of a reference series designated as (0). After that, 4 series of caffeine-resin composite samples designated as (1), (2), (3) and (4) were made, resulting in 12 samples designated from 1 to 12. The manufacturing parameters are listed in Table 3. The samples were cleaned in alcohol in a UW-01 Creality machine (Shenzen Creality 3D Technologyu Co., Ltd., Shenzen, China) in quick mode for 4 min and then dried with coupled air. The samples prepared in this way were further cured in XYZ UV Curing Chamber (New Kinpo Group, New Taipei City, Taiwan) for 10 min.

Table 3. DLP manufacturing parameters.

Parameter	Value
Layer height	0.05 mm
Bottom exposure time	35 s
Transition layer count (TL)	6
Bottom layer count	6
Exposure time	10.5 s
Transition type	linear
Rest time before lift	0 s
Rest time after lift	0 s
Rest time after retract	4 s
Bottom lift distance	6 mm
Lifting distance	6 mm
Bottom lift speed	150 mm/min
Bottom retract speed	150 mm/min
Bottom layer compensation	a = 0 mm, b = −0.04 mm

2.4. Mechanical Testing

The obtained samples were tested in static tensile test on Zwick/Roell Z020 machine with the load capacity of 20 kN. During the tensile test, the tensile rate was 1 mm/min in accordance with ISO 527. The mechanical parameters of tensile strength (σ_m), relative elongation (A) and Young's modulus (E) were determined.

2.5. *Caffeine Release Test*

2.5.1. Gravimetric Method

To evaluate the caffeine release from the composite system, the prepared samples were weighed (m_0) and dried in a drying apparatus at 50 °C for 24 h. Each sample was then weighed again (m_1) and placed in a beaker with 100 mL of distilled water and annealed at 37 °C for 12 h, simulating release conditions similar to contact with the human body. After caffeine release, the samples were weighed (m_2) and again annealed at 50 °C for 24 h to remove water. After the final drying treatment, the samples were weighed (m_3). Evaluation of the reduction in caffeine mass during the release process was performed by calculating the caffeine release rate (CR) from the following Formula (1):

$$CR = \frac{(m1 - m3)}{m1} \times 100\ \% \tag{1}$$

2.5.2. Global Threshold Method

Before the release process, microscopic images of the samples were taken with an Optika SZO-5 stereo microscope (OPTIKA S.r.l., Ponteranica, Italy) using a UCMOS05100KPA ToupTek camera (ToupTek Photonics, Hangzhou, China). Surface images were used to determine the approximate caffeine concentration of the sample by global image thresholding. Images of the sample structure were greyscale binarized with a fixed threshold of 200 and 50 for the caffeine located closest to the sample surface and inside the structure, respectively. The operations performed were due to the presence of black color for particles located on the surface of the sample and grey tones for particles located under the resin layer (in the volume of the material) for the original images. The result of the thresholding was histograms with the number of pixels, the counting of which made it possible to determine the percentage of caffeine in the sample. Sample images before and after thresholding are shown in Figure 1.

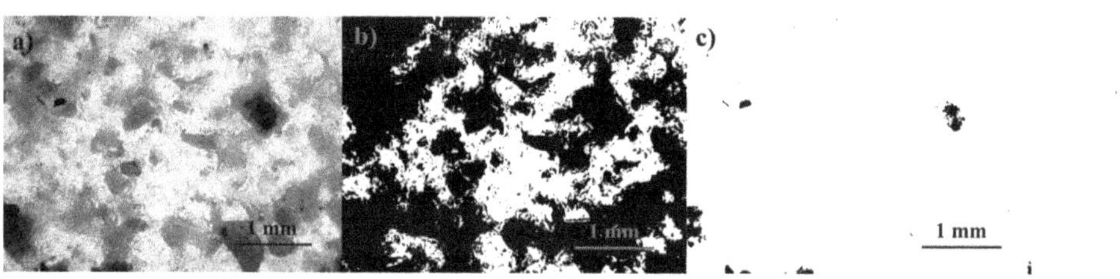

Figure 1. Images of 2nd sample (**a**) raw image from stereo microscope observations, (**b**) image after binarization with threshold of 50, and (**c**) image after binarization with threshold of 200.

The caffeine content of the samples after release was determined based on the initial caffeine weight determined by thresholding (m_K), and the difference in weight of the samples was determined by the weight method, as included in Equation (2) below:

$$CTCR = \frac{(m1 - m3)}{mK} \times 100\ \% \tag{2}$$

where CTCR is the coefficient of caffeine release, which is based on initial caffeine mass obtained from global threshold.

3. Results and Discussion

3.1. *Evaluation of Manufacturing Process and Samples*

Caffeine-resin composite samples are shown in Figure 2. Immediately after manufacturing, all samples were flexible and easily bent in the hands, which was not the case with pure resin prints. Samples printed with Aqua Clear Resin with the same geometry immediately after manufacturing were stiff. This could result from the presence of alcohol in the structure, which, being a caffeine solvent, is also a solvent for the acrylic resin. The addition of alcohol may have disrupted the photopolymerization process by preventing the structure from fully crosslinking, extending the photopolymerization time [22].

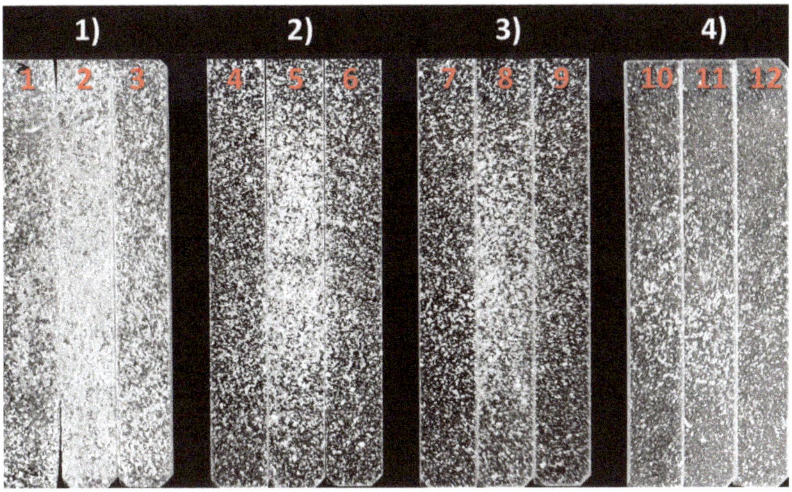

Figure 2. An illustrative image of 12 composite samples from series: (1), (2), (3), (4).

During the (1) series manufacturing, the first sample peeled off the work table and adhered to the film in the resin tank, which was caused by too much caffeine in the resin resulted in too weak adhesion of the sample to the work table. Despite this, the printing process for the composite samples using the printing parameters suggested by the resin manufacturer proceeded without any major problems for the other samples. The caffeine content of about 4% is therefore probably the limiting content of the caffeine filler that can be added to the resin without the appearance of defects in the samples resulting from peeling off the table. The dimensions of the samples, except for sample 1, were satisfactory and ranged from 2.06 to 2.08 and from 10.00 to 10.02 for thickness and width, respectively.

According to Figure 2, it can be seen that the distribution of caffeine along the length of the sample is not uniform; there is less caffeine at the ends of the sample than in the central part. With each successive manufacturing process, there is less and less caffeine in the samples; moreover, the centre sample in each batch contains more caffeine than the outer samples. Furthermore, the less undissolved caffeine in the form of agglomerates in the resin, the more homogeneous its distribution is, as can be seen for samples 10, 11, and 12. Between manufacturing processes sedimentation of caffeine could be seen in the resin tank so that its concentration in the samples is inhomogeneous not only on the surface but throughout the whole volume.

3.2. Mechanical Testing

The tensile properties of the samples are shown in Table 4. The values of tensile strength and Young's modulus have about 10% measurement error, while elongation has as much as 25% error. This may be explained by differences in the properties of the middle sample relative to the outer samples printed within 1 series, which can also be seen in the case of the caffeine content of the composite samples, for which the middle sample contained more caffeine than the outer samples. Despite this, the results are similar to those obtained for other commercially available resins [23].

Table 4. Mechanical parameters of resin and composite samples.

Series	Sample	σ_m (MPa)	E (MPa)	ε_B (%)
(0)	0.1–0.9	18.2 ± 1.9	834 ± 101	16.1 ± 4.0
(1)	1	12.6	1100	2.1
	2	7.9	503	5.6
	3	8.5	438	9.6
(2)	4	10.7	507	10.7
	5	10.3	530	10.3
	6	9.9	480	10.0
(3)	7	13.9	703	9.6
	8	12.9	682	8.8
	9	12.3	612	11.9
(4)	10	18.6	958	9.00
	11	15.6	831	7.0
	12	16.3	857	7.0

The caffeine content of the samples affects the strength properties of the composites by decreasing the strength and increasing the Young's modulus as the caffeine content increases. Among the reasons for this is the lack of interfacial adhesion between caffeine and resin due to the hydrophilic nature of the filler and the hydrophobic nature of the matrix, which causes a disruption in stress transfer in the structure [24]. Samples printed in the fourth series, 10, 11, and 12, have tensile strengths in the range of 16–18 MPa and Young's modulus on the order of 830–950 MPa, so they have similar properties to pure resin samples. Thus, it can be seen that the caffeine content corresponding to samples 10–12 is a limit for the preservation of the initial mechanical properties of the samples. Sample 1 deviates in properties from the other samples due to its peeling off the table during manufacturing. Lying on the resin tank foil, the sample did not grow in thickness for a constant distance from the UV source, resulting in multiple exposures of the same layer of material so that its stiffness increased by about 30%, while the stiffness of composite samples was usually lower than that of pure resin samples.

The relative elongation values of the samples, due to the very large measurement error for resin samples, are difficult to evaluate, but a certain dependence can be seen. All composite samples exhibit lower elongation than pure resin samples. Again, the reasons can be found in the lack of interfacial adhesion between caffeine and resin and the addition of alcohol, which during manufacturing probably disturbed the photopolymerization process causing defects in the crosslinked structure of the acrylic resin reducing its strength properties [22,25].

3.3. *Caffeine Release*

3.3.1. Gravimetric Analysis

Due to the great difficulty of precisely determining the initial and final caffeine content of the printed samples before and after the release test, a gravimetric method was used to determine the CR rate, which is limited to calculating the weight difference of the whole samples rather than the weight difference of the caffeine itself. The results of the weight measurements of the samples are given in Table 5.

The initial mass of the samples is 1.88–1.98 g depending on the series. Due to the highest caffeine content, which reduces the density of the structure as a result of the lack of interfacial interactions in the filler-matrix system, the lower initial mass is found in samples of the (1) series. Sample 1 as described earlier has the lowest mass as a result of the smallest number of printed layers.

Table 5. Masses of composite samples: initial—m_0, after drying—m_1, after release—m_2, and subsequent drying—m_3.

Series	Sample	m_0 (g)	m_1 (g)	m_2 (g)	m_3 (g)	CR (%)
(1)	1	0.57	0.53	0.58	0.48	10.0
	2	1.90	1.88	2.02	1.76	5.9
	3	1.88	1.86	2.01	1.76	5.1
(2)	4	1.96	1.94	2.12	1.87	3.5
	5	1.97	1.95	2.13	1.87	3.8
	6	1.95	1.92	2.10	1.86	3.6
(3)	7	1.97	1.95	2.13	1.90	2.2
	8	1.98	1.96	2.13	1.91	2.3
	9	1.96	1.94	2.12	1.89	2.4
(4)	10	1.98	1.96	2.14	1.94	0.8
	11	1.98	1.96	2.13	1.93	1.2
	12	1.96	1.94	2.11	1.92	1.1

The value of the calculated CR helps determine the system's ability to release caffeine at 37 °C for 12 h in an aqueous environment. The largest CR values can be observed for samples 1, 2 and 3, on the order of 5–10%, while the smallest for 10, 11 and 12, on the order of 1%. This is probably due to the decrease in the amount of caffeine present in the samples with each series. The less caffeine there is in the sample, the denser the structure is, making it more difficult for polar water molecules to penetrate the structure, reaching the caffeine particles and allowing them to dissolve and diffuse through the resin volume. Presumably, the more caffeine there is in the composite, the more disrupted the resin structure is, and more pores are between the caffeine and resin, improving the accessibility of the caffeine particles to water and allowing for easier release of caffeine from the system. In the case of CR, the values of the coefficient are similar for the middle sample and the outer samples in series (3) and (4), while for series (1) and (2), a slightly higher coefficient can be observed for the middle samples. The differences in the concentration and distribution of caffeine in the sample seen in Figure 1 are therefore relevant to the caffeine release process.

3.3.2. Global Threshold Analysis

Microscopic images of the surfaces of all composite samples were used to determine the degree of caffeine release based on global thresholding. Table 6 shows the numerical results obtained.

The total caffeine percentage m_K determined by thresholding decreases with each series and reaches values that are larger for the middle samples than for the outer samples within each series, which agrees with previous observations. The CTCR coefficient values differ from the CR values and are larger for all samples. As a general rule, no significant improvement in release can be seen for the middle samples relative to the outer samples when the release is evaluated on the basis of the CTCR coefficient.

The global thresholding method refers to operations based on a 2D image [26] and has other limitations such as choosing the right threshold, the quality of the initial image, ignoring the relationships between pixels [27–29] important due to the research performed. The sample is a 3D element and the distribution of caffeine over the height of the sample is not homogeneous, as can be seen by comparing the pure resin reference sample shown in Figure 3 with the composite samples shown in Figure 4. There is more caffeine in the surface layers of the composite samples than in the bottom layers, adjacent to the work table during manufacturing. The reason for this is the sedimentation mentioned above, whereby the undissolved caffeine macromolecules sank to the bottom during manufacturing. The vertical up-and-down movement of the working table was unable to ensure sufficient mixing of the resin-caffeine solution during manufacturing.

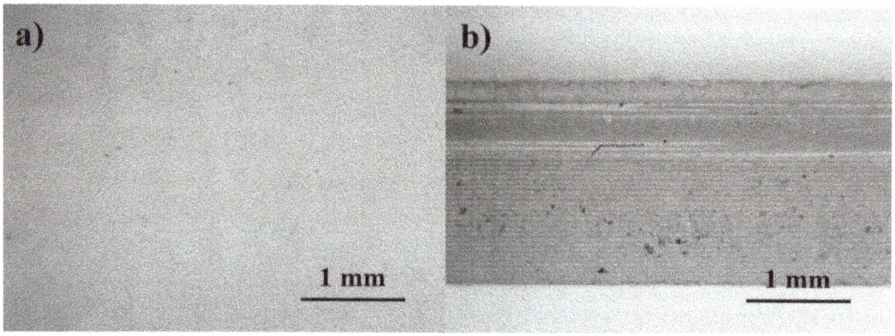

Figure 3. Images of the surface (**a**) and side (**b**) of the resin sample.

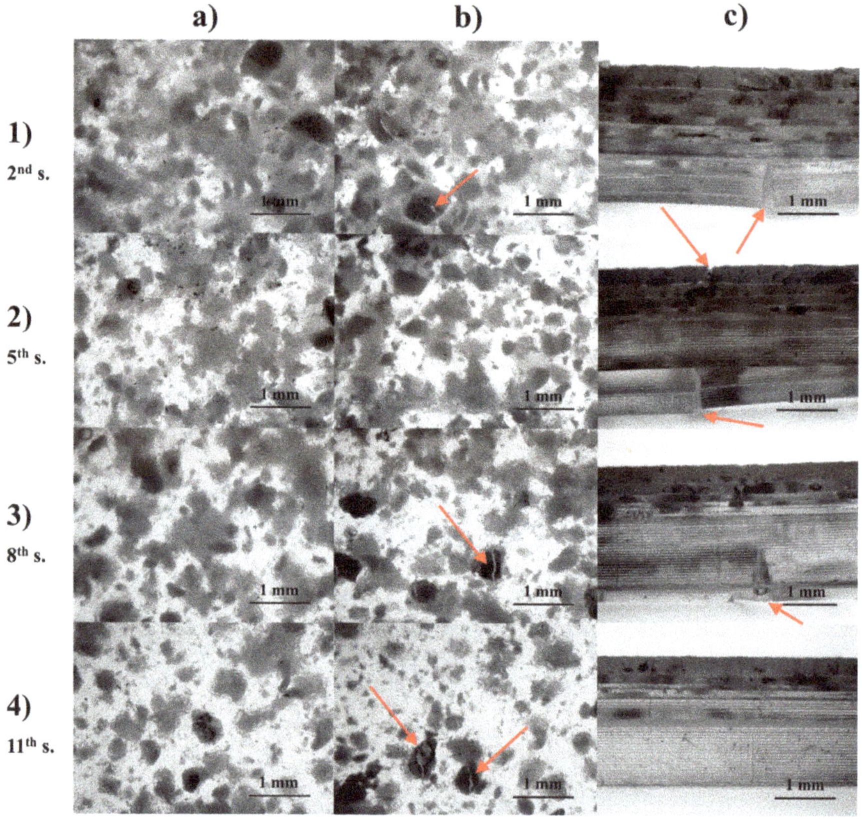

Figure 4. Images of the 2nd, 5th, 8th, and 11th composite sample (**a**) surfaces before releasing, (**b**) surfaces after releasing, and (**c**) sides after releasing. Red arrows indicate cracks.

Table 6. Percentage of caffeine on the surface and in the volume of the sample, and caffeine mass determined from thresholding.

Series	Sample	Surface Caffeine—I_{KS} (%)	Internal Caffeine—I_{KI} (%)	All Caffeine—$I_K = I_{KS} + I_{KI}$ (%)	$m_K = m_1 \times I_K$ (g)	CTCR (%)
(1)	1	1.2	63.0	64.2	0.34	15.6
	2	0.6	89.6	90.2	1.69	6.5
	3	0.1	80.4	80.5	1.50	6.3
(2)	4	0.1	64.0	64.1	1.24	5.4
	5	0.5	69.8	70.3	1.37	5.4
	6	0.2	58.5	58.8	1.13	6.1
(3)	7	0.0	57.4	57.4	1.12	3.8
	8	0.0	71.4	71.4	1.40	3.2
	9	0.2	56.8	57.0	1.10	4.1
(4)	10	0.4	57.0	57.4	1.12	1.5
	11	0.2	60.1	60.3	1.18	2.1
	12	0.1	58.0	58.1	1.13	1.9

In relation to the volumetric distribution of caffeine particles in the resin, the global thresholding method has some limitations. If the caffeine particles occur one below the other, they are visible as one particle in the surface image by which, after binarization of the image, the two particles are read by the software as one set of pixels. In addition, the arrangement of the particles also matters—for example, a flake-shaped particle angled will be determined to be smaller in volume. Therefore, global thresholding is an indicative method to assess caffeine release.

Figure 4b,c show numerous cracks in the samples in both the top layers, where more caffeine was present, and the bottom layers. Cracks were not present anywhere in the samples before release and probably appeared as a result of the accelerated release of caffeine at elevated temperatures. Furthermore, the caffeine, as an opaque medium, may have acted as a barrier to UV radiation, so the resin surrounding the caffeine may have been underexposed causing a reduction in the strength of the structure at these locations.

As the amount of caffeine in the sample decreases, a change in the trend of cracks can be seen; for the samples in the fourth series, cracks occurred mainly on the surface of the sample and not mainly in the bottom layers as before, as represented by the photos of sample 11. This is probably related to the presence of the largest caffeine agglomerates just on the surface due to sedimentation in previous sample series. Furthermore, the particles inside samples 10, 11, and 12 were small enough not to cause the samples to crack at the base during release. It is noteworthy that the samples with the smallest caffeine content underwent the smallest deformation during release and visually most closely resemble the reference sample in Figure 3.

The cracks in samples resulting from the caffeine release had a significant effect on the sample surface microscopic images, what made it impossible to assess the after release caffeine content using the global thresholding method. For this reason, the difference in the masses of the samples before and after release determined by the gravimetric method was used to calculate the CTCR coefficient rather than the caffeine mass after release determined from thresholding.

Based on the results of CTCR and CR, it can be evaluated whether the thresholding method can give results comparable to the gravimetric method. The values of the individual coefficients are presented in the graph in Figure 5. The CTCR coefficient obtains similar values to CR for most samples except for the first sample, which, due to the incorrect manufacturing process, was characterized by the worst structure properties: roughness, delamination, numerous caffeine particles protruding beyond the resin surface, insufficient thickness and under-crosslinked structure, which can be seen in Figure 6 in the form of numerous cavities and cracks in the sample surface layer. The quality of sample 1 determines the limiting value of caffeine that can be added to the resin to carry out the correct manufacturing process. Caffeine concentration of 4% or higher resulted in none of the samples in the series were printed correctly. In the case of sample 1, where caffeine was present throughout the sample, the m_K value determined by global thresholding was underestimated due to the presence of particles in the entire volume of the structure, which are less visible in microscopic images of the structure, making the CTCR ratio as much as 5.6 percentage points higher than the CR ratio.

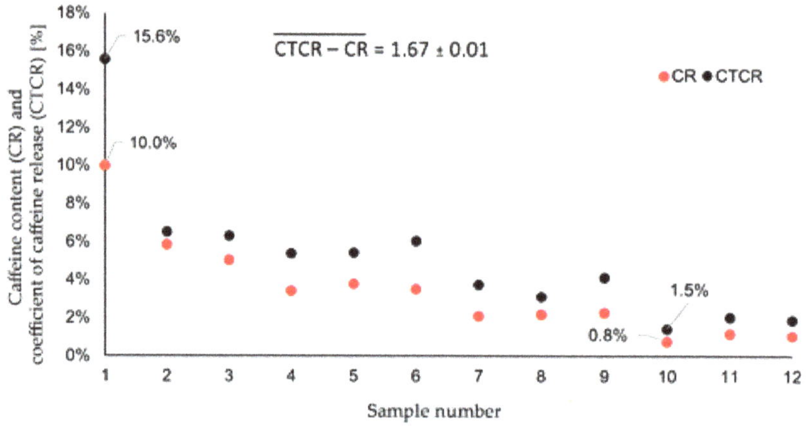

Figure 5. Graph of caffeine content and coefficient of caffeine release of the samples.

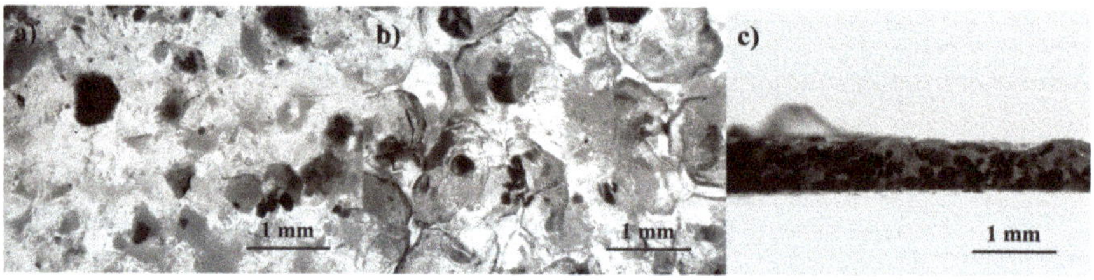

Figure 6. Images of 1st sample (**a**) surface before releasing, (**b**) surface after releasing, (**c**) side after releasing.

In the remaining samples with the proper geometry, based on Figure 4c, the caffeine particles in the sample are estimated to occur at a depth of around 0.5–1 mm from the surface. This contributed to the small differences in CTCR versus CR values due to the distribution of caffeine at shallow depth, a situation similar to the occurrence of caffeine only at the surface of the sample, which reduced the influence of particle distribution in space, increasing the correctness of the global thresholding method.

4. Conclusions

The present study succeeded in obtaining bio-composite samples made of photocurable acrylic resin and caffeine. On the basis of visual and strength tests, it can be concluded that the maximum weight content of caffeine in the resin to produce samples with acceptable performance is approximately 4%, which met the assumptions. A very important aspect of samples with undissolved filler particles is their dispersion in the matrix material, which in the case of the samples obtained was increasingly favourable as the concentration of caffeine decreased. Nevertheless, the manufacturing process of the composite samples was successful for a filler content not exceeding 4%, which meets the authors' goals.

The static tensile testing of the samples shows that the strength properties of the composites deteriorate with an increase in filler content, which is in line with the literature knowledge on the weakening of composites with a decrease in interactions between the hydrophilic filler and hydrophobic matrix, as was the case with the samples obtained.

However, weak interfacial interactions proved beneficial from the perspective of caffeine release from the system as expected. According to the release studies conducted, the effect of the amount of caffeine on the course of release can be observed. The more caffeine in the sample, probably, the easier it is for the caffeine to escape from the photopolymer structure due to disruption of the resin crosslinking process, which also affects the strength properties of the samples. The dynamic release of caffeine by increasing the temperature of the water during the release caused cracking of the samples in both the top and bottom layers of the sample, while increasing the caffeine content and the larger particle size intensifies this effect.

The global thresholding method proved to be helpful in evaluating the caffeine content of the samples before the release study, but it cannot replace the gravimetric method, which was proven by the significant differences in the caffeine content results of the composites obtained from the thresholding method and the gravimetric method. In the case of composites where particles are mainly distributed near the surface thanks to the thresholding method similar release results can be determined but for samples where particle dispersion runs throughout the volume of the samples, the results obtained were unreliable. Despite this, the thresholding method can be considered helpful for evaluating the presence of macromolecules in composites with limited filler solubility in the matrix, as was the case with caffeine and acrylic resin.

The DLP process is a promising method for obtaining composite systems of hydrophilic pharmaceutically active substances with hydrophobic resins used in additive manufacturing methods. This method, despite some limitations, is a potential method for obtaining composite transdermal systems, especially as a result of the high precision of the obtained parts. Thus, it is important to conduct further work on improving the technique for obtaining bio-composites with constituent materials of different polarity.

Author Contributions: Conceptualization, D.T. and W.K.; Methodology, D.T. and R.W.; Validation, R.W.; Formal analysis, D.T.; Investigation, D.T.; Resources, R.W. and W.K.; Writing—original draft, D.T.; Writing—review and editing, D.T. and R.W.; Funding acquisition, W.K. All authors have read and agreed to the published version of the manuscript.

Funding: This research was funded by statutory activity financed by the Polish Ministry of Science and Higher Education, grant number (0613/SBAD/4770).

Data Availability Statement: Not applicable.

Conflicts of Interest: The authors declare no conflict of interest.

References

1. El Aita, I.; Ponsar, H.; Quodbach, J. A critical review on 3D-printed dosage forms. *Curr. Pharm. Des.* **2018**, *24*, 4957–4978. [CrossRef] [PubMed]
2. Herrada-Manchón, H.; Rodríguez-González, D.; Alejandro Fernández, M.; Suñé-Pou, M.; Pérez-Lozano, P.; García-Montoya, E.; Aguilar, E. 3D printed gummies: Personalized drug dosage in a safe and appealing way. *Int. J. Pharm.* **2020**, *587*, 119687. [CrossRef] [PubMed]
3. Charoo, N.A.; Barakh Ali, S.F.; Mohamed, E.M.; Kuttolamadom, M.A.; Ozkan, T.; Khan, M.A.; Rahman, Z. Selective Laser Sintering 3D Printing—An Overview of the Technology and Pharmaceutical Applications. *Drug Dev. Ind. Pharm.* **2020**, *46*, 869–877. [CrossRef]
4. Scoutaris, N.; Ross, S.A.; Douroumis, D. 3D Printed "Starmix" Drug Loaded Dosage Forms for Paediatric Applications. *Pharm. Res.* **2018**, *35*, 34. [CrossRef]
5. Yu, D.G.; Yang, X.L.; Huang, W.D.; Liu, J.; Wang, Y.G.; Xu, H. Tablets with Material Gradients Fabricated by Three-Dimensional Printing. *J. Pharm. Sci.* **2007**, *96*, 2446–2456. [CrossRef] [PubMed]
6. Goole, J.; Amighi, K. 3D printing in pharmaceutics: A new tool for designing customized drug delivery systems. *Int. J. Pharm.* **2016**, *499*, 376–394. [CrossRef]

7. Melchels, F.P.W.; Feijen, J.; Grijpma, D.W. A review on stereolithography and its applications in biomedical engineering. *Biomaterials* **2010**, *31*, 6121–6130. [CrossRef] [PubMed]
8. Maines, E.M.; Porwal, M.K.; Ellison, C.J.; Reineke, T.M. Sustainable advances in SLA/DLP 3D printing materials and processes. *Green Chem.* **2021**, *23*, 6863–6897. [CrossRef]
9. Wang, J.; Goyanes, A.; Gaisford, S.; Basit, A.W. Stereolithographic (SLA) 3D printing of oral modified-release dosage forms. *Int. J. Pharm.* **2016**, *503*, 207–212. [CrossRef]
10. Palo, M.; Holländer, J.; Suominen, J.; Yliruusi, J.; Sandler, N. 3D printed drug delivery devices: Perspectives and technical challenges. *Expert Rev. Med. Devices* **2017**, *14*, 685–696. [CrossRef]
11. Kadry, H.; Wadnap, S.; Xu, C.; Ahsan, F. Digital light processing (DLP) 3D-printing technology and photoreactive polymers in fabrication of modified-release tablets. *Eur. J. Pharm. Sci.* **2019**, *135*, 60–67. [CrossRef] [PubMed]
12. Robles-Martinez, P.; Xu, X.; Trenfield, S.J.; Awad, A.; Goyanes, A.; Telford, R.; Basti, A.W.; Gaisford, S. 3D Printing of a Multi-Layered Polypill Containing Six Drugs Using a Novel Stereolithographic Method. *Pharmaceutics* **2019**, *11*, 274. [CrossRef] [PubMed]
13. Kundu, A.; Arnett, P.; Bagde, A.; Azim, N.; Kouagou, E.; Singh, M.; Rajaraman, S. DLP 3D Printed "Intelligent" Microneedle Array (iμNA) for Stimuli Responsive Release of Drugs and Its in Vitro and ex Vivo Characterization. *J. Microelectromechanical Syst.* **2020**, *29*, 685–691. [CrossRef]
14. Lim, S.H.; Kathuria, H.; Amir, M.H.B.; Zhang, X.; Duong, H.T.T.; Chi-Lui Ho, P.; Kang, L. High resolution photopolymer for 3D printing of personalised microneedle for transdermal delivery of anti-wrinkle small peptide. *J. Control. Release* **2021**, *329*, 907–918. [CrossRef]
15. Badruddoza, A.Z.M.; Godfrin, P.D.; Myerson, A.S.; Trout, B.L.; Doyle, P.S. Core–Shell Composite Hydrogels for Controlled Nanocrystal Formation and Release of Hydrophobic Active Pharmaceutical Ingredients. *Adv. Healthc. Mater.* **2016**, *5*, 1960–1968. [CrossRef] [PubMed]
16. Himawan, A.; Anjani, Q.K.; Detamornrat, U.; Vora, L.K.; Permana, A.D.; Ghanma, R.; Naser, Y.; Rahmawanty, D.; Scott, C.J.; Donnelly, R.F. Multifunctional low temperature-cured PVA/PVP/citric acid-based hydrogel forming microarray patches: Physicochemical characteristics and hydrophilic drug interaction. *Eur. Polym. J.* **2023**, *186*, 111836. [CrossRef]
17. Huang, L.; Himawan, E.; Belhadj, S.; Oswaldo Pérez García, R.; Durand, F.P.; Schipper, N.; Buzgo, M.; Simaite, A.; Marigo, V. Efficient Delivery of Hydrophilic Small Molecules to Retinal Cell Lines Using Gel Core-Containing Solid Lipid Nanoparticles. *Pharmaceutics* **2022**, *14*, 74. [CrossRef]
18. Roy, A.; Ray, P.G.; Bose, A.; Dhara, S.; Pal, S. pH-Responsive Copolymeric Network Gel Using Methacrylated β-Cyclodextrin for Controlled Codelivery of Hydrophilic and Hydrophobic Drugs. *ACS Appl. Bio Mater.* **2022**, *5*, 3530–3543. [CrossRef]
19. Chandran, R.; Tohit, M.R.E.; Stanslas, J.; Salim, N.; Mazlelaa, T.M.T. Investigation and Optimization of Hydrogel Microneedles for Transdermal Delivery of Caffeine. *Tissue Eng. Part C Methods* **2022**, *10*, 545–556. [CrossRef]
20. Safety Data Sheet for Aqua Clear Resin by Phrozen. Available online: https://drive.google.com/file/d/1NR46rNJ1z3yM8TRqyZ5TOedubqhtr4Il/view (accessed on 27 February 2023).
21. Safety Data Sheet for Caffeine. Available online: https://www.sigmaaldrich.com/PL/en/product/sial/c0750 (accessed on 27 February 2023).
22. Ye, Q.; Spencer, P.; Wang, Y.; Misra, A. Relationship of solvent to the photopolymerization process, properties, and structure in model dentin adhesives. *J. Biomed. Mater. Res. Part A* **2007**, *80*, 342–350. [CrossRef]
23. Bochnia, J. A Study of the Mechanical Properties of Naturally Aged Photopolymers Printed Using the PJM Technology. *Materials* **2023**, *16*, 400. [CrossRef]
24. Zeng, Q.H.; Xu, W.; Yu, A.B.; Paul, D.R. Quantification of the Interface Interactions in Polymer Nanocomposites. *Mater. Sci. Forum* **2010**, *654–656*, 2608–2611. [CrossRef]
25. Hu, K.; Kong, W.; Fu, X.; Zhou, C.; Lei, L. Resistivity optimization and properties of silver nanoparticles-filled alcohol-soluble conductive coating based on acrylic resin. *High Perform. Polym.* **2015**, *27*, 930–938. [CrossRef]
26. Padmasini, N.; Umamaheswari, R.; Sikkandar, M.Y. Chapter 10-State-of-the-Art of Level-Set Methods in Segmentation and Registration of Spectral Domain Optical Coherence Tomographic Retinal Images. In *Soft Computing Based Medical Image Analysis*, 1st ed.; Dey, N., Ashour, A.S., Shi, F., Balas, V.E., Eds.; Academic Press: Cambridge, MA, USA, 2018; pp. 163–181. [CrossRef]
27. Davies, E.R. The Role of thresholding. In *Computer Vision*, 5th ed.; Davies, E.R., Ed.; Academic Press: Cambridge, MA, USA, 2018. [CrossRef]
28. Rashidi, B.; Rashidi, B. Implementation of a High Speed Technique for Character Segmentation of License Plate Based on Thresholding Algorithm. *Int. J. Image Graph. Signal Process.* **2012**, *4*, 9–18. [CrossRef]
29. Chaubey, A.K. Comparison of the local and global thresholding methods in image segmentation. *J. Res. Rev.* **2016**, *2*, 1–4.

designs

Article

Dimensional Accuracy of Electron Beam Powder Bed Fusion with Ti-6Al-4V

Eric Bol and Mamidala Ramulu *

Mechanical Engineering Department, College of Engineering, University of Washington, Seattle, WA 98195, USA
* Correspondence: ramulum@uw.edu

Abstract: While much of additive manufacturing (AM) research is focused on microstructure, material properties, and defects, there is much less research in regards to understanding how well the part coming out of the machine matches the 3D model it is based on, as well as what are the key process parameters an engineer needs to care about when they are optimizing for AM. The purpose of this study was to understand the dimensional accuracy of the electron beam powder bed fusion (EB-PBF) process using specimens of different length scales from Ti-6Al-4V. Metrology of the specimens produced was performed using fringe projection, or laser scanning, to characterize the as-built geometry. At the meso-scale, specimen geometry and hatching history play a critical role in dimensional deviation. The effect of hatching history was further witnessed at the macro-scale while also demonstrating the effects of thermal expansion in EB-PBF. These results make the case for further process optimization in terms of dimensional accuracy in order to reduce post-processing costs and flow time.

Keywords: electron beam; powder bed fusion; dimensional accuracy; titanium

Citation: Bol, E.; Ramulu, M. Dimensional Accuracy of Electron Beam Powder Bed Fusion with Ti-6Al-4V. *Designs* **2023**, *7*, 53. https://doi.org/10.3390/designs7020053

Academic Editor: Obeidi Muhannad

Received: 28 February 2023
Revised: 31 March 2023
Accepted: 31 March 2023
Published: 6 April 2023

1. Introduction

Titanium alloy Ti6AL4V is the workhorse of the aerospace and medical industries due to its strength, density, durability, and corrosion resistance. These industries place a high priority on product safety, so dimensional accuracy must be well understood and high-quality manufacturing repeatability must be proven. When Ti-6Al-4V is converted into a powder through atomization processes and used as an additive manufacturing feedstock for electron beam powder bed fusion (EB-PBF) or laser beam powder bed fusion (LB-PBF), fully dense titanium hardware can be produced. This method of AM uses an electron beam as the energy source to selectively melt powder that is uniformly distributed within a build chamber in a layer-by-layer process. With this technology, increasingly complex structures can be designed and manufactured that could not be cost-effectively produced any other way [1].

In-situ process monitoring of the EB-PBF process is extremely difficult because the build chamber environment is a vacuum and heated to high temperatures, approximately half of the material melt temperature (~650 °C for titanium) [2–5]. In addition, the electron beam scan speed is very fast and is typically split into many points to maintain simultaneous melt pools [6]. Thus, it remains challenging and an active area of research to identify the nuances of what is happening with the beam power and the interaction with the melt pool. Key features of as-built EB-PBF specimens are often the surface morphology, or surface roughness, and their microstructure. Typically, EB-PBF uses a larger powder size distribution (PSD) than laser PBF machines, ranging from 45 to 106 μm in diameter, and it is one of the primary contributing factors to surface roughness [7]. The larger particles are used because they are more resistant to "smoking", or powder spreading, which is the buildup of excessive negative charge energy that forces the displacement of particles in all directions from the powder bed [8]. Process parameters, such as the beam speed, beam current, beam focus, scan strategy, etc., can have a significant effect on the resulting surface roughness, such that optimization can provide improvements to dimensional accuracy [9].

The as-built surface roughness of electron beam and laser PBF parts has been well documented, with numerous methods of metrology being implemented in attempts to characterize it [10–20]. Most focus on the surface-connected pores and partially melted attached particles that are difficult to describe by areal texture parameters defined in ISO 25178-2. The surface roughness of the top layer during the build has not been looked at in detail as it is subsequently remelted by the following layer. However, the recently melted topography affects how that next layer of particles is distributed, as well as the overall dimensional variation.

For traditional manufacturing, geometric dimensioning and tolerance (GD&T) standards have been well established based on ISO, ASME, and ASTM standards. The highly optimized and complex 3D structures that PBF can produce pose a challenge for traditional inspection and quality assurance methods. In the early stages of AM development, not much distinction was made between accuracy and resolution, particularly for prototype fabrication; however, with the expectation that AM will deliver high-quality finished production parts, there is a need for new or modified methods of measuring dimensional accuracy [21–23]. Various methods of optical 3D scanning are typically used to generate point clouds and then fit surfaces to the data to be compared with the CAD model.

Smith et al. used 3D scanning of optimized truss structures to measure dimensional deviation, yet found it difficult to determine if the variation seen in the measurements was true or if the specimens had been deformed when being manipulated for scanning [24]. In a study performed by Ameta et al., the authors introduced the concept of derived supplemental surfaces (DSS) based on "theoretical supplemental surfaces" (TSS) in ASME Y14.46 [25]. They measured a lattice structure using a laser confocal microscope, which generated a point cloud, and fitted the DSS using the Chebyshev and least-squares approaches. Gruber et al. used a fringe projection 3D scanning system and CT scans to analyze benchmark samples for geometric accuracy, and they found both were well suited for determining external feature part quality [26]. However, while most studies have focused on the accuracy of the particular metrological method employed, few have used multiple methods in tandem to characterize the dimensional accuracy of an AM process.

Previous work that revealed the dimensional deviation of a single beam width line melt and the optimization of an assembled structure were presented to ASME [27,28]. The purpose of this paper was to quantify the dimensional accuracy at different length scales in the EB-PBF process and identify sources of deviation for further improvement. Two experiments were conducted using Ti-6Al-4V powder, where the meso-scale experiment produced step and ramp specimens that measured variation based on layer topography, thickness, and build location, and the macro-scale experiment produced tapered box beams that had been topology optimized and used a 3D laser scanner to measure the as-built geometric dimensional deviation. These experiments demonstrated how important process optimization is for controlling variation from the initial layers to the final component quality.

2. Materials and Methods

2.1. EB-PBF Process

For the experiments described in this work, all samples were manufactured utilizing reused gas-atomized Grade 5 Ti-6Al-4V powder, originally obtained from the EB-PBF machine manufacturer, with a powder size distribution (PSD) of 45–106 μm in diameter. The EB-PBF process was performed with an Arcam A2X, shown in Figure 1, utilizing the default parameters provided by the machine manufacturer for Ti-6Al-4V that are organized into themes. The A2X emits electrons from an electrically heated tungsten filament with 60 keV of power acceleration and a current range from 0 to 50 mA. Because electrons will interact with atoms in the atmosphere, the process must take place within a high vacuum (normally 1×10^{-5} mbar) environment; however, a small amount of helium pressure is applied (2×10^{-3} mbar) to avoid oxygen contamination as well as reduce the amount of electrostatic charge buildup in the powder material. Three electromagnetic lenses that affect the spot shape, size (typically 250 μm in diameter), and position guide the beam to the powder bed. The method of energy transfer from the electrons to the powder bed material is by way of kinetic energy with elastic and inelastic collisions. In order to further reduce the risk of powder overcharge resulting in "smoking," an unfocused beam is employed to preheat the start plate and the powder bed prior to using a focused beam for melting [2,8,29,30].

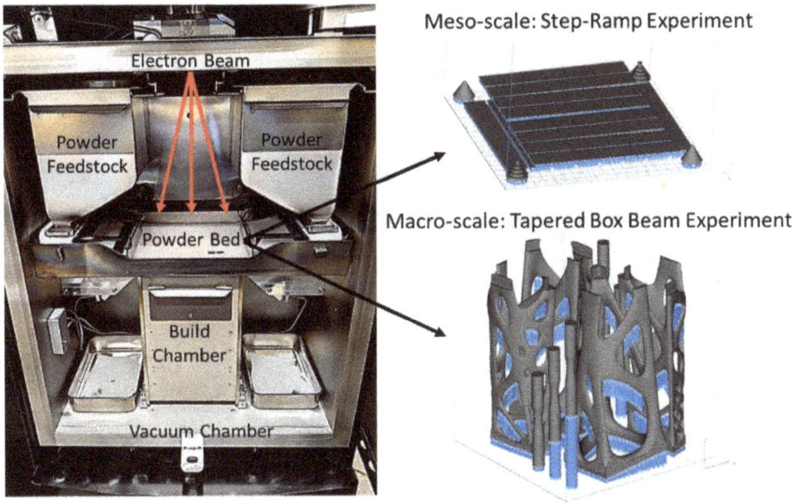

Figure 1. Image of the Arcam A2X EB-PBF experimental setup along with the build volume input models of the meso-scale step-ramp experiment and the macro-scale tapered box beam experiment. The experimental specimens are colored gray, while support features are shown in blue.

Process parameters for the experiments were grouped into four themes defined as start plate preheat, powder layer preheat, melt, and wafer (for support features). During the machine setup, the start plate preheat and powder layer preheat themes are set for the entire build, while the melt or wafer themes are assigned to individual parts, including supports. For example, the wafer theme parameters create a lower energy-dense beam to reduce the melt consolidation between layers, so that support features would be easier to remove in post-processing. A layer thickness of 50 μm was specified for all experiments, and a 210 mm × 210 mm × 10 mm SS316L start plate was utilized.

The printing process began with the start plate preheating to 730 °C, as measured by a thermocouple located directly beneath the start plate. Once the target temperature was reached, a layer of powder was fetched from the hoppers and distributed evenly across the build area by a rake. Every new layer proceeds with a preheat conducted in two stages defined by the Powder Layer Preheat theme. Preheat 1 scans the entire powder bed to bring the powder temperature to a "jump-safe" level, and then preheat 2 rescans the powder where melting is to occur, plus an offset, to further raise the powder temperature to a "melt-safe" level. Preheat 1 and 2 sinter the powder bed to prevent the powder from spreading when it is impacted by a focused beam for melting.

Following the preheat phase, the specimen outlines were melted using the contouring parameters and a spot-melt scan strategy. That was followed by an infill hatching scan, which rastered the beam back and forth. While contours are performed at a constant beam power, the hatching scan is continuously varied by four proprietary functions designed to keep the melt pool properties consistent. The current compensation function sets the beam current based on the length of the hatch line, while the speed function varies the velocity based on the beam current for a consistent melt pool size. The turning point function reduces the beam energy near the beginning of each hatch line, and the thickness function reduces the beam energy when melting close to an overhanging surface [24]. All of these parameters reside within the definition of the melt theme.

2.2. Experimental Design

Some level of distortion in welding and metal PBF is expected, and melting thin sheet metal was anticipated to amplify the dimension deviation. Therefore, for the meso-scale experiment, thin step and ramp specimens were designed as shown in Figure 2. The thinnest specimens started at 0.1 mm (2 layers) and gradually increased in thickness to 0.7 mm (14 layers), while the thicker specimens began at 0.2 mm (4 layers) and increased to 1.4 mm (28 layers). Each specimen was 140 mm long and 20 mm wide. Along the length, step specimens were divided into 7 steps of constant thickness, increasing at either 0.1 mm or 0.2 mm increments, while ramp specimens continuously increased from the starting thickness until reaching the final thickness and maintaining that constant for the final 20 mm. For the step specimens, the thickness-to-length ratio ranged from 0.5% to 3.5% and 1% to 7% for the thin and thick designs, respectively. Ramp specimens were designed to have a slope of 0.005 ($\theta \approx 0.29°$) and 0.01 ($\theta \approx 0.57°$) for the thin and thick designs, respectively. At these angles, the stair-case effect of the layering process will reveal the melt topology at every layer.

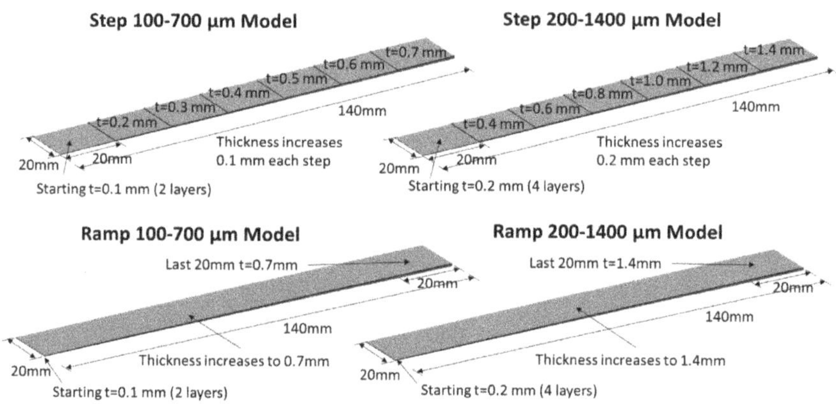

Figure 2. Meso-scale experiment specimen design illustrating the dimensions of the thin and thick CAD models of step and ramp.

The build for this experiment manufactured four sets of the four specimens from Figure 2, and the build volume configuration is presented in Figure 3. The first set of four specimens were lined up, side by side, 5 mm above the start plate, with skin surface supports extending to the plate. That same model setup was repeated adjacently but started at 7 mm above the start plate with supports extending to the start plate. Then that pattern of eight specimens was repeated at a higher location in the build volume, starting at 18.4 mm, oriented perpendicular to the previous set. Each upper specimen was provided with supports matching the first set, extending 5 mm below into the sintered powder but not extending to any other structures or the start plate. The lower (L) specimens were numbered L1 through L8, while the upper (U) specimens were numbered U1 through U8. The odd numbers are ramp models, while the even numbers are step models.

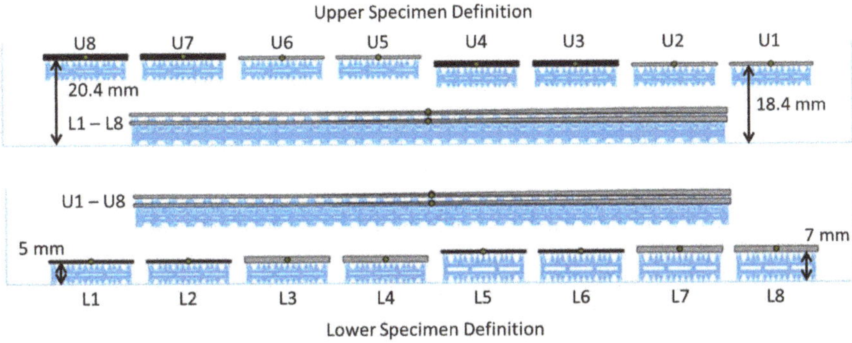

Figure 3. Meso-scale experimental design showing the side views of the build volume with specimen labels and build height dimensions of the upper and lower specimen sets. The experimental specimens are colored gray, while support features are shown in blue.

For the macro-scale dimensional accuracy experiment, a topology-optimized tapered box beam was selected as the specimen. The box beam design is based on the forward strut box of a fan case-mounted turbofan engine pylon and scaled down to a size just slightly larger than the Arcam A2X volume, such that it was necessary to be split into halves and assembled. Commercially available topology optimization software that contains a proprietary implementation of the SIMP (solid isotropic material with penalty) method [31–34] was employed as a guide for where to place material using mass-minimization criteria. The result guided follow-on design refinements based on maximum principal stress minimization criteria under bending and torsional loads. Further redesign efforts focused on elements of Design for Additive Manufacturing (DfAM) criteria such as build orientation, support minimization, powder removal, and method of assembly. The final design of the topology-optimized tapered box beam is displayed in Figure 4.

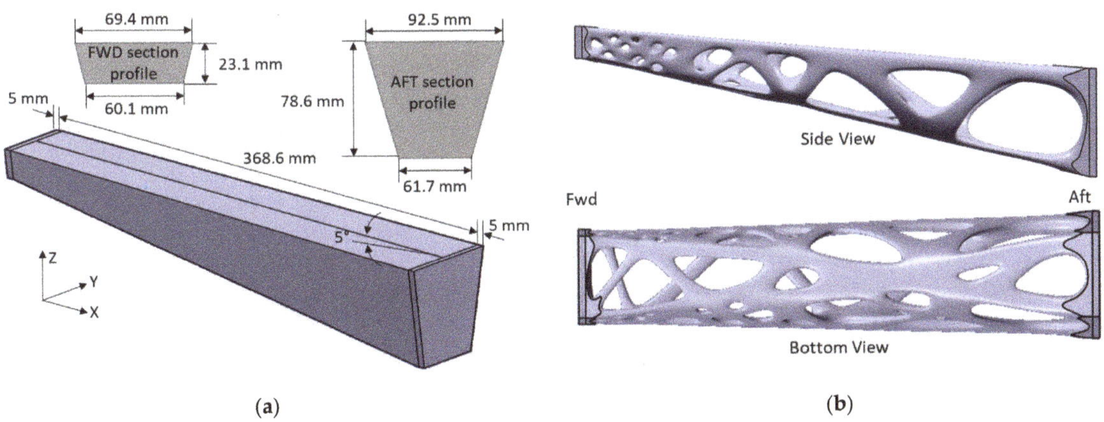

Figure 4. Macro-scale dimensional accuracy specimen design illustrating the: (**a**) design space dimensions; and (**b**) the final overall shape of the tapered box beam design showing the side view and bottom view.

Six beam specimens were produced over three builds, two per build, as shown in Figure 5. The beams are labeled with the nomenclature "Front" (F) and "Back" to distinguish their location within the build chamber, with the front being closest to the build chamber door. By placing the mid-span join furthest from the build plate in the z-direction, the goal was to magnify any dimensional deviation present in the buildup of the layer-by-layer process. The interfacing surfaces at the three joint locations were designed such that the two halves would slide together axially with a surface clearance of 0.1 mm. This was done to create a tight fit based on the mechanical-chemical bonding strategy, where the interfacing surfaces were solution cleaned with acetone, sanded with a diamond rotary tool, then solution cleaned again, before being bonded using JB Weld and curing while clamped for >24 h.

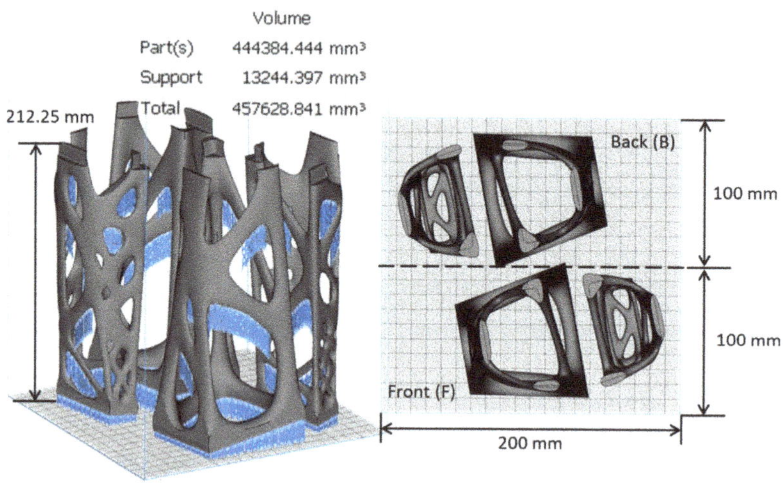

Figure 5. Macro-scale tapered box beam experiment: build support strategy and build volume layout identifying the front and back specimen locations. The experimental specimens are colored gray, while support features are shown in blue.

2.3. Metrology

The topology of the step and ramp specimens was measured using a Keyence VR-3100 three-dimensional profile measurement macroscope, which utilizes high-intensity LED light and a 4-megapixel monochrome CMOS to obtain a one-shot fringe projection image. Fringe projection enables the instantaneous measurement of the length, height, volume, angle, etc. of a portion of any object. Prior to examination, the specimens were post-processed using the Arcam Powder Recovery System (PRS). Due to the differing materials between the steel start plate and the titanium specimens, during cooling the interfacing welds cracked such that the specimens were removed by hand without cutting. The down-skin support features were not removed in order to provide a more stable orientation on the imaging platform, and for each specimen, several images were taken at 40× magnification and stitched together using the system software in order to capture the entire length.

The geometric accuracy of the assembled topology-optimized tapered box beams was determined by a 3D laser scan. A Kreon Ace measuring arm, mounted to a calibrated table, was combined with a Skyline 3D scanner that specifies an accuracy of up to 9 μm. Each specimen was placed on the table in various orientations while the operator manually manipulated the scanner by hand, sweeping along the surface of the part without contact in multiple passes. The scanner utilizes an HD camera and blue laser so that it can measure reflective materials like titanium, and it generates a dense point cloud to capture the surface topology with a scan speed of up to 600,000 points per second. Software assembled the point clouds into a single topological surface per beam and then converted them to an STL file. Metrology analysis was accomplished using GOM Inspect software, which examined the parameters of total length and total volume based on an overall best fit compared to the nominal CAD geometry.

3. Results

In this section, the results of the multi-scale manufacturing experiments and metrological analysis of the step-ramp and tapered box beam specimens are presented. In addition, witness coupons were also produced to examine the microstructure and tensile properties of the additively produced samples. These aspects and others are important to consider for any sort of part or process qualification and are briefly reported in Figure 6. From the SEM image of the microstructure taken from specimen L2, the dual phase of fine lamellar α and β that appears as a basket weave (Widmanstätten) is expected from Grade 5 Ti-6Al-4V [35]. The tensile properties reflect a more brittle Ti-6Al-4V, which can be expected from heavily reused powder feedstock, with only about 4% elongation but very high yield strength (0.2% offset) of 1097 MPa and an elastic modulus of 103 GPa.

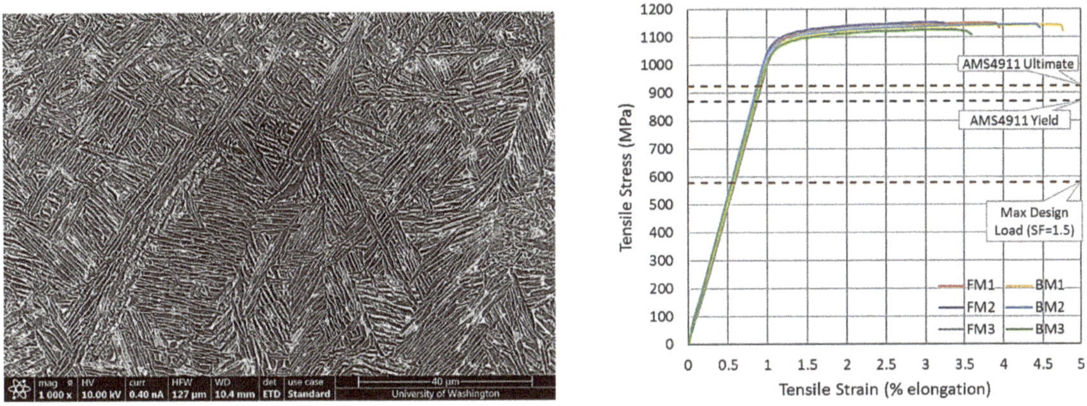

(a) (b)

Figure 6. Properties of the additively produced specimens: (**a**) SEM image of specimen L2 microstructure at the 500 µm design thickness (up is the build direction); and (**b**) tensile properties taken from tapered box beam specimen builds 1–3, while also displaying comparisons for AMS 4911 and the maximum design load criteria.

3.1. Meso-Scale Step-Ramp Experiment

The top surfaces of the fringe projection images of step and ramp specimens, L1 through L4, are displayed in Figure 7, with the designed thickness of the layers identified. From the ramp specimens, L1 and L3, the staircase effect of the layering process is clearly shown, which also provides the hatching pattern and angle that occurred on every layer. step specimens L2 and L4 provide a good contrast with large 20 mm × 20 mm melt areas. On the left-hand side of the specimens L1 and L2, the thinnest 100 µm thickness (two layers) shows an enormous quantity of pores, many of which contain partially melted or sintered powder inside. The thinnest layers of L3 and L4 have only a few pores because they started at four layers thick and can be directly compared with the 200 µm levels in L1 and L2. Progressing upward in designed thickness, random pores were still present in various specimens and layers. Typically, these pores were found at the intersection of the contour and hatch scans, or at what appeared to be the stop or start of a hatch scan.

As observed due to the variation in contrast, there was a lack of uniformity in the hatching melt at every layer on every specimen. Some hatch lines appeared wide and well melted, while in some cases, a few millimeters away, the lines are thin and raised such that they appear to not be touching the adjacent lines. Several hatch lines also appear to meander along the scan direction. These optical differences are more clearly visualized in three dimensions, as shown in Figure 8, using an example from L3 at the 550 µm design thickness. Optically, the shading reveals that very few layers appear to have a completely uniform melt despite a constant thermal mass. The 450 μm layer of L1 and L3 appears to have the most uniform melt, whereas on other layers there exists a significant contrast in the hatch scan melt, particularly in the ramp specimens L1 and L3.

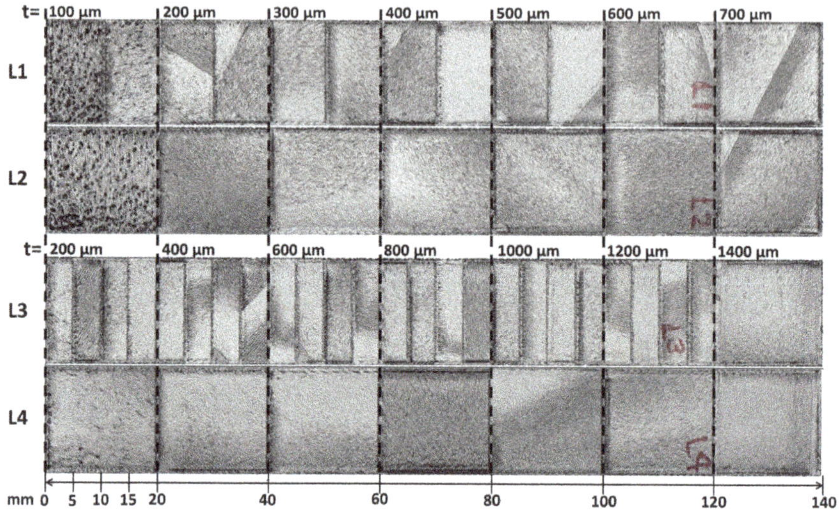

Figure 7. Set 1 of 4 step and ramp specimens, L1 through L4, from top to bottom, with the designed thickness indicated in black text starting at the dashed lines.

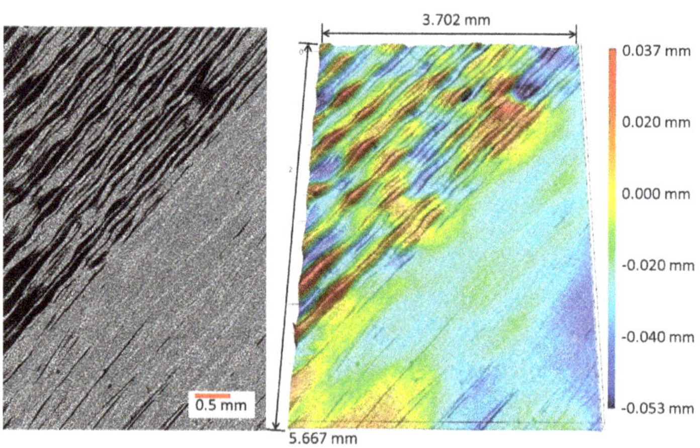

Figure 8. Specimen L3, Layer 8 (550 μm design thickness), 40× optical image (**left**), and 3D height data image (**right**). This image displays a transition in the scan lines melt quality and surface roughness within a single layer.

Height data was analyzed to extract a topological centerline profile of the specimens along their length. This is shown in Figure 9 compared to the CAD file dimensions to demonstrate the level of dimensional accuracy in the thin and thick specimens. As there were no perfectly flat regions with which to calibrate a zero plane, the top layer, being the most consistent, established the zero-reference plane for specimens L1 through L8. The contour scan created raised ridges around the periphery of each layer, and these are depicted as sharp spikes in the plot, occurring at regular 5 mm, 10 mm, or 20 mm intervals depending on the specimen geometry.

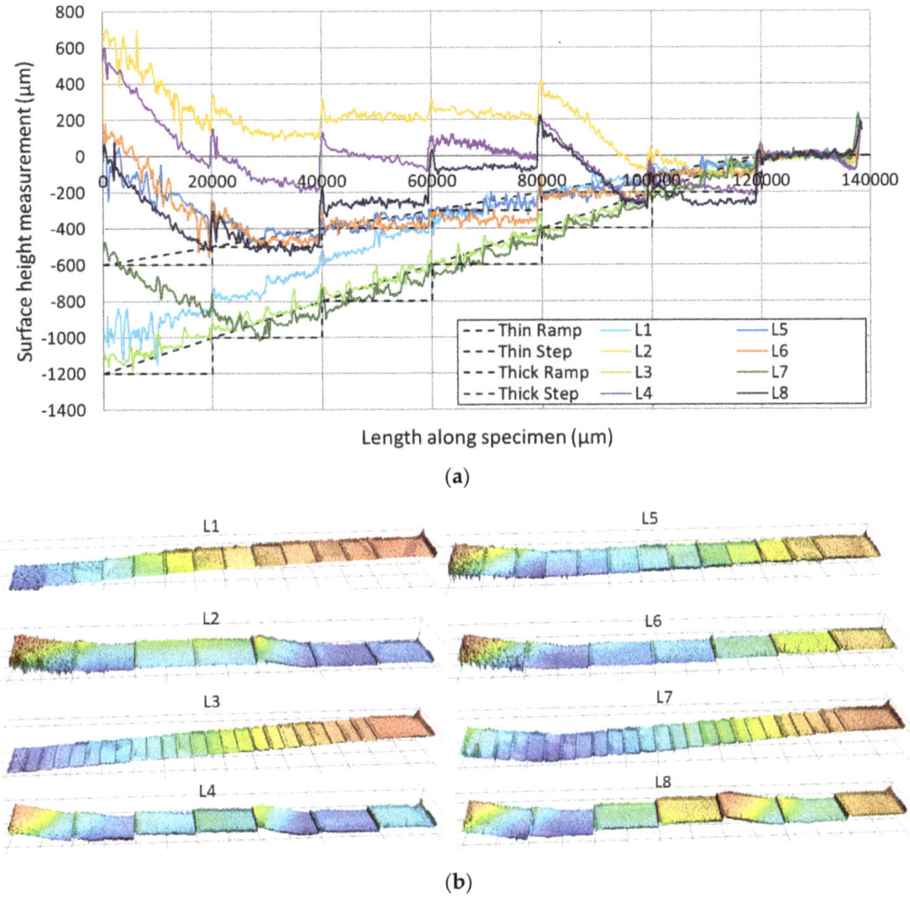

Figure 9. Upper surface measurements of the lower thin and thick ramp and step specimens: (**a**) centerline profile versus the expected CAD geometry; (**b**) three-dimensional surface height plots provided as graphical representations of specimens L1 through L8, where the desired behavior is to gradually transition from blue to red as demonstrated by L3 (colors are not set to the same scale).

As anticipated, these thin titanium specimens were distorted from their CAD geometry. The only exception was the thicker ramp specimen, L3, which most closely matched the CAD profile, only deviating by about 50 µm in the thinnest 200 µm region. Specimens L5, L6, and L7 also showed good agreement with the CAD profile beyond the thinnest 30 mm of the span length, but prior to that point, veered sharply upwards by a total of 600–700 µm. Specimen L1 still resembles a ramp, but distorts downward a total of 300 µm, from the mid-span to the thinner regions, at a rate of 10 µm in height per mm.

Other than L6, the step specimens (L2, L4, and L8) all displayed a different mode of distortion from the ramp specimens, indicating the level of dimensional accuracy in EB-PBF is affected by the geometry being manufactured. Relative to the top step surface, the thin and thick step specimens deviated sharply upwards at the 5th step, and only a few steps appear flat. Most of the steps appear to have a different slope, and the total deviation for L2, L4, and L8 was approximately 1.25 mm, 1.8 mm, and 1.3 mm, respectively, along the 140 mm length.

The remaining upper specimens U1–U8, that were manufactured off of the build plate with floating supports, showed significantly higher levels of dimensional deviation, particularly at the extremities, ranging from 2 mm to 4 mm. Those eight specimens were distorted to the point that an adequate reference plane could not be established from which to make direct comparisons. However, the middle thickness regions of the upper specimens agreed with the CAD geometry, similar to what L3 and L7 demonstrated. Overall, dimensional accuracy is typically worse when associated with large flat surfaces that are parallel to the build plane. No two specimens of the same CAD geometry demonstrated the same distortion behavior.

Shading differences in the upper surface topography in Figure 7 clearly show differences in hatching scan angles for different layers. It is well known that AM machines will rotate the hatching angle between layers as a method to reduce anisotropy in the material microstructures and prevent large grain growth in the build direction. Because the ramp specimens revealed the hatch topography for every layer, a closer inspection of the hatch angles was performed with the expectation that the hatching angle would be the same within any given layer across all specimens. Figure 10 plots the exposed hatch angles for each specimen for the given design thickness layers.

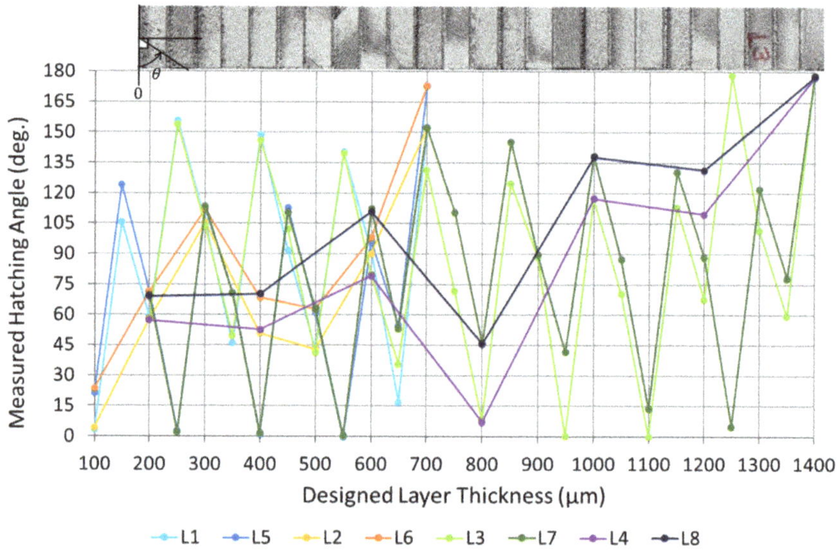

Figure 10. Hatching scan angles for specimens L1-L8 as measured by visual inspection of the optimal images.

For example, at the 200 µm level, specimens L1-L4 are all hatched with a 58-degree scan angle, while L5-8 were hatched with a 70-degree scan angle. However, at 400 µm, L1 and L3 were hatched at a 147-degree scan angle, while the immediately adjacent specimens L2 and L4 were hatched at a 52-degree angle. While a few degree measurement error may exist in the accuracy of the visual inspections, it cannot account for a 95 degree difference. At that particular layer, the machine software chose to melt the ramp specimens, which were not next to each other, at a vastly different angle than the step specimens. Figure 10 reveals that no two specimens, regardless of the geometry or build location, have the same hatching angle history, which would explain the variations in dimensional accuracy.

3.2. Macro-Scale Beam Experiment

Following fabrication, the topology-optimized tapered box beam specimens exhibited a rough as-built surface condition that is typical of the powder bed fusion process. Surface curvature and space-frame member transitions of the parts in this experiment were smooth, without facets, and surfaces opposite the build direction that were not supported had a rougher surface than parallel or top surfaces. Since the components were at an angle such that no final surface was parallel with the build plane, a fairly uniform surface condition was obtained. Support features were manually removed with dykes and smoothed using a tungsten carbide rotary tool, while both the forward and aft ends of the beam were ground flat with a belt sander. Figure 11 shows a fully bonded beam specimen as well as a portion of the 3D scan data showing surface features and missing surface regions. All six beams were successfully assembled, which demonstrated the geometric accuracy based on a CAD interface clearance of only 0.1 mm.

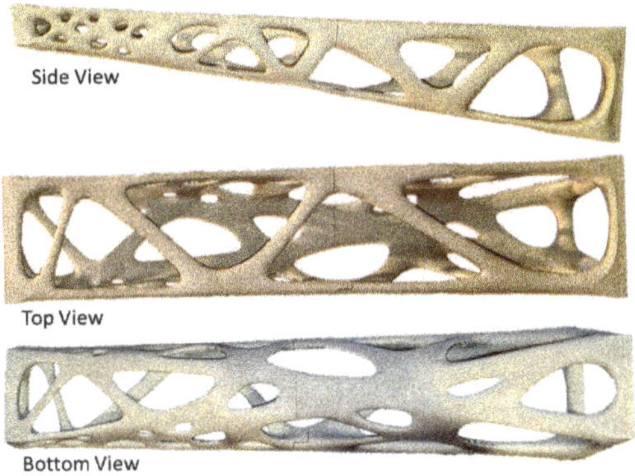

(a)

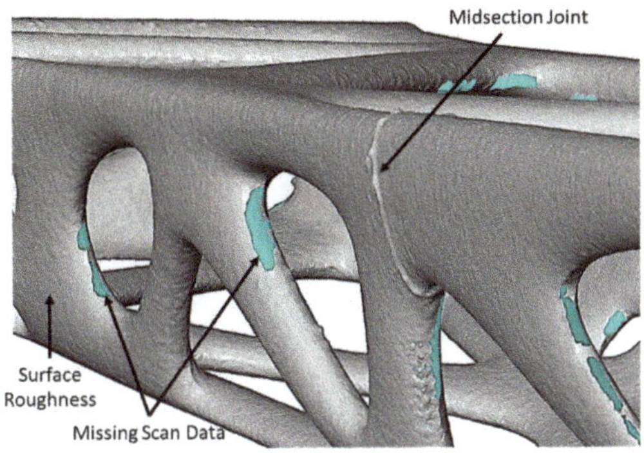

(b)

Figure 11. Macro-scale specimen of a topology-optimized taper box beam manufactured from Ti-6Al-4V via the EB-PBF process: (**a**) Side, top, and bottom view images, respectively, of one assembled specimen; (**b**) Example section of scanned STL data from specimen 1F.

A surface-best fit comparison of the six sets of scanned geometry to the CAD model had good results with alignment despite the absence of datum features. The alignment primarily matched the middle of the beams, such that the largest deviations in accuracy were revealed at the forward and aft extremities. Generally, all of the beam specimens were smaller than the CAD model. When overlaid with the nominal geometry, the outer surfaces of the scanned data were either equal to or inside the nominal surface, implying the printed beams had all thermally contracted as they cooled. The contraction is most notable at the forward and aft ends of the beams, as can be seen in the side views of Figure 12, where the transparent gray CAD surface protrudes beyond the colored scan surface. Scanned surfaces were only observed protruding from the CAD model on the inner surfaces of the beam that are depicted in yellow and red.

Figure 13 displays the results of the length and volume measurements of each of the six beam specimens. The designed nominal length (L_0) from the aft plane to the forward plane was 378.59 mm; however, the longest specimens, 3B (build: 3, location: back) and 1B (build: 1, location: back), were measured at 376.4 mm, 2.19 mm shorter than nominal. Specimen 3F (build: 3, location: front) was the shortest of the 6 beams, which measured 376.14 mm, 2.45 mm shorter than nominal. This provides a deviation range across the six beams of only 0.26 mm. The average length of the six beams was 376.29 mm, which is 2.30 mm, or 0.6%, shorter than nominal, but with a standard deviation of only 0.036%, or 0.10 mm (0.004 inches).

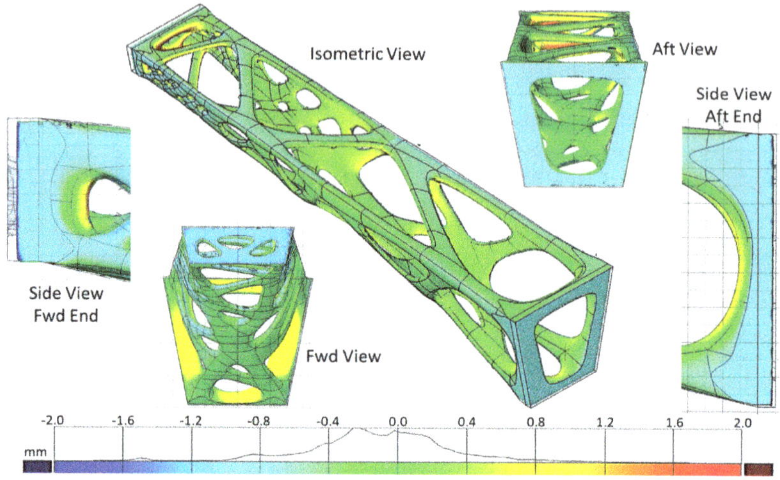

Figure 12. Example of geometric best-fit surface analysis comparison to nominal CAD using scan results of specimen 1F. The scanned surface is overlaid with a transparent CAD model, where blue indicates that the scanned surface resides inside the CAD geometry and yellow indicates that it is protruding from the CAD geometry.

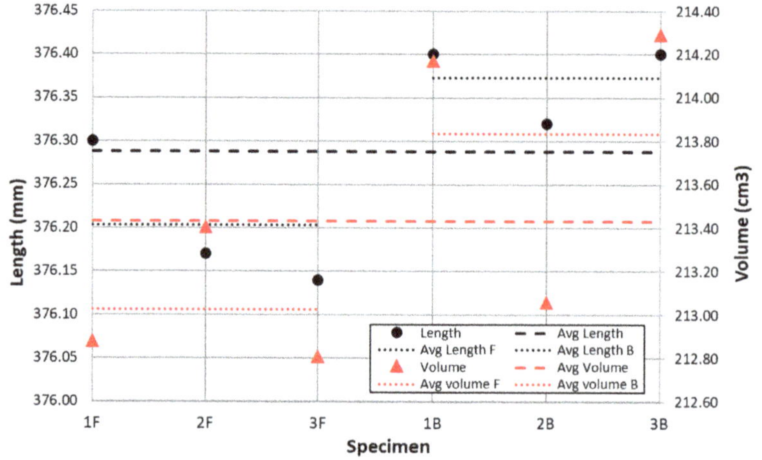

Figure 13. Plot of the six beam specimens' total length and total volume as measured from 3D scan STL files, grouped by build location. For reference, the designed nominal length (L_0) is 378.59 mm.

Despite the laser scanner not being able to completely scan all surfaces, a total volume measurement was still possible and was compared with the nominal CAD volume of 222.11 cm^3. Similar to the length parameters, the largest and smallest volume specimens were 3B and 3F, respectively, where the volume of 3B measured 214.29 cm^3 and that of 3F measured 212.80 cm^3, giving a range of 1.49 cm^3 (0.09 in^3). At an assumed titanium density of 4.43 g/cc, that reveals a weight variation of 6.6 g (0.015 lbs) amongst the 6 specimens. The average volume was 213.43 cm^3, which is a difference of 8.68 cm^3 from the nominal, or a 3.9% reduction in volume, and the standard deviation was very small at only 0.59 cm^3.

Table 1 displays the average and standard deviation of the specimens grouped by build location. Among those in the same location, the variation is smaller than the overall range. In Figure 13, the specimens were also grouped by location within the build chamber instead of by build number, displaying the beams printed in the front of the chamber followed by those printed in the back. From this perspective, with the same part geometry, there is more significant dimensional deviation based on location within a build than from a series of builds. With respect to length, all the front specimens were shorter than those in the back, and while there was one back beam that had a volume measurement similar to the front beams, the average volume of the back beams was higher.

Table 1. Average length and volume measurements of the six topology-optimized tapered box beam specimens grouped by build location and the CAD nominal values for comparison.

Specimen Group	Length (mm)	Volume (cm^3)
Avg. Front (F)	376.20	213.03
Range Front (F)	0.16	0.60
Std. Dev. Front (F)	0.07	0.27
Avg. Back (B)	376.37	213.83
Range Back (B)	0.08	1.24
Std. Dev. Back (B)	0.04	0.55
Delta Avg. (F−B)	−0.17	−0.81
Delta Std. Dev. (F−B)	0.03	−0.29
CAD Nominal	378.59	222.11

4. Discussion

In the meso-scale step-ramp experiment, the anticipated porosity from sparsely supported melt regions was revealed in the thinnest regions of the specimens. High amounts of top-layer surface-connected porosity were observed in design thicknesses below 400 µm. This experiment revealed that the intra-layer melt topology varied with a dynamically changing hatch angle. Arcam states that their EB-PBF machine control software performs hatch optimization, such that it will rotate the scan angle to optimize scan lengths. This is tied into the current compensation function process parameter for the purpose of increasing productivity by reducing the number of scan lines [24]. It also involves merging scan lines from different regions and reordering part scans.

However, different hatch angles between components of the same geometry demonstrated a source of variability. The same model, printed at different z-heights and locations within the build chamber, will result in differing dimensional deviations if the scan path of the parts throughout the layers is not identical [7,10]. The hatching history is therefore akin to weld sequencing for a large weldment assembly. A proper weld sequence will minimize distortion of the overall assembly, whereas a change in the weld sequence will change how the weld assembly distorts. Furthermore, a randomized welding sequence will result in a chaotic dimensional deviation such that no two assemblies will be identical. That was the case for the step and ramp specimens in this experiment, which is undesirable for dimensional stability, quality, and repeatability.

From the macro-scale topology-optimized tapered box beam experiment, the most significant result was that there was more variability from part location within the build chamber than there was from build to build. As exhibited by the step-ramp specimens, this was due to each part of the build having a different hatching history. When parts with identical hatching histories are compared from build to build, the dimensional variability is reduced. However, the beams were all significantly shorter than the CAD model, such that the deviation due to thermal contraction dwarfed the level of part-to-part variation by an order of magnitude. This result indicates that the EB-PBF process is repeatable, and the beams could have more closely matched the CAD model if they had been compensated for thermal expansion at the elevated build chamber temperature [21,23].

The thermal contraction of the titanium beams can be quantified by the overall length measured from the scanned surfaces. The standard equation for linear thermal expansion is shown in Equation (1):

$$\Delta L = \alpha L_0 \Delta T \quad (1)$$

assuming a Ti-6Al-4V linear coefficient of thermal expansion (CTE) (α) to be 9.70 µm/m-°C up to 650 °C, the predicted change in length (ΔL) at an approximate 625 °C change in temperature (ΔT) is 2.30 mm over the beam span, matching the average 2.30 mm deviation from nominal. This also led to a non-trivial volume deviation of 3.9% and further stresses the need for compensating CAD models for thermal expansion prior to converting them to an STL file to be manufactured in the EB-PBF process. While only Ti-6Al-4V was utilized in this work, the demonstrated mechanisms of dimensional deviation discussed here would also be seen with other materials, to varying degrees. Improved accuracy by means of implementing the same scan path for the same geometry or by compensating the model for CTE (625 °C for Ti-6Al-4V) would not impact the material properties or microstructure.

5. Conclusions

Two experiments were conducted at different length scales, meso- and macro-, to quantify the dimension accuracy of the EB-PBF process using Ti-6Al-4V and to identify the sources of variability. The meso-scale experiment used thin step and ramp specimens to capture deviation as thickness increased, while the macro-scale experiment used a topology-optimized tapered box beam to ascertain the deviation across the build chamber and multiple builds. The following conclusions were reached:

- for thin samples of the same geometry, different hatching histories reduced the dimensional accuracy, causing random deviations many times the thickness, up to 4 mm;
- for large models of the same geometry, the range of dimensional deviation across three builds was 0.26 mm;
- the average 2.3 mm deviation from the nominal geometry could have been mitigated if the tapered box beam models had been compensated for thermal expansion using a ΔT of 625 °C;
- a high degree of geometric accuracy at the assembly level was demonstrated with a modeled interface clearance of 0.1 mm.

A limitation of this work is that it was only conducted using an EB-PBF system with Ti-6-Al-4V, and results would likely vary if the same experiments were performed on a LB-PBF system due to the differences in beam energy and build temperature. Furthermore, the lack of a datum scheme for the specimen models complicates the repeatability of the metrological measurements. This work is directly applicable to any metal fusion processing, which would include laser-based PBF. Future work will involve the study of changes in microstructure and microhardness with respect to thermal mass as well as the characterization of internal porosity utilizing µCT.

Author Contributions: Conceptualization, M.R. and E.B.; methodology, M.R. and E.B.; software, E.B.; validation M.R. and E.B.; formal analysis, E.B.; investigation, M.R. and E.B.; resources, M.R.; data curation, M.R. and E.B.; writing—original draft preparation, E.B.; writing—review and editing, M.R. and E.B.; visualization, E.B.; supervision, M.R.; project administration, M.R.; funding acquisition, M.R. All authors have read and agreed to the published version of the manuscript.

Funding: This work was partially accomplished using facilities funded by the Joint Center for Deployment and Research in Earth Abundant Materials (JCDREAM) in Washington State. The authors sincerely acknowledge the Boeing Pennell Professorship and the Boeing Learning Together Program for educational funding in support of this work.

Data Availability Statement: Data from these experiments may be made available upon request to the corresponding author; however, any data deemed proprietary or private cannot be transferred.

Acknowledgments: The authors would like to acknowledge Bill Kuykendall of Mechanical Engineering for his assistance on the ARCAM machine during this investigation.

Conflicts of Interest: The authors declare no conflict of interest.

References

1. Seepersad, C.C. Challenges and Opportunities in Design for Additive Manufacturing. *3D Print. Addit. Manuf.* **2014**, *1*, 10–13. [CrossRef]
2. Persenot, T.; Burr, A.; Martin, G.; Buffiere, J.-Y.; Dendievel, R.; Maire, E. Effect of build orientation on the fatigue properties of as-built Electron Beam Melted Ti-6Al-4V alloy. *Int. J. Fatigue* **2019**, *118*, 65–76. [CrossRef]
3. Riedlbauer, D.; Scharowsky, T.; Singer, R.F.; Steinmann, P.; Körner, C.; Mergheim, J. Macroscopic simulation and experimental measurement of melt pool characteristics in selective electron beam melting of Ti-6Al-4V. *Int. J. Adv. Manuf. Technol.* **2017**, *88*, 1309–1317. [CrossRef]
4. Boone, N.; Zhu, C.; Smith, C.; Todd, I.; Willmott, J. Thermal near infrared monitoring system for electron beam melting with emissivity tracking. *Addit. Manuf.* **2018**, *22*, 601–605. [CrossRef]
5. Landau, E.; Tiferet, E.; Ganor, Y.; Ganeriwala, R.; Matthews, M.; Braun, D.; Chonin, M.; Ziskind, G. Thermal characterization of the build chamber in electron beam melting. *Addit. Manuf.* **2020**, *36*, 101535. [CrossRef]
6. Wojtuszewski, R.; Banas, A.; Oliwa, M. Additive manufacturing of titanium alloys. In *Annual Forum Proceedings*; AHS International: Fairfax, VA, USA, 2018; pp. 1–18.
7. Sames, W.J.; List, F.A.; Pannala, S.; Dehoff, R.R.; Babu, S.S. The metallurgy and processing science of metal additive manufacturing. *Int. Mater. Rev.* **2016**, *61*, 315–360. [CrossRef]
8. Cordero, Z.C.; Meyer, H.M.; Nandwana, P.; Dehoff, R.R. Powder bed charging during electron-beam additive manufacturing. *Acta Mater.* **2017**, *124*, 437–445. [CrossRef]
9. Zhao, X.; Dadbakhsh, S.; Rashid, A. Contouring strategies to improve the tensile properties and quality of EBM printed Inconel 625 parts. *J. Manuf. Process.* **2021**, *62*, 418–429. [CrossRef]
10. Körner, C. Additive manufacturing of metallic components by selective electron beam melting—A review. *Int. Mater. Rev.* **2016**, *61*, 361–377. [CrossRef]
11. Townsend, A.; Senin, N.; Blunt, L.; Leach, R.K.; Taylor, J.S. Surface texture metrology for metal additive manufacturing: A review. *Precis. Eng.* **2016**, *46*, 34–47. [CrossRef]
12. Newton, L.; Senin, N.; Chatzivagiannis, E.; Smith, B.; Leach, R. Feature-based characterisation of Ti6Al4V electron beam powder bed fusion surfaces fabricated at different surface orientations. *Addit. Manuf.* **2020**, *35*, 101273. [CrossRef]
13. Lee, S.; Rasoolian, B.; Silva, D.F.; Pegues, J.W.; Shamsaei, N. Surface roughness parameter and modeling for fatigue behavior of additive manufactured parts: A non-destructive data-driven approach. *Addit. Manuf.* **2021**, *46*, 102094. [CrossRef]
14. Flys, O.; Berglund, J.; Rosén, B.-G. Using confocal fusion for measurement of metal AM surface texture. *Surf. Topogr. Metrol. Prop.* **2020**, *8*, 024003. [CrossRef]
15. Lou, S.; Zhu, Z.; Zeng, W.; Majewski, C.; Scott, P.J.; Jiang, X. Material ratio curve of 3D surface topography of additively manufactured parts: An attempt to characterise open surface pores. *Surf. Topogr. Metrol. Prop.* **2021**, *9*, 015029. [CrossRef]
16. Thompson, A.; Senin, N.; Giusca, C.; Leach, R. Topography of selectively laser melted surfaces: A comparison of different measurement methods. *CIRP Ann. Manuf. Technol.* **2017**, *66*, 543–546. [CrossRef]
17. Wang, P.; Sin, W.J.; Nai, M.L.S.; Wei, J. Effects of Processing Parameters on Surface Roughness of Additive Manufactured Ti-6Al-4V via Electron Beam Melting. *Materials* **2017**, *10*, 1121. [CrossRef] [PubMed]
18. Seewig, J.; Scott, P.J.; Eifler, M.; Barwick, B.; Hüser, D. Crossing-The-Line Segmentation as a Basis for Rsm and Rc Evaluation. *Surf. Topogr. Metrol. Prop.* **2020**, *8*, 024010. [CrossRef]
19. Bukkapatnam, S.T.; Iquebal, A.S.; Kumara, S.R. Planar random graph representations of spatiotemporal surface morphology: Application to finishing of 3-D printed components. *CIRP Ann.* **2018**, *67*, 495–498. [CrossRef]
20. Du Plessis, A.; Tshibalanganda, M.; Yadroitsava, I.; Yadroitsev, I. On the Evaluation of Surface Roughness: X-ray Tomography Reveals Hidden Details. In *Progress in Additive Manufacturing 2020*; Shamsaei, N., Seifi, M., Eds.; ASTM International: West Conshohocken, PA, USA, 2022; pp. 208–222.
21. Gao, W.; Zhang, Y.; Ramanujan, D.; Ramani, K.; Chen, Y.; Williams, C.B.; Wang, C.C.L.; Shin, Y.C.; Zhang, S.; Zavattieri, P.D. The status, challenges, and future of additive manufacturing in engineering. *Comput.-Aided Des.* **2015**, *69*, 65–89. [CrossRef]
22. Marrugo, A.G.; Gao, F.; Zhang, S. State-of-the-art active optical techniques for three-dimensional surface metrology: A review. *J. Opt. Soc. Am. A* **2020**, *37*, B60. [CrossRef]
23. Leach, R.; Bourell, D.; Carmignato, S.; Donmez, A.; Senin, N.; Dewulf, W. Geometrical metrology for metal additive manufacturing. *CIRP Ann.* **2019**, *68*, 677–700. [CrossRef]
24. Smith, C.; Derguti, F.; Nava, E.H.; Thomas, M.; Tammas-Williams, S.; Gulizia, S.; Fraser, D.; Todd, I. Dimensional accuracy of Electron Beam Melting (EBM) additive manufacture with regard to weight optimized truss structures. *J. Mater. Process. Technol.* **2016**, *229*, 128–138. [CrossRef]
25. Ameta, G.; Fox, J.; Witherell, P. Tolerancing and Verification of Additive Manufactured Lattice with Supplemental Surfaces. *Procedia CIRP* **2018**, *75*, 69–74. [CrossRef]

26. Gruber, S.; Grunert, C.; Riede, M.; López, E.; Marquardt, A.; Brueckner, F.; Leyens, C. Comparison of dimensional accuracy and tolerances of powder bed based and nozzle based additive manufacturing processes. *J. Laser Appl.* **2020**, *32*, 032016. [CrossRef]
27. Bol, E.; Doyle, C.; Ramulu, M. Process Parameter Effects on Melt Topology and Dimensional Deviation in Electron Beam Melted Ti-6Al-4V. In Proceedings of the ASME 2021 International Mechanical Engineering Congress and Exposition, Virtual, 1–5 November 2021; American Society of Mechanical Engineers: New York, NY, USA, 2021; Volume 85550, p. V02AT02A011.
28. Bol, E.; Ramulu, M. Repeatability of a Topology Optimized Tapered Box Beam Additively Manufactured with Electron Beam Melted Ti-6Al-4V. In Proceedings of the ASME 2020 International Mechanical Engineering Congress and Exposition, Virtual, 16–19 November 2020; American Society of Mechanical Engineers: New York, NY, USA, 2020; Volume 84485, p. V02AT02A049.
29. Zhang, X.; Leary, M.; Tang, H.; Song, T.; Qian, M. Selective electron beam manufactured Ti-6Al-4V lattice structures for orthopedic implant applications: Current status and outstanding challenges. *Curr. Opin. Solid State Mater. Sci.* **2018**, *22*, 75–99. [CrossRef]
30. Damri, E.; Tiferet, E.; Braun, D.; Ganor, Y.; Chonin, M.; Orion, I. Effects of Gas Pressure during Electron Beam Energy Deposition in the EBM Additive Manufacturing Process. *Metals* **2021**, *11*, 601. [CrossRef]
31. Bendsøe, M.P. Optimal shape design as a material distribution problem. *Struct. Optim.* **1989**, *1*, 193. [CrossRef]
32. Zhou, M.; Rozvany, G. The COC algorithm, Part II: Topological, geometrical and generalized shape optimization. *Comput. Methods Appl. Mech. Eng.* **1991**, *89*, 309–336. [CrossRef]
33. Mlejnek, H.P. Some aspects of the genesis of structures. *Struct. Optim.* **1992**, *5*, 64–69. [CrossRef]
34. Muñoz, D.; Nadal, E.; Albelda, J.; Chinesta, F.; Ródenas, J. Allying topology and shape optimization through machine learning algorithms. *Finite Elem. Anal. Des.* **2022**, *204*, 103719. [CrossRef]
35. Zhang, L.-C.; Wang, J.; Liu, Y.; Jia, Z.; Liang, S.-X. Additive manufacturing of titanium alloys. In *Reference Module in Materials Science and Materials Engineering*; Elsevier: Amsterdam, The Netherlands, 2020.

designs

MDPI

Article

By Visualizing the Deformation with Mechanoluminescent Particles, Additive Manufacturing Offers a Practical Alternative to Stress and Strain Simulation

Ernests Einbergs *, Agnese Spustaka, Virginija Vitola and Aleksejs Zolotarjovs

Institute of Solid State Physics, University of Latvia, Kengaraga Str. 8, LV 1063 Riga, Latvia
* Correspondence: ernests.einbergs@cfi.lu.lv

Abstract: The use of stress–strain analysis in structural design or mechanical components is critical for avoiding or investigating structural failures. In the case of complicated designs, mathematical full-field stress modeling produces imprecise predictions. Experimental analysis can be used as a replacement for mathematical modeling, but with the use of currently available strain gauges, it is cumbersome and impossible in the case of moving parts. Mechanoluminescent materials transform mechanical energy into visible light and can be used as a replacement for strain gauges to monitor strain/stress. Three-dimensional printing technology has made major advances in terms of additive manufacturing. In this article, we describe a method to produce an ML 3D print. The fabricated samples are precise and versatile and satisfy the need for easy and non-destructible spatial stress analysis. A 3D printed photopolymer sample with $SrAl_2O_4$: Eu, Dy particle addition only to the final layers was tested, and the number of layers was optimized. It was determined that the optimal number of layers for easy detection is in the range of 10 to 20 layers. It opens the possibility for the real-time evaluation of complex uneven forces on complex parts, thus having a good potential for commercialization.

Keywords: mechanoluminescence; additive manufacturing; 3D printing; stress analysis; mechanical components; non-destructive; full-field strain measurements

Citation: Einbergs, E.; Spustaka, A.; Vitola, V.; Zolotarjovs, A. By Visualizing the Deformation with Mechanoluminescent Particles, Additive Manufacturing Offers a Practical Alternative to Stress and Strain Simulation. *Designs* **2023**, *7*, 54. https://doi.org/10.3390/designs7020054

Academic Editor: Ali Zolfagharian

Received: 1 March 2023
Revised: 31 March 2023
Accepted: 4 April 2023
Published: 7 April 2023

1. Introduction

Three-dimensional printing technology is an additive manufacturing technique for producing a variety of complex structures from three-dimensional models. Manufacturing and logistics operations can be improved by 3D printing, which has evolved over the years and includes a wide range of methods, materials, and tools. In recent years, it has made significant advancements, as the expiration of earlier patents, which allowed manufacturers to create new 3D printing devices, is one of the key factors contributing to this technology's accessibility. Recent innovations have decreased the price of 3D printers, extending their use in laboratories, schools, and homes. The technology has enabled the production of functional prototypes, lightweight components, and customized tools, among other things. New applications are constantly appearing as a result of the constant development of cutting-edge materials and additive manufacturing techniques, significantly advancing a number of industries, including engineering and healthcare [1–3]. A need for specialized stress sensors has also arisen from the use of stimulus-responsive materials in 3D printing, where printed objects can change shape or qualities in reaction to external stimuli, such as heat [4], light [5], electricity [6], magnetism [7], water [8], mechanical stress, etc. Stress, pressure, and capacitive touch sensing have already been extensively used in a variety of fields, ranging from mobile phones to electronic touch screens, automobiles, and aircraft.

Mechanoluminescence (ML) is the non-thermal emission of light when a material is subjected to stress; thus, ML materials are capable of transforming mechanical energy into visible light [9,10]. This method of strain/stress monitoring with ML materials is currently under development, and the intensity of ML exhibits a strong correlation with the applied stress, making it suitable for stress sensing and even mapping the stress distribution.

ML materials have been studied for their potential use in various applications, including stress sensing, damage detection, and structural health monitoring [11–15]. They have been used in the development of smart materials, such as self-healing materials and sensors for detecting structural changes in buildings, bridges, and other structures. The method of active strain/stress monitoring with ML materials is currently being developed [16,17]. ML layers are a powerful addition to 3D prototyping, coupled with an optical sensor that can be used as a real-time stress sensor in a 3D-printed part.

While numerical analysis, such as that performed with COMSOL, can provide a simulated approach to stress distribution in parts [18,19], there are often significant differences between a computer model and an experiment that can be created by effects that are unaccounted for, such as sample warping during printing. This can lead to imprecise results, and the extensive testing of parts is usually required before they can be implemented in final assemblies. Conventional stress sensors, such as piezoresistive stress sensors, are commonly used for this purpose. However, they are not always suitable, particularly in cases where the part is in motion, such as rotating gear. Furthermore, conventional stress sensors only provide information about the stress at the point, line, or interface where the sensor is applied, and do not provide a full picture of the stress distribution within the mechanical part. Therefore, alternative methods of stress sensing have been sought. The incorporation of mechanoluminescent layers into 3D-printed parts, along with an optical sensor, offers a promising solution for real-time stress sensing and mapping in 3D-printed parts. This technology provides a non-destructive method for detecting stress in various parts of a complex design, including moving parts, and can offer valuable insights into the stress distribution within the printed structure.

This article explores the possibility of adding ML powder ($SrAl_2O_4$: Eu, Dy) to the final layers of a 3D-printed mechanical part, along with a method for real-time stress mapping in the printed part. The number of layers with ML particles that are most efficient is deduced in terms of the ML intensity and the spatial resolution. This research has the potential to advance the field of 3D printing and provide valuable insights into the stress distribution of printed structures.

2. Materials and Methods

While calculating the stress–strain analysis in uniform, elementary objects are relatively simple, obtaining an analytical solution for more complicated geometric shapes becomes difficult. Our method is preferred for determining structural weak points in components with non-uniform stress distribution. A model was developed with the goal of generating non-linear stress distribution during deflection. A dog-bone similar shape was chosen with the dimensions of $100 \times 30 \times 7$ mm. The center is 12 mm wide, a spline function is employed to form the desired shape, and 3 circular cutouts (radius of 5 mm) are added to create structural weaknesses. The aforementioned shape is shown in Figure 1. Large cutouts at both ends of the designed shape serve no purpose other than reducing the overall amount of photopolymer used during LCD stereolithography.

The model was created with Fusion 360 and sliced with Chitubox v1.9.4. The printing parameters were as follows: layer height of 0.01 mm, bottom exposure of 20 s, exposure of 5 s, movement speed of 110 mm/s, lift distance of 5 mm, and rest time of 5 s. It is important to keep the layer height low while printing with a polymer and ML particle mixture, as the particles absorb a considerable part of UV illumination. Overall movement speed was irrelevant, but a pause (rest time) between layers had to be implemented due to the mixture having a higher viscosity than a pure photopolymer and taking longer to fill the void left by the previous layer.

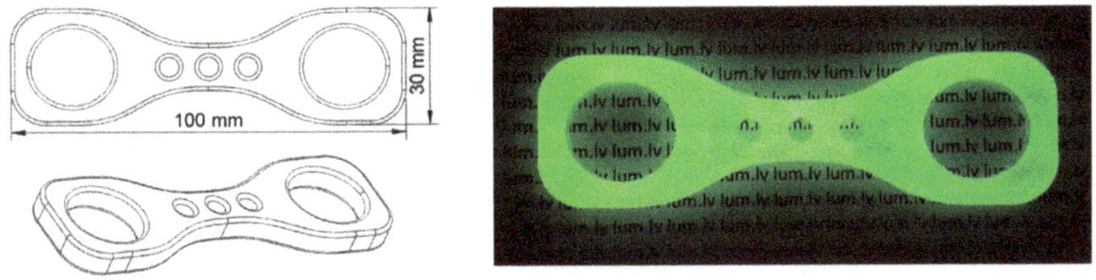

(**a**) (**b**)

Figure 1. (**a**) Used dog-bone-like design as seen in Fusion360; (**b**) sample with 20 luminescent layers 30 s after UV excitation.

Multiple research groups have extensively examined strontium aluminate [12,16,20,21], which is among the most studied materials with long-lasting and predictable, repeatable afterglow and ML properties. Therefore, this material was chosen as the ML powder for this research. Prints were made with a conventional and affordable stereolithography (SLA) printer Elegoo Mars 2 Pro. Before application, the polymer and ML powder mixture was stirred with Heathrow Scientific Vortex for 30 min, after which it was poured into the reservoir. Transparent ABS-like photopolymer purchased from the printer's manufacturer Elegoo was mixed with monoclinic $SrAl_2O_4$ powder (purchased from Sigma Aldrich) activated with 1 at% Eu^{2+} and 2 at% Dy^{3+} at a ratio of 20 to 1. A preliminary study was performed that determined that the ratio produces samples reliably without meaningful surface defects. Considerably higher concentrations do not suspend for long enough within the photopolymer during printing and settle faster than is desirable. The bulk of the sample was printed with pure photopolymer because, during a bending flexural test, only the outermost layers are subjected to stress. The inner structural layers can be pure photopolymer for cost-saving purposes. For the final N layers (2, 5, 10, 20, 30), the holding vat was removed, cleansed with pure ethanol, and filled with the polymer and ML particle mixture, resulting in a 0.02–0.30 mm thick ML layer. For samples that were meant to contain more than 5 layers of the luminescent material, the print was paused every 5 successful layers and stirred to facilitate the highest achievable particle distribution homogeneity. If the sample displayed any surface defects or fractures during the rest time of the print, the procedure was stopped and restarted from the beginning.

A custom-built measurement apparatus that is capable of deformation and the collection of the emitted light and can provide the precise measurement of displacement, the load applied to the sample as well as the sensitivity to light was employed to measure ML. The setup and working principles of this system are described in detail in [22]. The system is designed around a leadscrew-driven cart system. Three-point flexural tests [23] were performed with the aid of a Nema 23 stepper motor coupled with a 1 to 10 planetary gear set to amplify the force generated. An SFU2005 ballscrew is connected to the gearset shaft with a rigid coupler. A 3D-printed mount connects the ball-bearing cart and the screw nut on the lead screw. The mount is equipped with a load pin with a diameter of 15 mm and a height of 30 mm. The constructed system moves with an accuracy of 6 ± 2 μm per step. Measurements were performed by capturing 16-bit pixel depth images with CMOS Blackfly BFLY–U3–23S6N camera at 2 fps (frames per second) with all built-in image processing functions disabled. Although the camera supports up to 41 fps, a higher framerate is undesirable due to the decrease in the dynamic range under low light conditions and decreased signal-to-noise ratio with the built-in amplifier enabled. Samples were excited for 60 s with a wide spectrum UV light source (450 nm peak).

3. Results

Both afterglow and ML exhibit the same emission spectrum, indicating that Eu^{2+} is the luminescence center involved in both temperature and stress-induced recombination. The emission spectrum of the material is a typical Eu^{2+} emission spectrum [16,24–26]—broad emission band with a maximum at 530 nm (green light) under UV excitation arising from the luminescent $4f^65d \rightarrow 4f^7$ transition. The Eu^{2+} luminescence centers are involved in both temperature and stress-induced recombination because both the afterglow and mechanoluminescence have the same spectrum. The sample exhibits afterglow that is comprised of two components—fast decay and slow decay. The afterglow is very predictable in the same circumstances (mainly temperature). The fast decay arises from charge carrier thermal release from the trapping centers, and the slow decay arises from the emptying of the energetically deeper trapping centers booth by thermal energy and the tunneling of the charge carriers (electrons) to the luminescence center; it is believed that it might also be the mechanism that is responsible for the observed ML during deformation [12,20,27–30] because the probability of tunneling is distance dependent.

A delay between the sample illumination and spatial intensity acquisition was implemented to avoid over-exposure of the CMOS camera during intensive afterglow [21]. The fast component of Eu^{2+} center decay has a significantly higher intensity than that of ML observed during deformation, and it became hard to distinguish ML from the background signal. Therefore, measurements were initiated 3 min after the end of UV irradiation. A shorter delay will yield a lessened ML signal when corrected by subtracting the background signal (further contrast), and a longer delay will make the ML signal more distinguishable at the expense of overall signal intensity.

Luminescence intensity was recorded during one loading-unloading cycle. A region of interest was cut out of the acquired images spanning the length and width of the sample and was integrated to display the change in luminescence intensity. Figure 2a shows the light emission increase as a result of applied stress; for illustrative purposes, afterglow emission was subtracted to highlight the ML. Figure 2b shows the corrected signal and deflection of the sample. High response to even small deformations (starting from 1.6 mm deflection) was observed. No plastic deformation was observed in the range of deflections used. The resulting increase in luminescence emission is clearly distinguishable and visible.

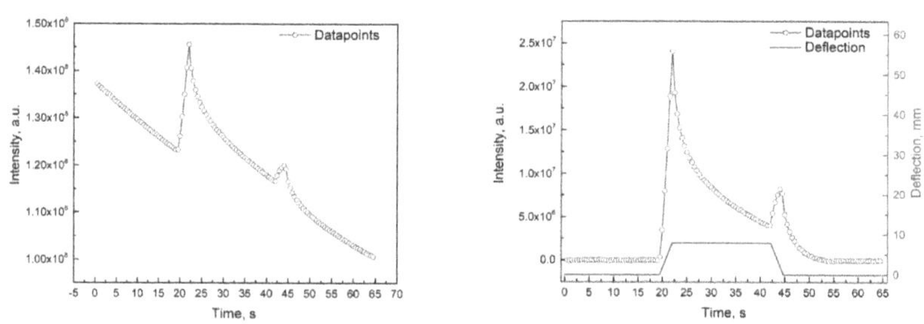

(a) (b)

Figure 2. (**a**) Integral luminescent emission in one loading–unloading cycle of the 3D-printed sample with 10 luminescent layers; (**b**) deflection of sample and illustration of ML without background afterglow of $SrAl_2O_4$: Eu, Dy.

The measurement error was estimated by repeatedly performing the same measurement under identical conditions. To evaluate measurement repeatability, samples were deformed by 6 mm in an identical experimental environment. It was assumed that the degradation of the photopolymer is statistically insignificant for a population of 10 measurements and the expected value is constant. As no operator input was necessary during repeated tests, it was assumed that the deviation from the average value directly represents the inconsistency of the system itself. The residual from the calculated average contrast is shown in Figure 3. The coefficient of variation was determined to be 4.98%. It was assumed that the overall error of the constructed system would not exceed 10% of the absolute measured value and was added to Figures 4 and 5.

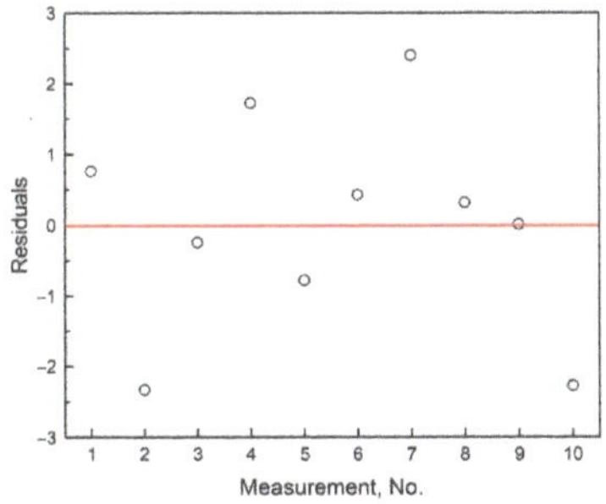

Figure 3. The estimation of the contrast measurement error is 4.98%, based on calculating the relative standard deviation after numerous repetitions of a single loading–unloading cycle.

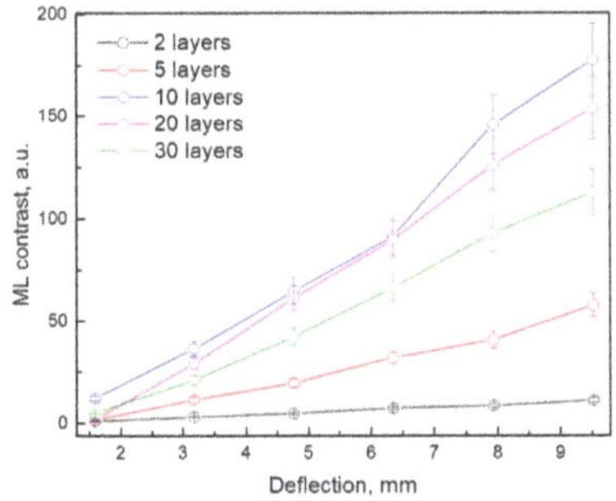

Figure 4. Contrast of ML for samples with N (2, 5, 10, 20, 30) layers changing maximum deflection with an upper estimate of the measurement error of 10%.

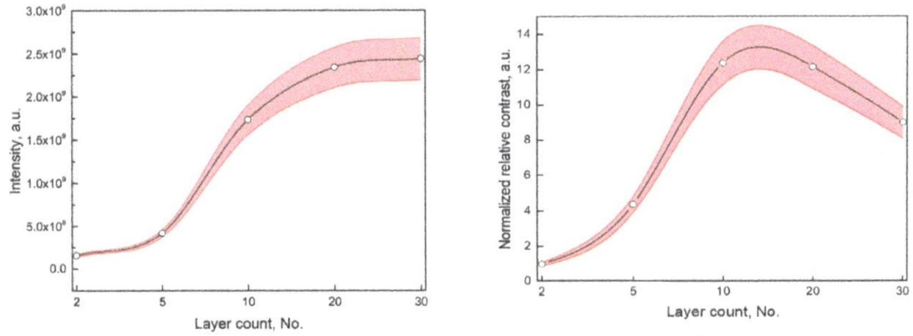

(**a**) (**b**)

Figure 5. (**a**) The integral luminescence intensity of 3D-printed samples with different counts of ML layers; (**b**) the normalized contrast for 8 mm deflection; points are connected with a spline function for illustrative purposes; the red region represents the estimated 10% error.

For rapid development purposes, it is important to determine the optimal number of layers for stress distribution visualization. Since ML additives are multiple times more expensive compared to photopolymers, a cost-saving measure was implemented, where only the outer layers during a 3D print are filled with ML particles. There are several reasons as to why this was deemed not to impact the final result. The layers that are closer to the neutral axis are subjected to smaller strain and in turn, their contribution to the overall ML signal is proportionally smaller. As the particles will settle eventually, fewer layers will require less interruption of the print procedure and manual mixing. Therefore, a few layers close to the surface of the sample will provide the highest contrast due to being as far from the neutral axis as possible, but due to the absence of luminescent material, the amount of light emitted during ML is harder to perceive and distinguish from the background signal. During the determination of the optimal number of layers, the contrast was calculated by subtracting the afterglow intensity from the peak observed during deflection. Results are shown in Figure 4. Samples were not mutually normalized to focus on the easiest-to-detect ML signal. If measurements are normalized, the highest contrast is observed for the samples with a smaller amount of ML layers, but it requires a sensitive registration device, which would prove to be inaccurate if less sensitive equipment is used. Figure 5 displays the difference between overall intensity and the determined contrast during ML.

From experimental observations, the highest cost performance is achieved by implementing no fewer than 10 surface layers printed from a mixture of $SrAl_2O_4$: Eu, Dy and photopolymer at a ratio of 1:20. Samples containing 10 luminescent layers displayed almost the same ML contrast as the sample with 20 layers, though slightly higher, but it can be disregarded within the determined margin of error. Samples with 5 and 2 layers have drastically weaker overall luminescence intensity, and the contrast decreases around 3 times when compared to the achievable luminescence with 10 and 20 layers. The sample with 30 layers has the brightest afterglow and, as mentioned previously, has a reduced contrast due to the elevated baseline from layers weakly contributing to the effect of ML. Due to the relatively high cost of ML powder and the fact that the ML signal is strong enough to be noticeable, using fewer than 10 layers of the ML powder and photopolymer mixture is feasible.

The main point of interest in this research, however, is not the integral ML intensities response to strain, as that was already proven as an indication of deformation amplitude by other researchers. Here, we set the main focus on the spatial real-time visualization of mechanical stress. Spatial stress distribution can be deduced in real time by employing the same method described in one of our previous works [23]. For samples with a low contrast, it is possible to rely on the linearity of ML. When a sample is subjected to a constant stress gradient (deformed with constant speed), the rise in intensity is linear, which is observable in Figure 2b as well as in the research by other groups [12,16,24]. By taking each pixel of the image or a region of interest and performing a linear fit over time, the slope in the linear regression formula represents the intensity change during deformation. A simpler approach can be employed in cases with high contrast. ML is a form of stimulated luminescence; any photon released during mechanical interaction is produced at the expense of afterglow. By observing the change in light intensity between a frame during deflection and the beginning of the fast decay right after the end of mechanical interaction, depleted regions can be observed. Due to the fact that ML intensity is related to the absolute stress to which the sample is subjected, it is reasonable to anticipate that the depleted regions are directly linked to surface regions subjected to higher stress. Figure 6a shows the described difference in registered light intensity by subtracting the two frames. Minor image inconsistency is expected due to the reflective surface of the load pin. Figure 6b shows the theoretically calculated von Mises stress distribution. The calculated values are not applicable to the printed samples and are not provided since distribution is geometry related and the absolute value is dependent on the material property.

The distribution that was theoretically calculated and the obtained image match well. Theoretical calculations were performed with COMSOL Multiphysics. The geometry used for calculations was imported from the same file used for the 3D printing of the samples. To minimize the degrees of freedom, a number of simplifications were introduced. It is impossible to create a true-to-life model with no limitations since the calculation will never converge. The supporting pins were considered to be rigid bodies that could be replaced by a fixed-end constraint. The loading pin was modeled as an ideal rigid cylinder. The sample consists of layers. The anisotropic mechanical properties were ignored, and the sample was assumed to be isotropic. The surface layers are a composite structure with anisotropic properties, which may be dependent on print parameters. The resolution of acquired frames was reduced to decrease the noise of the image. The reduction was performed by adding together values of 9 pixels arranged in a 3 by 3 square region. If further resolution reduction was applied, the overall image noise would decrease, but the surface features of the sample would become harder to distinguish. If a good print can be achieved with an ML particle-to-polymer ratio higher than 1:20, it would lower the image noise and the number of layers necessary to estimate surface stress distribution.

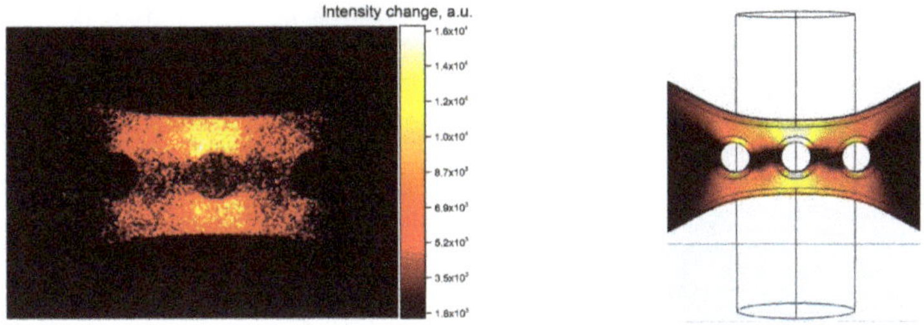

(a) (b)

Figure 6. (**a**) Obtained image after subtracting two consecutive frames; (**b**) numerical model of stress distribution.

4. Discussion

A novel method for visualizing structural weak points is tested and proven to be viable. The additive manufacture of ML scaled-down models in rapid development settings could provide unforeseen benefits. Three-dimensional printing of a prototype design and its observation with a camera is incomparably faster than performing theoretical calculations with the same accuracy required for a reasonable conclusion. The experimentally acquired stress distribution map is comparable to theoretically calculated models and encourages further methods' refinement for applications in practical circumstances. Further studies should focus on optimizing the ratio between photopolymer and ML material. The most effective excitation required for $SrAl_2O_4$: Eu, Dy luminescence is within the same spectral range as the photopolymer hardening; therefore, complementary research might determine the deterioration of the photopolymer. In the scope of this investigation, no meaningful change in mechanical properties was observed for the tested samples.

To make stress visualization with additive manufacturing more compelling for industrial applications, a combination of ML particles and polymers could be sought. Long-term UV exposure might cause the current combination to degrade, and ML particles further from the surface layers might not be excited and consecutively would not contribute to the observable ML.

Discrepancies between theoretical calculations and empirical observations might be caused by anisotropic mechanical properties. In Figure 6b, the local maxima in the von Mises stress can be observed around the outermost circular cutouts. The absence of intensity change in Figure 6a might be caused by the nonlinear force transfer between the layers and the fact that the layers observed have ceramic particles embedded. The observation of ML implies that either the polymer has adhered to the particles or a strong enough compressive force acts upon them.

Theoretical calculations were performed with the finite element method. This study focuses on von Mises stress. Similar research in the field has shown that a correlation between ML and maximal tensile stress and shear stress exists [12]. The von Mises stress is a measure of the equivalent stress in a material, taking into account tensile and shear stresses. While principal stress could be used as well, as it displays the stress component acting along the principal axis, it is a measure of the maximum and minimum normal stress on a material, while the von Mises directly visualizes the surface disruption magnitude while subjected to external forces and is not material property dependent.

5. Conclusions

A method for additively manufacturing mechanoluminescent samples by employing stereolithographic 3D printing was developed. Stress visualization and determination with an empirical method have immense potential in the rapid development phase of any production or construction process. The elimination of structural flaws in components or structural elements intended for industrial or civil applications has the potential to prevent design errors that would otherwise go unnoticed, given the complex nature of theoretical calculations. A complex 3D model was created, and theoretical stress distribution was calculated by COMSOL. The theoretical stress distribution during loading was compared to empirical measurements—the model was 3D printed with several finishing layers of ML powder, and the ML intensity was mapped during deformation. It was concluded that the empirical stress distribution obtained from a direct measurement method is in accordance with the calculated von Mises stress distribution and can therefore be improved for more complex technological uses. It was discovered that it is feasible to just apply ML powder to the top layers of the print for the sake of material conservation. The number of ML layers for the best ML contrast was found to be 10 layers. Future research into the appropriate ratio of ML particles to photopolymer has the potential to significantly reduce production costs even more.

This approach has the potential to be a valuable contribution to computational stress–strain analysis. It enables the real-time measurement of complicated unequal stresses on complex parts, resulting in a high commercialization potential.

Author Contributions: Conceptualization, A.S. and V.V.; methodology, E.E.; validation, A.Z. and V.V.; formal analysis, A.S. and E.E.; investigation, E.E. and A.S.; data curation, A.Z. and E.E.; writing—original draft preparation, E.E.; writing—review and editing, V.V., E.E. and A.S.; visualization, A.S. and E.E.; supervision, A.Z. All authors have read and agreed to the published version of the manuscript.

Funding: This work was supported by the European Regional Development Fund, grant number 1.1.1.1/20/A/138, 2021–2023. Institute of Solid State Physics, University of Latvia as the Center of Excellence has received funding from the European Union's Horizon 2020 Framework Programme H2020-WIDESPREAD-01-2016-2017-TeamingPhase2 under grant agreement No.739508, project CAMART2.

Data Availability Statement: The data presented in this study are available on request from the corresponding author.

Conflicts of Interest: The authors declare no conflict of interest.

References

1. Siripongpreda, T.; Hoven, V.P.; Narupai, B.; Rodthongku, N. Emerging 3D Printing Based on Polymers and Nanomaterial Additives: Enhancement of Properties and Potential Applications. *Eur. Polym. J.* **2023**, *184*, 111806. [CrossRef]
2. Ravi, P.; Burch, M.B.; Farahani, S.; Chepelev, L.L.; Yang, D.; Ali, A.; Joyce, J.R.; Lawera, N.; Stringer, J.; Morris, J.M.; et al. Utility and Costs During the Initial Year of 3D Printing in an Academic Hospital. *J. Am. Coll. Radiol.* **2022**, *20*, 193–204. [CrossRef] [PubMed]
3. Cheng, P.; Peng, Y.; Li, S.; Rao, Y.; le Duigou, A.; Wang, K.; Ahzi, S. 3D Printed Continuous Fiber Reinforced Composite Lightweight Structures: A Review and Outlook. *Compos. B Eng.* **2023**, *250*, 110450. [CrossRef]
4. Zarek, M.; Layani, M.; Cooperstein, I.; Sachyani, E.; Cohn, D.; Magdassi, S. 3D Printing of Shape Memory Polymers for Flexible Electronic Devices. *Adv. Mater.* **2016**, *28*, 4449–4454. [CrossRef] [PubMed]
5. Yang, H.; Leow, W.R.; Wang, T.; Wang, J.; Yu, J.; He, K.; Qi, D.; Wan, C.; Chen, X. 3D Printed Photoresponsive Devices Based on Shape Memory Composites. *Adv. Mater.* **2017**, *29*, 1701627. [CrossRef]
6. Liu, Y.; Zhang, F.; Leng, J.; Fu, K.; Lu, X.L.; Wang, L.; Cotton, C.; Sun, B.; Gu, B.; Chou, T.W. Remotely and Sequentially Controlled Actuation of Electroactivated Carbon Nanotube/Shape Memory Polymer Composites. *Adv. Mater. Technol.* **2019**, *4*, 1900600. [CrossRef]
7. Ze, Q.; Kuang, X.; Wu, S.; Wong, J.; Montgomery, S.M.; Zhang, R.; Kovitz, J.M.; Yang, F.; Qi, H.J.; Zhao, R. Magnetic Shape Memory Polymers with Integrated Multifunctional Shape Manipulation. *Adv. Mater.* **2020**, *32*, 1906657. [CrossRef]
8. Zhao, Z.; Wu, J.; Mu, X.; Chen, H.; Qi, H.J.; Fang, D. Desolvation Induced Origami of Photocurable Polymers by Digit Light Processing. *Macromol. Rapid Commun.* **2017**, *38*, 1600625. [CrossRef]

9. Chandra, B.P.; Shrivastava, K.K. Dependence of Mechanoluminescence in Rochelle-salt Crystals on the Charge-produced During Their Fracture. *J. Phys. Chem. Solids* **1978**, *39*, 939–940. [CrossRef]
10. Bünzli, J.C.G.; Wong, K.L. Lanthanide Mechanoluminescence. *J. Rare Earths* **2018**, *36*, 1–41. [CrossRef]
11. Terasaki, N.; Xu, C.-N.; Li, C.; Zhang, L.; Li, C.; Ono, D.; Tsubai, M.; Adachi, Y.; Imai, Y.; Ueno, N.; et al. Visualization of Active Crack on Bridge in Use by Mechanoluminescent Sensor. In Proceedings of the Health Monitoring of Structural and Biological Systems 2012, San Diego, CA, USA, 11–15 March 2012; SPIE: Bellingham, WA, USA; Volume 8348, p. 83482D.
12. Feng, A.; Smet, P.F. A Review of Mechanoluminescence in Inorganic Solids: Compounds, Mechanisms, Models and Applications. *Materials* **2018**, *11*, 484. [CrossRef] [PubMed]
13. Fujio, Y.; Xu, C.N.; Terasawa, Y.; Sakata, Y.; Yamabe, J.; Ueno, N.; Terasaki, N.; Yoshida, A.; Watanabe, S.; Murakami, Y. Sheet Sensor Using $SrAl_2O_4$: Eu Mechanoluminescent Material for Visualizing Inner Crack of High-Pressure Hydrogen Vessel. *Int. J. Hydrogen Energy* **2016**, *41*, 1333–1340. [CrossRef]
14. Ahn, S.Y.; Timilsina, S.; Shin, H.G.; Lee, J.H.; Kim, S.H.; Sohn, K.S.; Kwon, Y.N.; Lee, K.H.; Kim, J.S. In Situ Health Monitoring of Multiscale Structures and Its Instantaneous Verification Using Mechanoluminescence and Dual Machine Learning. *iScience* **2023**, *26*, 105758. [CrossRef] [PubMed]
15. Shin, H.G.; Timilsina, S.; Sohn, K.S.; Kim, J.S. Digital Image Correlation Compatible Mechanoluminescent Skin for Structural Health Monitoring. *Adv. Sci.* **2022**, *9*, 2105889. [CrossRef] [PubMed]
16. Jia, Y.; Yei, M.; Jia, W. Stress-Induced Mechanoluminescence in $SrAl_2O_4$: Eu^{2+}, Dy^{3+}. *Opt. Mater.* **2006**, *28*, 974–979. [CrossRef]
17. Liu, L.; Xu, C.N.; Yoshida, A.; Tu, D.; Ueno, N.; Kainuma, S. Scalable Elasticoluminescent Strain Sensor for Precise Dynamic Stress Imaging and Onsite Infrastructure Diagnosis. *Adv. Mater. Technol.* **2019**, *4*, 1800336. [CrossRef]
18. Jafari, A.; Broumand, P.; Vahab, M.; Khalili, N. An EXtended Finite Element Method Implementation in COMSOL Multiphysics: Solid Mechanics. *Finite Elem. Anal. Des.* **2022**, *202*, 103707. [CrossRef]
19. Sivakumar, N.; Kanagasabapathy, H.; Srikanth, H.P. Static Multiple, Distributed Piezoelectric Actuator Structural Deformation and Bending Analysis Using Comsol. *Mater. Today Proc.* **2018**, *5*, 11516–11525. [CrossRef]
20. Zhuang, Y.; Xie, R.J. Mechanoluminescence Rebrightening the Prospects of Stress Sensing: A Review. *Adv. Mater.* **2021**, *33*, 2005925. [CrossRef]
21. Kim, J.S.; Kwon, Y.N.; Shin, N.; Sohn, K.S. Mechanoluminescent $SrAl_2O_4$: Eu, Dy Phosphor for Use in Visualization of Quasidynamic Crack Propagation. *Appl. Phys. Lett.* **2007**, *90*, 241916. [CrossRef]
22. Einbergs, E.; Zolotarjovs, A. Programmable Material Testing Device for Mechanoluminescence Measurements. *HardwareX* **2022**, *12*, e00349. [CrossRef] [PubMed]
23. Einbergs, E.; Zolotarjovs, A.; Bite, I.; Vītola, V.; Spustaka, A.; Tunēns, G.; Arnautov, A. A Mechanoluminescence Based Approach to Spatial Mechanical Stress Visualisation of Additively Manufactured (3D Printed) Parts. *Materialia* **2022**, *24*, 101516. [CrossRef]
24. Azad, A.I.; Rahimi, M.R.; Yun, G.J. Quantitative Full-Field Strain Measurements by SAOED ($SrAl_2O_4$: Eu^{2+}, Dy^{3+}) Mechanoluminescent Materials. *Smart Mater. Struct.* **2016**, *25*, 095032. [CrossRef]
25. Rojas-Hernandez, R.E.; Rubio-Marcos, F.; Rodriguez, M.Á.; Fernandez, J.F. Long Lasting Phosphors: SrAl2O4: Eu, Dy as the Most Studied Material. *Renew. Sustain. Energy Rev.* **2018**, *81*, 2759–2770. [CrossRef]
26. Vitola, V.; Bite, I.; Millers, D.; Zolotarjovs, A.; Laganovska, K.; Smits, K.; Spustaka, A. The Boron Effect on Low Temperature Luminescence of $SrAl_2O_4$: Eu, Dy. *Ceram. Int.* **2020**, *46*, 26377–26381. [CrossRef]
27. Botterman, J.; Smet, P.F. Persistent Phosphor $SrAl_2O_4$: Eu, Dy in Outdoor Conditions: Saved by the Trap Distribution. *Opt. Express* **2015**, *23*, A868. [CrossRef]
28. Vitola, V.; Millers, D.; Smits, K.; Bite, I.; Zolotarjovs, A. The Search for Defects in Undoped $SrAl_2O_4$ Material. *Opt. Mater.* **2019**, *87*, 48–52. [CrossRef]
29. Doke, G.; Krieke, G.; Antuzevics, A.; Sarakovskis, A.; Berzina, B. Optical Properties of Red-Emitting Long Afterglow Phosphor Mg_2Si_1-XGexO$_4$: Mn^{2+}/Mn^{4+}. *Opt. Mater.* **2023**, *137*, 113500. [CrossRef]
30. Zhang, J.C.; Wang, X.; Marriott, G.; Xu, C.N. Trap-Controlled Mechanoluminescent Materials. *Prog. Mater. Sci.* **2019**, *103*, 678–742. [CrossRef]

designs

Article

A Novel Feature-Based Manufacturability Assessment System for Evaluating Selective Laser Melting and Subtractive Manufacturing Injection Moulding Tool Inserts

Mennatallah F. El Kashouty [1,2,3], Allan E. W. Rennie [3,*] and Mootaz Ghazy [2]

[1] Al Fouad Co., for Automotive Spare Parts, Alexandria 23722, Egypt
[2] Industrial and Management Engineering Department, College of Engineering and Technology, Arab Academy for Science, Technology, and Maritime Transport, Abu Qir Campus, Alexandria 21913, Egypt
[3] School of Engineering, Lancaster University, Bailrigg, Lancaster LA1 4YW, UK
* Correspondence: a.rennie@lancaster.ac.uk

Abstract: Challenges caused by design complexities during the design stages of a product must be coordinated and overcome by the selection of a suitable manufacturing approach. Additive manufacturing (AM) is capable of fabricating complex shapes, yet there are limiting aspects to surface integrity, dimensional accuracy, and, in some instances, design restrictions. Therefore, the goal is essentially to establish the complex areas of a tool during the design stage to achieve the desired quality levels for the corresponding injection moulding tool insert. When adopting a manufacturing approach, it is essential to acknowledge limitations and restrictions. This paper presents the development of a feature-based manufacturability assessment system (FBMAS) to demonstrate the feasibility of integrating selective laser melting (SLM), a metal-based AM technology, with subtractive manufacturing for any given part. The areas on the tool inserts that hold the most geometrical complexities to manufacture are focused on the FBMAS and the design features that are critical for the FBMAS are defined. Furthermore, the structural approach used for developing the FBMAS graphical user interface is defined while explaining how it can be operated effectively and in a user-friendly approach. The systematic approach established is successful in capturing the benefits of SLM and subtractive methods of manufacturing, whilst defining design limitations of each manufacturing method. Finally, the FBMAS developed was validated and verified against the criteria set by experts in the field, and the system's logic was proven to be accurate when tested. The decision recommendations proved to correlate with the determined recommendations of the field experts in evaluating the feature manufacturability of the tool inserts.

Keywords: feature selection; decision-based system; additive manufacturing; selective laser melting; injection moulding; tool inserts; automotive industry

Citation: El Kashouty, M.F.; Rennie, A.E.W.; Ghazy, M. A Novel Feature-Based Manufacturability Assessment System for Evaluating Selective Laser Melting and Subtractive Manufacturing Injection Moulding Tool Inserts. *Designs* **2023**, 7, 68. https://doi.org/10.3390/designs7030068

Academic Editor: Obeidi Muhannad

Received: 1 March 2023
Revised: 2 May 2023
Accepted: 15 May 2023
Published: 22 May 2023

1. Introduction

Knowledge-based expert systems (KBES), or expert systems, as they are sometimes referred to, are interactive systems that require expert knowledge. KBES are computer systems that are capable of imitating intelligent human behaviour in problem solving [1]. The knowledge of an expert system is accumulated through the collective input of experience and expertise from numerous individual experts. Therefore, the collective experience of experts provides users with valued recommendations that can assist them in the decision making process. Expert systems are considered one type of KBES that denotes information in the form of 'IF-THEN' statements until a certain conclusion is reached [2]. Başak and Gülesin [3] stated that expert systems enhance quality and productivity and decrease costs. Furthermore, it is understandable that these types of systems are formulated in a step-by-step structure, where the user is led through the sequence of steps to reach a certain decision, whilst also comprehending how that decision has been made.

The purpose of the research presented here was to develop an expert system that contains expert data regarding the selection process that provides the user with decision making recommendations for manufacturing an injection moulding tool insert for the aftermarket automotive sector. SLM has shown promising potential in the fabrication of injection mould tooling inserts. Numerous researchers have directed their attention over recent years towards the development of decision-making systems or assessment methodologies to generally integrate the benefits of additive manufacturing (AM) technology in tooling processes with less focus on feature limitations. Therefore, more research must be oriented towards developing a systematic approach that evaluates the manufacturability feature limitations of SLM technology in comparison to conventional methods and presenting the challenging outcomes.

Pal et al. [4] discussed a methodology where quality function deployment (QFD) and analytic network process (ANP) are integrated to convert customer needs into product technical requirements. The second stage of the study was a decision-making tool used for prioritising the engineering requirements based on customer needs for selecting and evaluating an appropriate rapid prototyping (RP) approach for fabricating a rapid casting tool. Nagahanumaiah et al. [5] presented a systematic approach for manufacturability analysis of moulds produced by rapid tooling (RT) methods, the approach being founded in three phases: mould feature manufacturability, secondary elements compatibility, and cost effectiveness. The geometric features of the mould core and cavity were evaluated for manufacturability using a fuzzy analytic hierarchy process (Fuzzy-AHP) methodology, as geometric compatibility for manufacturing a feature is characterised by a 'pass' or 'fail' approach. Another study proposed an AM manufacturability assessment approach that incorporates the use of automated feature recognition from the original CAD model with established process knowledge to optimise process outputs for metal AM processes [6].

Additional work has been conducted by Nagahanumaiah et al. [7], where a computer-aided RT process selection and manufacturability evaluation methodology was presented for injection moulding. The process selection supports mould cost estimation models and process capability databases. The model is based on a QFD process capability mapping with a set of tooling requirements that are prioritised through a pairwise comparison using AHP. In the work of Nagahanumaiah and Ravi [8], a generic approach was investigated for using grey relational analysis to quantify the effect of different moulding process variables on selected quality parameters for parts produced from direct metal laser sintering (DMLS) moulds. Data for dimensional error and weight difference were normalised, often called grey relation generation, to define the relationship between the desired and actual experimental data.

Kerbrat et al. [9] developed a methodology that estimates the complexity of tools using manufacturability index calculations based on octree decomposition for machining and AM. In this approach, areas with the most complexity are focused on and the calculated indices indicate which areas are advantageous for machining or manufacturing via an AM process. In this case, tools are seen as separate single modules that are further assembled. Townsend and Urbanic [10] related AM with computer numerical control (CNC) machining in a holistic approach for design and manufacturing, which defines the strength and weaknesses of each process. Moreover, for any given criteria, one of the processes shows a distinct advantage over the other. The processes are mapped simultaneously to the geometry and function of the part with regard to process strength. In the referred study, modules were created to group part geometry, and process selection was determined to fabricate the modules. Functionality is associated with part geometry; hence, this systems approach proposed applying an AHP model that quantifies decision making for process selection.

Ponche et al. [11] proposed a numerical chain based on a new design for AM (DfAM) methodology detailing both design requirements and manufacturing specificities. The quality of the parts produced is significantly affected by the physical phenomena occurring during AM fabrication. Therefore, the methodology proposed in the work offers a new DfAM approach detailing design requirements and manufacturing specifications right from the part design stage, which allows for the optimisation of geometry for thin-walled metal parts. However, the work of the study conducted was restricted to extruded parts. Zhang et al. [12] proposed an evaluation framework in which quantitative indicators were defined according to the design needs of the specific AM process to convey information from the process planning for improving the design. Referring to the user's manufacturing requirements, the purpose of the framework is to check whether a designed part is suitable to be fabricated using AM processes.

Design and manufacturing are the key considerations for developing a product, and recently, combining additive and subtractive manufacturing technologies has gained much attention. In the design process, design rules are set and defined to take account of various manufacturing constraints. Different definitions have been recommended by previous research for machining features with different viewpoints. Başak and Gülesin [3] reviewed earlier studies concluding that a feature-based design involves defining all the necessary information in a database regarding part geometry, surface topology, dimensioning, and tolerances. Other studies considered a feature used in computer-aided design (CAD) as a geometric shape, and based on the type of application, it can be defined as geometric, manufacturing, or an assembly feature.

Sormaz and Khoshnevis [13] defined a machining feature as a volumetric geometry that is machinable in a single operation, but expressed concerns due to restricting the definition to the removal of material volume. Wang [14] proposed a machining feature definition that entails surface features, geometrical features, and volumetric features. Givehchi et al. [15] added to the definition the state of the feature boundary representation. Le et al. [16] adopted a definition that describes a machining feature as a geometrical shape with a set of specifications that can be acknowledged by at least one machining process. Zhang et al. [17] proposed a definition for AM features in the same manner as machining features for which at least one AM process is known. The definition is based on the characterisation of AM processes that has an impact on build orientation and PBF in particular, which can manifest important effects on surface roughness and mechanical properties. Therefore, in this work, the manufacturing feature definition from the work of Le et al. [16] was adopted, which refers to both AM and machining features.

For the purposes of this research, design features with relevance to the scope of work for additive and subtractive manufacturing technologies were defined and are presented in Table 1 [18–20].

Table 1. Design Feature Illustrations and Definitions.

Design Feature	Illustration	Definition
Hole	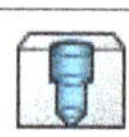	A hole feature originates from a rounded profile. Hole types include 'through', 'blind', and 'tapered'.
Slot	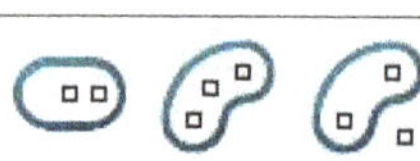	A slot is a perimeter that has a constant centre line and width. Slot types include 'blind', which are contoured with two ends, and 'through', which pass completely through the part.
Pocket		A pocket is a feature with an open or a closed perimeter often called an open pocket or a closed pocket. Pocket types include 'through' and 'blind'.
Boss Extrude		A boss extrude feature adds to the area of the surface through extrudes above the planar surface.
Freeform Pattern	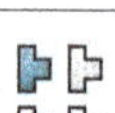	Any feature that has multiples that can be grouped together to create a pattern design. They can be machined as individual features or as a pattern.

Table 1. *Cont.*

Design Feature	Illustration	Definition
Fillet		Fillets are rounded corners. A curve created at the intersection of two or more faces.
Sharp Edge		A sharp edge on the external side of a body.
Undercut	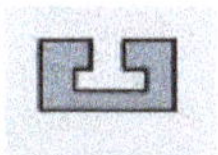	An undercut refers to a feature that is described as a non-visible recessed surface that is inaccessible using a straight tool.
Tapping		Tapping is responsible for creating screw threads in a hole.
Negative draft	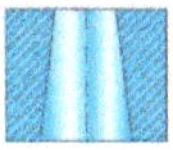	In a part viewed from a plan view, the side walls are tapered towards the bottom; the internal dimension at the bottom has a larger dimension compared to the top.

2. System Development

During the design stage, the designer is free to explore different "design for manufacturing" approaches, such as design for subtractive manufacturing (DfSM), DfAM, and design for hybrid manufacturing (DfHM), given that access to the manufacturing systems is available. Therefore, the first step after the CAD design of a tool insert is developed, is for the designer to analyse its manufacturability. When developing the FBMAS, the following specifications and limitations are considered for the GUI development.

2.1. Recognising System Specifications and Limitations

The following are the targeted system specifications:

- Applying the feature-based system to assist users in defining and evaluating manufacturability limitations of a given tool insert based on a set of predetermined features criteria;
- The system is feature-based, evaluating the tool insert as multiple features and providing recommendations according to rules in the "IF-THEN" format that are constructed in the knowledge base. The "IF" part includes the condition clauses and the "THEN" part includes the resulting sentences;
- Feature specifications of diameter to length ratios are derived from SECO [21,22];
- The separate feature recommendations are processed to provide the user with a generic part recommendation;
- The system is interactive in assisting the user to assess the feature-based manufacturability limitations and provide recommendations for which manufacturing technique to use.

The main limitations set for the developed feature-based system were:

- The only technologies that the FBMAS can be applied (i.e. will be limited) to are SLM for AM, CNC machining, die-sink electric discharge machining (EDM) and wire EDM for subtractive manufacturing;
- The rules set for the system were constructed on the basis of individualisation, with overlapping features being outside the scope of this research;
- The maximum part size allowed for this system is associated with the maximum volume of commercially acknowledged SLM machine systems (e.g., SLM Solutions [23]; 500 mm × 280 mm × 850 mm). The build platform wall allowance is understood based on technical user experience;

- The critical features identified for this study are limited to hole, slot, pocket, boss extrude, and freeform pattern (refer to Table 1);
- The maximum number of designs of a feature allowed for this system is five designs—this rule applies to each of the features individually;
- Economic cost factors were disregarded in this research.

2.2. Graphical User Interface

The FBMAS architecture is a fixed-inflexible system that can only provide the user with what the developer has predetermined for the system. The system was developed using Matlab (MathWorks, Matlab academic version R2017a). The logic of the system comprises fixed rules that define the design constraints provided by human experts. Those design rules was set for SLM and the subtractive manufacturing methods as focused by the FBMAS. Figure 1 provides a schematic illustration of the general FBMAS structure.

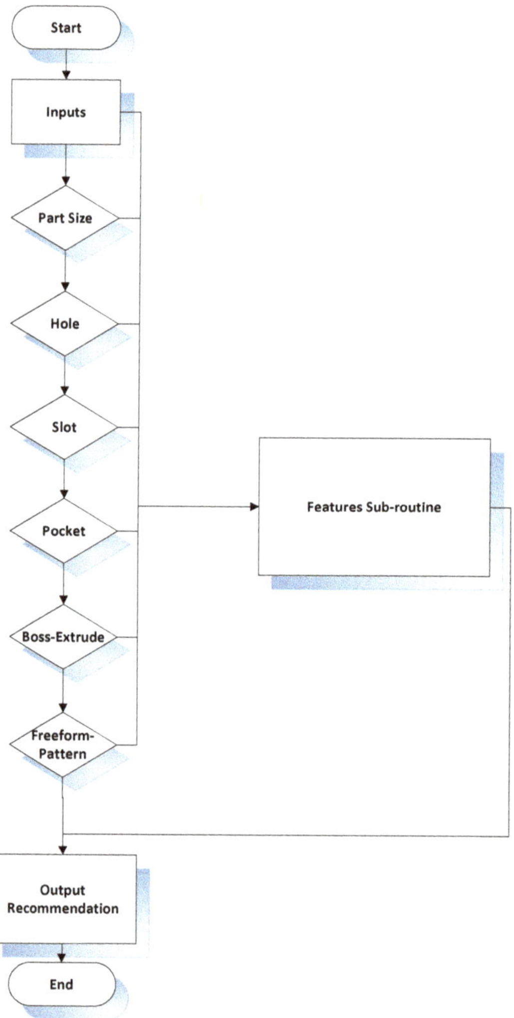

Figure 1. Schematic diagram of the FBMAS general structure.

The user is required to input the necessary information for each feature in the form of queries in the GUI for the different defined feature designs. The system then formulates the information and returns an output to the user with the decision recommendation for each feature design. After all identified features are assessed, the FBMAS displays a list of the individual feature decision recommendations and the overall recommendation for the part manufacturing. Figure 2 displays a graphical illustration of the FBMAS from the initialisation stage to displaying the recommendations.

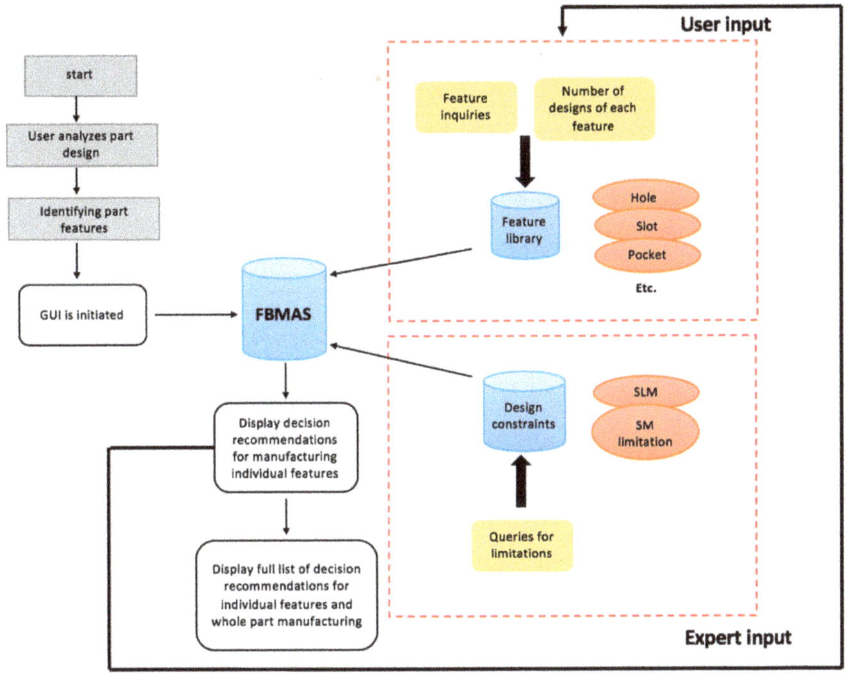

Figure 2. Graphical illustration of the FBMAS from initialisation stage to displaying recommendations.

2.2.1. FBMAS Main Initialisation

Figure 3 illustrates the primary screen that appears to the user when the FBMAS is initialised. In the first screen, there are two main panels, and the user is requested to input the necessary information for all fields in both panels.

The first panel comprises inquiries about the main part sizes. The maximum part length, width, and height were set to 500 mm, 280 mm, and 850 mm, respectively. The maximum part dimensions specified in this research were based on the maximum featured commercial SLM system on the market that is capable of efficiently producing large-volumetric-size metal components. The user must enter values for the three dimensions.

Depending on the rules and the constraints set for the maximum part size, the returned queries are checked with the design constraints as shown in the system logic in Figure 4. A decision recommendation is fed to the designer for the injection mould tool insert to be manufactured using subtractive technologies in the likelihood that the insert cannot be separated into smaller modules. If the insert design can be separated into individual modules, the user is recommended to separate the part before any further evaluation is conducted. After the recommendation message is displayed, the system terminates, and each module evaluated is treated as a separate entity.

Figure 3. Initial screen of the FBMAS.

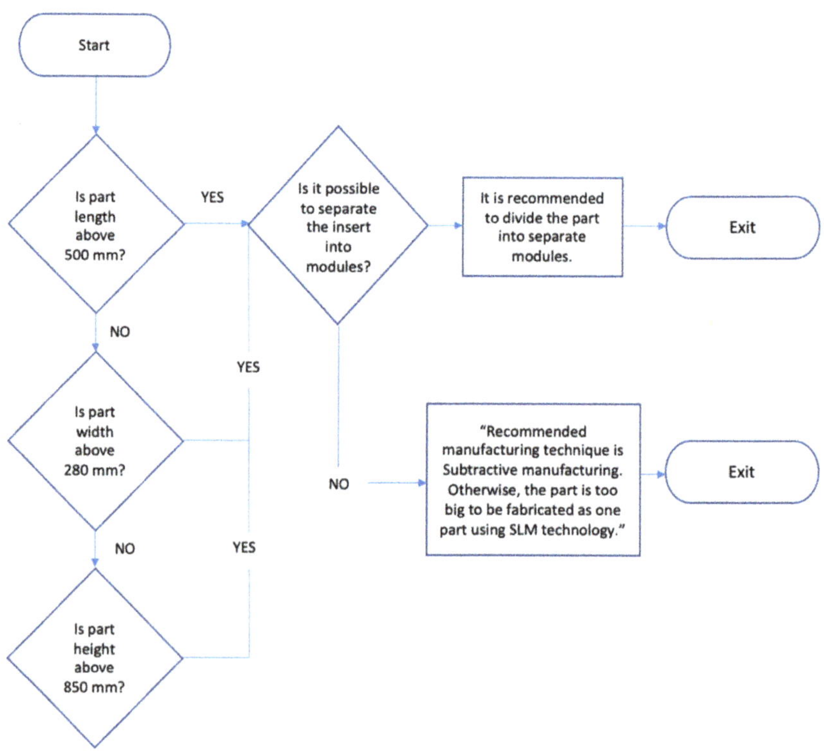

Figure 4. Size constraints.

As for the second panel, the user is obligated to specify the features that are identified for the given tool insert. For the purpose of clarification, each feature is accompanied with a sneak-peek descriptive 3D illustration providing affirmation of the user's feature selection. After the features are selected, the designated feature screens are activated for the user. At this stage, the next button activates the system to screen the constraints, and in case one of the constraints is met, the user is provided with a valid recommendation on how the insert should be manufactured. Otherwise, the next button activates the subsequent screens relying on the features selected from the third panel.

2.2.2. Design Features

In the likely event of a hole feature being selected, the user is approached with multiple inquiries. First, in a separate page, the user is prompted to input the number of different hole designs (as shown in Figure 5).

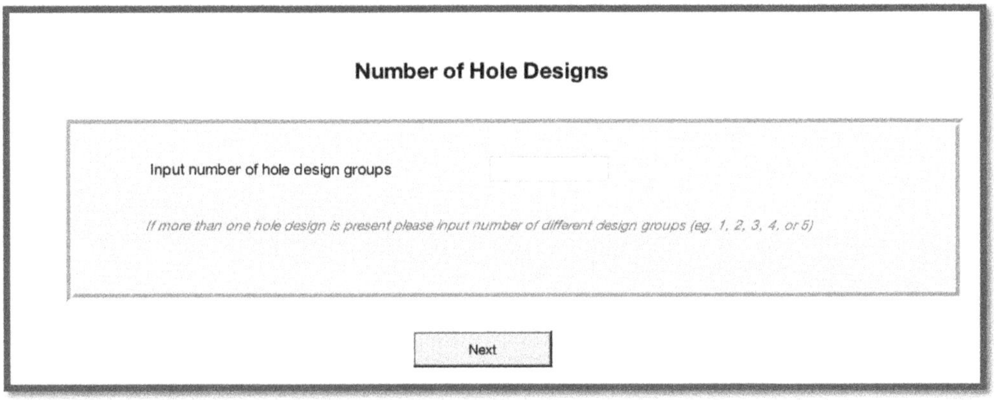

Figure 5. The user is prompted to enter the number of hole designs.

The maximum number of a given group of feature designs allowed for this system is five. For example, the maximum number of different designs for a hole feature is five—the same rule applies to all features of the FBMAS.

After the user inputs the number of hole designs, they are driven through a sequence of questions to identify the features' criteria and limitations. Those limitations are gauged through a set of logical rules that have an impact on the choice in manufacturing technology. The resulting recommendation decision for the hole feature page is saved, to be displayed in the recommendation list page. The recommendation list page is displayed at the end of the system after all the features of the insert are evaluated. If the user identified that there is more than one hole design, then the system is prompted to open the same number of design pages as specified by the user. Figure 6 displays the hole feature design page that appears to the user when a hole feature is selected in the initial page.

A set of questions are listed in the page, and the user has to provide an answer to each question. The questions generated are the result of the compiled design information obtained from inquiries and investigations conducted with experts in the automotive industry. First, the user is questioned to determine whether the hole has a negative draft. Therefore, if the user identifies that there is in fact a negative draft, a decision recommendation is displayed for the hole feature.

At this point, evaluating the rest of the feature criteria after a decision recommendation is made is unnecessary, because the outcome from the evaluation dominates any other outcome that may follow. If the user acknowledges this feature criterion, then a decision recommendation is displayed to indicate that a negative draft is not achievable using any subtractive method of manufacturing. Figure 7 is an example of the decision recommendation displayed.

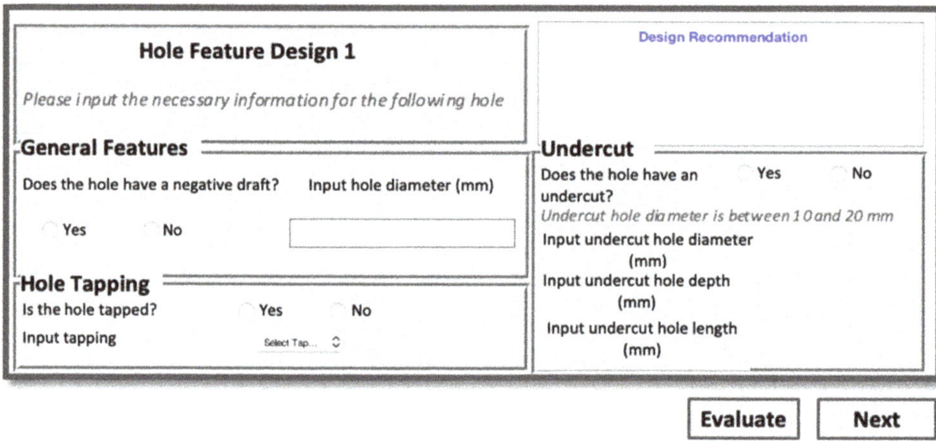

Figure 6. Hole feature main page displayed for the FBMAS.

Figure 7. Example of a displayed design feature recommendation.

However, if the abovementioned feature criterion is not present, then the logical flow of the system continues to evaluate the rest of the feature criteria. Further on, the user is required to answer a set of questions that inquire whether there is an undercut feature or not; if the user agrees that there is in fact an undercut feature, more questions have to be answered. First, the user is required to input the undercut hole diameter, followed by the undercut depth, and finally the length of the undercut feature.

Certain design limitations must be taken into account; for this, FBMAS design rules for SLM and subtractive technologies were the founding base for the selection system. It was found that the minimum diameter that SLM technology can accomplish for an open feature is 1 mm [24–26]. For subtractive technology, design guidelines were acquired and validated through experts in the automotive industry for the production of injection moulding tool inserts [27,28]. Creating an undercut feature requires the use of a T-slot cutting tool with specific diameters, as shown in Figure 8. The minimum hole diameter recommended is 10 mm, and the maximum is 20 mm; otherwise, it is considered a pocket. If the user inputs a value lower than 10 mm or greater than 20 mm, a message appears to direct them to input a valid undercut hole diameter. The length L of the undercut varies depending on the diameter that corresponds with the length of the T-slot cutting tool. The depth of the undercut is derived from the equation:

$$\text{Undercut depth} = \frac{(\text{Dc} - \text{Dm})}{2}$$

where Dc is the diameter of the cutter and Dm is the diameter of the tool shank.

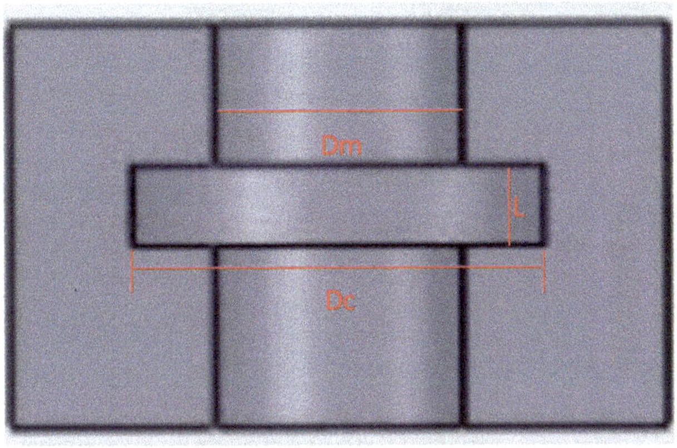

Figure 8. Illustrative undercut diagram.

Therefore, if the user assumes that there is an undercut feature and the required information is entered, then the system proceeds to analyse and assess the design rules that are defined for the undercut feature. Furthermore, a decision recommendation is displayed for the user identifying the proper manufacturing technology to seek. The system proceeds to enquire about other hole feature limitations. If no undercut is detected, the user is queried for the existence of hole tapping. If the user affirms, the following question examines the tapping size. According to the user's response, a decision recommendation is displayed if a limitation is detected; otherwise, the system resumes and enquires about additional limitations.

At this point, if none of the previous hole features presented a defined limitation, the decision system proceeds to enquire about the hole diameter. The minimum open feature diameter that can be accomplished by the SLM technology is 1 mm. However, for subtractive manufacturing, hole diameter and depth are associated with the cutting tool dimensions; therefore, it is important to signify the ratio of hole diameter to depth as a design limitation. The minimum permissible hole diameter is 1 mm, and the maximum is 20 mm. If the user enters a value outside the permissible range, a message appears, alerting the user to input a valid hole diameter. If the hole diameter exceeds 20 mm, a message is displayed for the user to refer to the pocket feature.

The user is required to enter the hole diameter, and the FBMAS system is responsible for assessing the information entered. Depending on the value submitted for the diameter, the system prompts the user to answer a question. For example, if the user enters a value of 1 mm for the hole diameter, the system proceeds to enquire whether the ratio of diameter to length is 2:1. If 'yes', a decision recommendation is displayed, but if 'no', another question appears to check if it is a blind hole. Figure 9 displays an example of the prompted question.

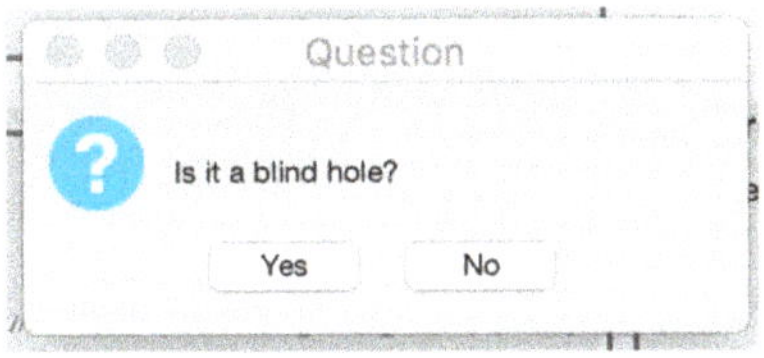

Figure 9. Example of question prompted by the system.

After the FBMAS enquires about all the defined hole feature limitations for each of the identified hole designs, the decision system proceeds to enquire about limitations that are detected in the subsequent design features in the same manner as the hole feature. The logical sequence in which the questions are arranged is dependent on the significance of each feature criteria to the decision making process. The slot feature is the next in the main logical flowchart.

During investigations for this research, it was shown that injection moulding tool inserts for automotive applications are most likely complex in design. In definition, the term complex means that multiple features are mutually integrated in one component, requiring the use of multiple manufacturing methods to achieve the desired design. Referring back to the slot feature design page, the user is asked whether the slot feature has sharp-edged corners. To create a sharp-edged corner, the user can either select the use of SLM technology or a subtractive manufacturing approach. To establish which subtractive manufacturing approach to use, the user has to determine whether the slot is through or blind. Then, EDM methods are attempted. Depending on the type of slot, for example if it is through, then wire EDM is recommended; otherwise, if the slot is blind, conventional die-sink EDM is recommended. Consequently, the system is guided to question the presence of an undercut in the slot feature. The slot feature follows the same design rules as the hole feature design. Moreover, the maximum undercut slot diameter is 25 mm; otherwise, the user is advised to refer to the pocket feature.

The pocket feature is the third feature in the system's logic and is assessed in the same manner as the previous design features discussed. The fourth feature in the FBMAS is the boss extrude. After the user selects the boss extrude feature from the main initialisation page, the feature design page is prompted. The FBMAS is executed to assess feature design limitations depending on the information entered by the user. First, the user is requested to enter the number of different boss extrude group features. According to the number of designs entered, design pages are opened subsequently. For each design page, the user is asked to fill out the enquiry fields. The user is asked whether there are any sharp-edged corners or corner fillets less than 1 mm in diameter.

If no constraints are identified, the FBMAS enquires about the spacing between the boss extrude feature and the nearest wall. Furthermore, the FBMAS checks if the height to width ratio of the boss extrude feature is more than 8:1; if the user confirms, then SLM technology is disqualified as a potential manufacturing technique.

A freeform pattern feature is simply multiple repetitions of an individual design feature. Significantly, the feature diameter is the key design criterion to query so as to assess design limitations for a freeform pattern feature. First, the user is asked about the number of freeform pattern designs. The same set of design rules for a feature diameter query was followed in the previous design features. The pattern diameter is directly associated with the feature's depth. Therefore, the design ratios were followed by the FBMAS to assess the adequate manufacturing technique for implementation. The minimum permissible diameter for CNC machining a freeform pattern design is 0.25 mm. The user is required to enter the diameter, and the system is responsible for assessing the input. Depending on the value submitted for the diameter, the system prompts the user to answer a question about the ratio, and according to the answer, a decision recommendation is displayed. If the user enters a value outside the permissible range, a message appears alerting the user to input a valid diameter.

2.2.3. Feature and Part Decision Recommendations

The decision recommendation page is the last stage of the FBMAS. At this point, the user has initiated all the necessary feature design pages that are of relevance to the part under consideration. In the recommendation page, each feature is displayed in the upper tab menu. When the user presses on one of the feature tabs, a display of the identified design groups of a given feature are displayed. For each design group specified by the user, a decision recommendation for the manufacturability of the given feature is presented, along with an explanation of limitations. Figure 10 shows the recommended decision for five hole feature design groups. These recommendations provide the user with an insight into the different capabilities and limitations of the defined manufacturing technologies in this system when it comes to design feature manufacturability.

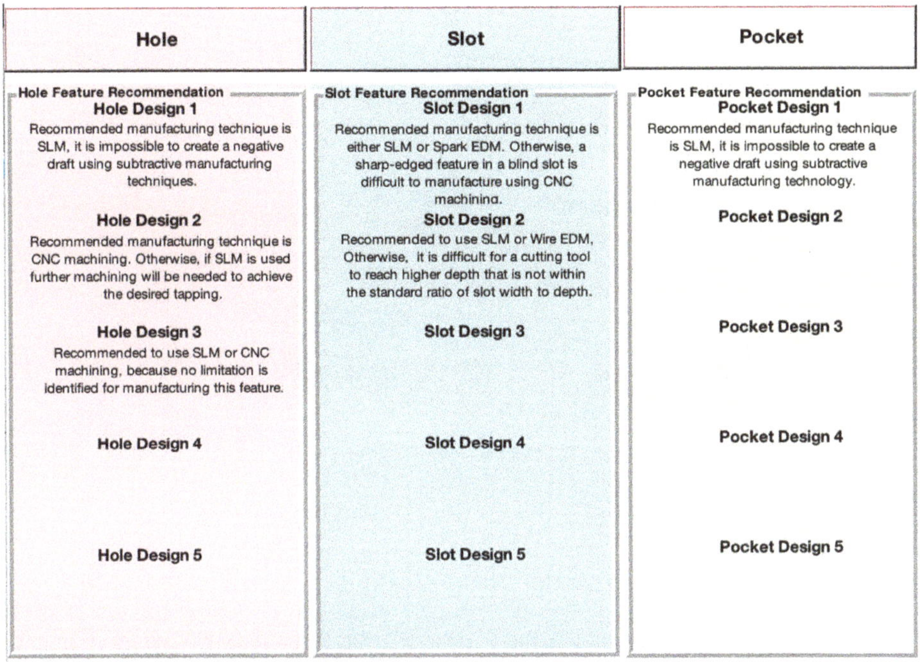

Figure 10. Design recommendation page of the FBMAS.

3. System Verification and Validation

The verification of the system was established by inspecting the logic of the developed FBMAS. To validate the system for the purpose of supporting the fabrication of injection moulding tool inserts for the automotive industry, the process for the case studies was conducted in consultation with a stakeholder from the industry. The output decision recommendations of the FBMAS were compared with actual decisions made by the experts consulted to assess how well the system works. Considering how verification and validation are related to the development process of the FBMAS, Figure 11 displays the modelling paradigm. The paradigm is adopted from the simplified version illustrated by Sargent [29] for the verification and validation of simulation models.

The principal knowledge was captured to correspond with the need of this study. The real-life design evaluation process is the problem entity that needs to be modelled. The logical depiction of the system is the conceptual model, and the programming of the conceptual model is the computerised model. To develop the conceptual model, extensive analysis and flowchart modelling was carried out to validate compliance with the actual system. Verifying the computerised model ensures that the computer programming and implementation is conducted with no faults. Operational validation was carried out with sufficient experimentation to ensure that the model's outcome provides accurate results as intended in actual situations. Finally, in this work, data validity throughout all stages of the verification and validation process was performed to ensure that the design feature limitations are correctly defined and represented. The two main verification and validation techniques acquired in this work are event and extreme condition tests. In the "event" test, the model was run to depict similarities with the real-life system. As for the "extreme condition" test, the outcome should be perceived as acceptable regardless of the extreme inputs to the system [29].

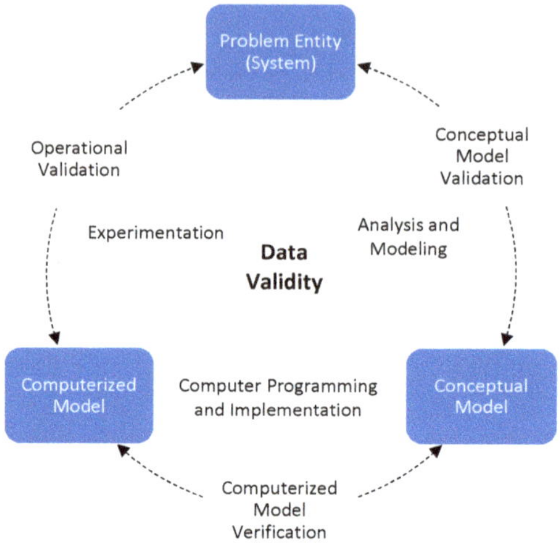

Figure 11. FBMAS verification and validation paradigm [29].

3.1. System Verification

Prior to utilising the developed FBMAS, the system had to be examined to verify that it operated accurately. The verification process was executed in several stages. Initially, the logic and interface of the system were verified through the different stages of system development. Furthermore, after completion of the system development, it was examined as a whole to ensure that it worked properly.

Different scenarios were established to examine the system's performance when subjected to different inputs and the effect of these variations on the system's outputs. Input variations were mainly set to the part size and presence of a given feature. To test the systems operation, variations in part size and feature existence were determined for the FBMAS to acquire the expected output. To verify the accuracy of the system's performance, the same criteria were tested manually to compare and ensure that the same results were acquired.

As an example for the verification process, the system was tested in various scenarios of entering different part size values. Firstly, the system was tested under extreme conditions where the input data provide plausible outputs for unlikely, extreme conditions. For example, if part size in any of the X, Y, and Z directions is zero, a message is displayed to state that a valid part size must be entered. Other event scenarios are were outwith part sizes above and below the SLM design limitations of 500 mm, 280 mm, and 850 mm, respectively. The results retrieved from the FBMAS were similar to those results determined from manually processing the system. Changing the inputs results in correspondingly altered outputs.

This system verification method was carried out multiple times to ensure the reliability of the FBMAS in accurately following the programmed logical design rules. Additionally, the same verification approach to test for feature manufacturability evaluation was used. Another event scenario was the system being fed with inputs that are known to provide a decision recommendation for manufacturability using SLM technology, and we checked that the output provided an accurate outcome. This method was followed to trace all the possible logical approaches of providing numerous inputs to the FBMAS and retrieve plausible outputs. We compared the outputs retrieved through the verification process of the FBMAS with the manual process at various stages through the development of the FBMAS. In continuously seeking to verify the system at all stages, errors were effortlessly detected and corrected instantaneously.

3.2. System Validation

The primary purpose for validating the FBMAS was to ensure that the system provides realistically feasible outcomes, assisting users in evaluating the manufacturability of the design features of injection moulding tool inserts for the aftermarket spare parts automotive industry. This approach ensures that the knowledge of experts for SLM technology and subtractive manufacturing techniques are accurately captured and constructed within the structure of the developed FBMAS. Design constraints were set to outline limitations that exist for the defined methods of manufacturing. Those constraints were defined by industrial experts and conform to the design constraints that do actually exist and cause manufacturing restrictions. Three industrial injection moulding core or cavity inserts were selected from the industry to validate the system. The case studies were selected by the experts to test the system's decision outcomes compared with the actual outcomes due to challenging limitations faced during manufacturing. The selection of the case studies was conducted under supervision and consultation of the industrial experts who have hands-on experience in the manufacturing of injection moulding tool inserts for the aftermarket automotive industry. El Kashouty [30] described the case studies comprehensively; nevertheless, in this work, the reflector study is discussed concisely.

4. Discussion

The study under review is for a headlamp reflector. The reflector is a standalone part that is not assembled to fit any other component. Figure 12 shows the product.

Figure 12. Reflector product, dimensions in (mm).

4.1. Manufacturability Assessment

The core tool insert for the reflector product was manufactured using two approaches, both subtractive and AM. Initially, the 3D CAD model (as shown in Figure 13) was prepared, and the core insert was CNC machined using a tapered-end mill with a 0.25 mm diameter and shank diameter of 3 mm to achieve the required sharp-edged freeform pattern design, although the required tip diameter was set to be sharp edged.

Figure 14a,b demonstrate a simplified design of the repetitive pattern of the core insert. Acquiring a tapered-end mill with a diameter of zero was impossible. Therefore, the CNC-machined insert did not deliver the stipulated results in accordance with part quality and accuracy. Furthermore, experts confirmed that if only subtractive methods are targeted, using die-sink EDM manufacturing techniques will deliver more satisfactory results.

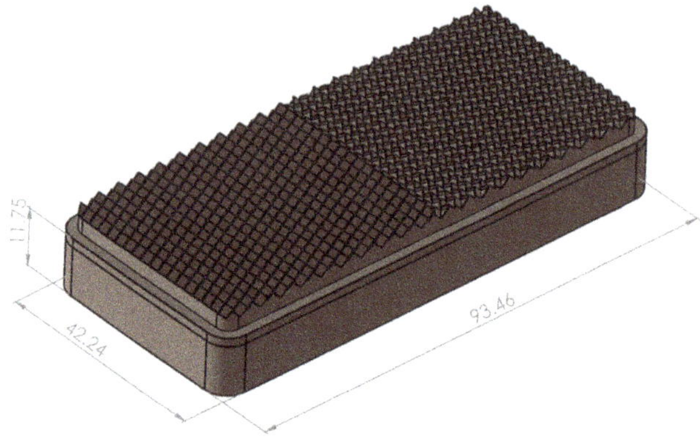

Figure 13. Reflector's core tool insert, dimensions in (mm).

(a) (b)

Figure 14. (**a**) A simplified view of the repetitive pattern of the core insert. (**b**) The reflector product with a simplified view of the repetitive pattern attached to the core.

The second approach for manufacturing was to use SLM technology. The core insert was successfully fabricated; moreover, minor postprocessing was required to achieve the desired surface finish. It was confirmed by experts that SLM offered positive results in fabricating the identified design features with no limitations.

4.2. Core Insert Features Evaluation Using FBMAS

The core's feature specifications were fed to the system and the possible decision recommendations were processed and displayed by the FBMAS. The acknowledged design features fed to the FBMAS were the minimum freeform pattern diameter. The identified pattern design requires that the base have a diameter of zero, as shown in Figure 15.

The user enters the identified features, and the freeform pattern design page is displayed. The user inputs the required information and presses the next button to display the decision recommendations, as shown in Figure 16. The FBMAS states that to manufacture the freeform pattern design of the core insert, it is recommended to use SLM technology as the ideal manufacturing technique, as opposed to subtractive manufacturing. The FBMAS recommendations conform to the recommendations indicated by the consulted experts, given the availability of the manufacturing systems.

Results obtained by Kashouty et al. [31] successfully demonstrate that employing SLM technology for producing tool inserts with complex surface topology proved to be an effective and efficient alternative to subtractive manufacturing. Significant benefits in terms of surface roughness, dimensional accuracy, and product functionality were achieved through the use of SLM technology for the fabricated tool inserts in comparison to their CNC counterparts.

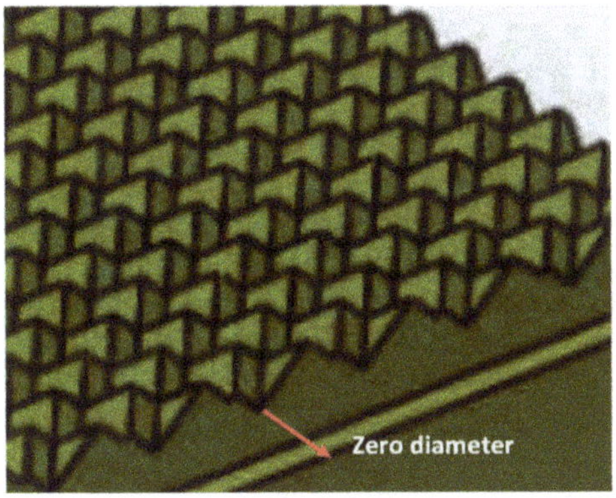

Figure 15. Detailed view of core insert.

Figure 16. Reflector's decision recommendation.

5. Conclusions

The work presented in this paper discussed the systematic approach for developing the feature-based manufacturability assessment system. The areas on the tool inserts that hold the most geometrical complexities to manufacture are focused on each manufacturing method, whilst defining their design limitations. The methodical description of the system's logical operations was clearly recognised through the presented segments of the flowchart and applied through the GUI. The main logic which the system follows is "IF-THEN" rules, used to define design limitations that assist users in determining the proper manufacturing method for the tool insert under consideration. The conditions of the "IF-THEN" are based on constraints set by the operations of SLM technology and the defined methods of subtractive manufacturing in the system. The system focuses on identifying the outcome through decision recommendations for the individual design features as well as the whole part in question. It was noted that the developed FBMAS decision recommendations proved to be in correspondence with the decision recommendations of the field experts in evaluating the feature manufacturability of the tool inserts. The developed FBMAS was self-verified against the criteria set by the field experts. The system's logic was proven to be accurate when tested. Selected tool inserts assisted in the validation process, exhibiting variability in the type of design feature validated for each study.

The developed FBMAS was verified, and the system's logic was proven to be accurate when tested. For the reflector's core insert, it was shown that the required base pattern must have a sharp-edge tip; therefore, the FBMAS recommended that the core insert be manufactured using SLM technology. Addressing the system's specifications and limitations provided the user with a focused insight into the positive outcomes of evaluating the tool insert's feature manufacturability, although there are other aspects to consider when selecting the adequate methods of manufacturing a tool insert. The FBMAS decision recommendations proved to correspond with the decision recommendations of the field experts in evaluating the feature manufacturability of the appointed tool inserts.

Author Contributions: Conceptualisation, M.F.E.K., A.E.W.R. and M.G.; Formal analysis, M.F.E.K.; Funding acquisition, A.E.W.R.; Investigation, M.F.E.K.; Methodology, M.F.E.K.; Project administration, A.E.W.R. and M.G.; Supervision, A.E.W.R. and M.G.; Validation, M.F.E.K., A.E.W.R. and M.G.; Writing—original draft, M.F.E.K.; Writing—review and editing, M.F.E.K., A.E.W.R. and M.G. All authors have read and agreed to the published version of the manuscript.

Funding: This research was funded by the UK Engineering and Physical Sciences Research Council (EPSRC; Grant Reference EP/R511560/1) and Al Fouad Co., for Automotive Spare Parts.

Data Availability Statement: Not applicable.

Acknowledgments: The authors are appreciative of the technical support provided by Croft Additive Manufacturing Ltd. (UK), Al Fouad Co., for Automotive Spare Parts (Egypt), the Department of Manufacturing Engineering at Alexandria University (Egypt), and the Yousef Jameel Science and Technology Research Center at The American University of Cairo (Egypt) for the assistance provided during the preparation of this paper.

Conflicts of Interest: The authors declare that there are no conflict of interest.

References

1. Kumar, S.P.L. Knowledge-based expert system in manufacturing planning: State-of-the-art review. *Int. J. Prod. Res.* **2018**, *57*, 4766–4790. [CrossRef]
2. Chen, Y.; Hsu, C.Y.; Liu, L.; Yang, S. Constructing a nutrition diagnosis expert system. *Expert Syst. Appl.* **2012**, *39*, 2132–2156. [CrossRef]
3. Başak, H.; Gülesin, M. A feature based parametric design program and expert system for design. *Math. Comput. Appl.* **2004**, *9*, 359–370. [CrossRef]
4. Pal, D.K.; Ravi, B.; Bhargava, L.S. Rapid tooling route selection for metal casting using QFD-ANP methodology. *Int. J. Comput. Integr. Manuf.* **2007**, *20*, 338–354. [CrossRef]
5. Nagahanumaiah; Ravi, B.; Mukherjee, N.P. Rapid tooling manufacturability evaluation using fuzzy-AHP methodology. *Int. J. Prod. Res.* **2007**, *45*, 1161–1181. [CrossRef]
6. Stavropoulos, P.; Tzimanis, K.; Souflas, T.; Bikas, H. Knowledge-based manufacturability assessment for optimization of additive manufacturing processes based on automated feature recognition from CAD models. *Int. J. Adv. Manuf. Technol.* **2022**, *122*, 993–1007. [CrossRef]
7. Nagahanumaiah; Subburaj, K.; Ravi, B. Computer aided rapid tooling process selection and manufacturability evaluation for injection mold development. *Comput. Ind.* **2008**, *59*, 262–276. [CrossRef]
8. Nagahanumaiah; Ravi, B. Effects of injection molding parameters on shrinkage and weight of plastic part—Produced by DMLS mold. *Rapid Prototyp. J.* **2009**, *15*, 179–186. [CrossRef]
9. Kerbrat, O.; Mognol, P.; Hascoet, J.Y. Manufacturing complexity evaluation at the design stage for both machining and layered manufacturing. *CIRP J. Manuf. Sci. Technol.* **2010**, *2*, 208–215. [CrossRef]
10. Townsend, V.; Urbanic, J. Relating additive and subtractive processes in a teleological and modular approach. *Rapid Prototyp. J.* **2012**, *18*, 324–338. [CrossRef]
11. Ponche, R.; Kerbrat, O.; Mognol, P.; Hascoet, J.-Y.Y. A novel methodology of design for Additive Manufacturing applied to Additive Laser Manufacturing process. *Robot. Comput. Integr. Manuf.* **2014**, *30*, 389–398. [CrossRef]
12. Zhang, Y.; Bernard, A.; Gupta, R.K.; Harik, R. Evaluating the Design for Additive Manufacturing: A Process Planning Perspective. *Procedia CIRP* **2014**, *21*, 144–150. [CrossRef]
13. Sormaz, D.N.; Khoshnevis, B. Modeling of manufacturing feature interactions for automated process planning. *J. Manuf. Syst.* **2000**, *19*, 28–45. [CrossRef]
14. Wang, L. An overview of function block enabled adaptive process planning for machining. *J. Manuf. Syst.* **2015**, *35*, 10–25. [CrossRef]
15. Givehchi, M.; Haghighi, A.; Wang, L. Generic machining process sequencing through a revised enriched machining feature concept. *J. Manuf. Syst.* **2015**, *37*, 564–575. [CrossRef]
16. Le, V.T.; Paris, H.; Mandil, G. Extracting features for manufacture of parts from existing components based on combining additive and subtractive technologies. *Int. J. Interact. Des. Manuf.* **2018**, *12*, 525–536. [CrossRef]
17. Zhang, Y.; Bernard, A.; Gupta, R.K.; Harik, R. Feature based building orientation optimization for additive manufacturing. *Rapid Prototyp. J.* **2016**, *22*, 358–376. [CrossRef]
18. LaCourse, D. *What is Feature Detection Machining*; MecSoft Corporation: Irvine, CA, USA, 2017.

19. Thornton, V. Glossary of CAD Terms. In *Material, Reference*; Goodheart-Willcox Co., Inc.: Tinley Park, IL, USA, 2017.
20. *Solidworks Essential Manual*; Dassault Systémes SolidWorks Corporation: Waltham, MA, USA, 2012; pp. 1–512.
21. SECO. *Catalog & Technical Guide 2019.2 Solid End Mills*; SECO: Fagersta, Sweden, 2019.
22. SECO. *Catalog & Technical Guide 2019.2 Holemaking*; SECO: Fagersta, Sweden, 2019.
23. SLM Solutions Group. *SLM 800—Large Format Selective Laser Melting*; SLM Solutions Group: Lübeck, Germany, 2017.
24. EPMA. *Introduction to Additive Manufacturing Technology*; EPMA: Chantilly, France, 2013.
25. Renishaw PLC. *Educational Article—Design for Metal AM—A Beginner's Guide*; Renishaw PLC: Gloucestershire, UK, 2017; Volume 44.
26. Diegel, O.; Nordin, A.; Motte, D. *Design for Metal AM*; VTT Technical Research Centre of Finland Ltd.: Espoo, Finland, 2017.
27. Drake, P.J., Jr. *Dimensioning and Tolerancing Handbook*; McGraw-Hill: New York, NY, USA, 1999.
28. Henzold, G. *Geometrical Dimensioning and Tolerancing for Design, Manufacturing and Inspection*, 2nd ed.; Elsevier Ltd.: Amsterdam, The Netherlands, 2006; ISBN 0-7506-6738-9.
29. Sargent, R.G. Advanced Tutorials: Verification and Validation of Simulation Models. In Proceedings of the 2011 Winter Simulation Conference, Phoenix, AZ, USA, 11–14 December 2011; pp. 183–198.
30. El Kashouty, M.F. Development of a Novel Feature Based Manufacturability Assessment System for High-Volume Injection Moulding Tool Inserts. Ph.D. Thesis, Lancaster University, Lancaster, UK, 2020.
31. Kashouty, M.F.E.; Rennie, A.E.W.; Ghazy, M.; Aziz, A.A.E. Selective laser melting for improving quality characteristics of a prism shaped topology injection mould tool insert for the automotive industry. *Proc. Inst. Mech. Eng. Part C J. Mech. Eng. Sci.* **2021**, *235*, 7021–7032. [CrossRef]

designs MDPI

Article

A Structural and Thermal Comparative Review of 3D-Printed Wall Shapes

Nicholas D. Bello [1,*] and Ali M. Memari [2]

1 Civil Engineering Master of Science, The Pennsylvania State University, University Park, PA 16802, USA
2 Architectural Engineering and Civil and Environmental Engineering Departments, The Pennsylvania State University, University Park, PA 16802, USA; amm7@psu.edu
* Correspondence: ndb5283@psu.edu

Abstract: This article explores several aspects of the three-dimensional concrete printing (3DCP) industry. More specifically, it begins with a literature review discussing the background of this technology. This literature review also explores several of the challenges that the industry is currently facing. In this way, a knowledge gap is identified. More specifically, there are few studies that have explored the structural and thermal performance of typical walls printed in this industry. Therefore, we used the simulation tool in SolidWorks to examine the structural behavior of several different wall types when pressure was applied to the exterior face. In addition to this, the thermal performance of different wall types was also studied in SolidWorks by applying a temperature difference between the exterior and interior faces of each wall. For example, one wall shape in this study had minimum factor of safety of approximately 100 due when a load was applied, and the same wall lost approximately 212 W due to the temperature difference applied in this study. Finally, SolidWorks was used to calculate the moment of inertia of the cross sections of several of these walls, which helped to provide a better understanding of each wall's structural rigidity.

Keywords: 3D printing concrete; structural analysis; thermal analysis; SolidWorks

Citation: Bello, N.D.; Memari, A.M. A Structural and Thermal Comparative Review of 3D-Printed Wall Shapes. *Designs* **2023**, *7*, 80. https://doi.org/10.3390/designs7030080

Academic Editor: Obeidi Muhannad

Received: 17 May 2023
Revised: 4 June 2023
Accepted: 7 June 2023
Published: 19 June 2023

1. Introduction

The construction industry is constantly aiming to improve, and as a result, new innovations are constantly being created. One of these innovations introduced in recent years is three-dimensional (3D) concrete printing (3DCP). The 3D concrete printers act in a similar manner to other forms of 3D printers that are commonly used throughout society [1]. Instead of printing plastic, 3D concrete printers extrude concrete from their nozzles, following the designated path of the programmer [1]. More specifically, the programmer inputs the desired path of the nozzle using a computer-aided design (CAD) software, which the machine then translates into a path so as to print the given shape by stacking concrete layers on top of each other [1]. More specifically, companies that utilize 3DCP extrude concrete to create a wall outline and cavities (similar to formwork) [1]. The cavities present in these printed shapes are then filled with concrete and/or insulation [1]. Nevertheless, there are several companies at the forefront of this new technology, each utilizing a different arrangement of concrete and insulation inside the wall.

With 3DCP being such a new technology, there has not been substantial research into the effects of each company's different wall configurations. Thus, the main objective of this study was to determine the structural and thermal performance of several different wall configurations. An examination of online resources, the literature, and other sources helped us to determine the wall shapes that are used in the industry.

2. Literature Review

2.1. Background of the 3DCP Industry

According to Zhang et al. [2], a certain construction industry issue that has increased in severity in recent years is related to the migration of individuals to urbanized areas [2]. Of the current issues, none are more severe than the inability to find qualified workers, the overuse of construction materials, and even the safety of construction workers [2].

As a result, researchers across the world have developed possible solutions to help the construction industry to overcome these problems. In 1997, Joseph Pegna originally developed the idea of concrete that could quickly be cured [3]. Then, in 2002, Behrokh Khoshnevis developed the idea of extruding concrete layer by layer to create a formwork which can later be filled with cast concrete [2,3]. This process is now known as contour crafting [2,3]. In more detail, Elfatah discussed several advantages that are possible within the 3DCP industry, including the improvement of the construction time, the creation of more elaborate designs, improvement of the accuracy of construction, and even reductions in construction waste [1]. Since these ideas were initially developed by Pegna and Khoshnevis, concrete printing technology has grown significantly. More specifically, universities have begun testing and research using this technology [3]. Additionally, several companies throughout the world have started to use 3D printing technology to construct buildings.

2.2. State of 3DCP Technology

As mentioned above, there are universities and companies currently engaged either in R&D related to 3DCP or in the actual construction of buildings with printed walls [3]. As for the process of 3DCP, initially, the users of the technology must start by mixing their concrete [4] using water, and then they must pump the concrete into the printer's head [4]. As it stands, there are two main printer types: the robotic arm style and gantry style [1]. In a gantry-style printer, the printer head sits on a metal frame, which has the capability of moving the printer head on all three axes [1]. On the other hand, a robotic-arm-style printer places the printer head at the end of an arm, which is connected to the main podium of the printer [1]. This printer has the ability to extend and contract its arm and rotate in order to print concrete [1]. After the concrete is pumped into the printer's head, it will flow through the nozzle, which deposits the concrete onto the printing surface [4]. Though this is how a typical gantry-style printer works; other printer types follow a similar process [4]. As previously mentioned, the concrete is extruded onto the desired surface and built up layer by layer to create a formwork in contour crafting [2,4]. However, through the work of researchers, many challenges have already been addressed in an attempt to improve 3DCP. For example, researchers and users of 3DCP have been able to identify several printing characteristics that affect the effectiveness of the extruded concrete.

First, pumpability is a characteristic that has been extensively studied in 3DCP [5]. Verian et al. defined pumpability as "the ease and reliability with which material is moved through the delivery system" [5]. Additionally, printability is another characteristic that affects the quality of printed concrete [5]. This term refers to the ease with which concrete flows out of the nozzle of the printer [5]. Buildability is another key characteristic of printed concrete, which is defined as the printed concrete's "resistance to deformation under load when it is not fully cured" [5]. Clearly, this is an essential characteristic of printed concrete, as extruded concrete is not fully set by the time additional layers are extruded on top of the previously printed layers [5]. In summary, researchers have highlighted these characteristics, which are important for construction quality using this technology [5].

Additionally, bonding between subsequently printed layers has been a point of emphasis for researchers. For example, Verian et al. [5] stated that they developed a means to reduce cold joint forming between subsequent layers, regardless of the time between the printing of subsequent layers [5]. To increase the bond between layers, Verian et al. [5] suggested applying a primer to the previously printed layer directly before the next concrete layer is printed, which, accordingly, increases the bonding between layers. Developments such as these are key to the advancement of this technology and allow the material to behave even more effectively. In fact, Buswell et al. [6] outlined some key developments, such as achieving a density as strong or even higher compared to that of cast concrete.

The 3DCP industry also faces the challenge of improving the durability of printed shapes [6]. For example, Buswell et al. stated that cracking can propagate through printed structures due to shrinkage [6]. In other words, after concrete is extruded from the printer's nozzle, it hardens and shrinks, which leads to cracking in the structure [6]. Buswell et al. also stated that this is due to the elimination of forms during the curing phase of the concrete [6]. As a result, the printed concrete is exposed to its surroundings while curing [6]. Although the removal of concrete forms is beneficial for the construction industry, shrinkage and durability concerns must be addressed.

Additionally, researchers have examined the effect of incorporating insulation into these concrete printed structures [7]. Dey et al., in particular, completed an experiment showing the effect of including insulation in 3DCP structures [7]. In this experiment, printed concrete slabs were heated from one side, and then the thermal inertia of the slab was measured with and without insulation [7]. In conclusion, it was found that the presence of insulation increased the thermal inertia of the slab, meaning that the temperature difference between the exterior and interior surfaces was higher when insulation was present versus when no insulation was present [7]. Pessoa et al. introduced several hypotheses, such as that of placing insulation inside cavities to further enhance the thermal performance of these walls and, at the same time, limit thermal bridging [8].

As previously mentioned, the 3DCP industry is attempting to improve sustainability in construction and reduce material waste, which currently plagues this industry [2]. Reducing construction waste is especially important for the 3DCP industry, as concrete and brick are the two materials that contribute most to construction waste [9]. In this regard, a study conducted by Wu et al. discussed the use of construction waste as a replacement for cement [9]. Aligning with the overarching goals of the 3DCP industry, this new type of mixture has the potential to be used in the 3DCP industry [9]. During the study conducted by Wu et al., it was found that ground brick and concrete waste are viable replacements for part of the cement and sand that is typically used in a concrete mixture [9]. More specifically, Wu et al. concluded that it is best to use ground concrete and brick waste to replace 10% of the concrete's cement and 10% of the concrete's sand [9]. Innovations such as these are increasingly being studied in the construction industry as a whole and have potential to be used in the 3DCP industry and to further one of its overarching goals (sustainability).

2.3. Knowledge Gap in the 3DCP Industry

Despite all of this development, there are still many unknowns in the concrete printing industry. As previously mentioned, there are several companies currently using the concrete printing techniques discussed above, and many of these companies utilize different wall shapes. However, it remains unknown how the shapes of the walls change their structural and thermal behaviors when filled with cast concrete and insulation. More specifically, many studies have been conducted throughout this industry in regard to the consistency, strength, and other parameters of concrete that is used for 3DCP. However, very little research has been conducted on the effect that each unique shape has on the wall's structural and thermal behaviors. As a result, the goal of this study is to determine the structural and thermal behaviors of different walls according to the shapes that they take. In this way, the wall shapes can be compared and contrasted.

3. Study Setup in SolidWorks

3.1. Structural Study Setup in SolidWorks

In order to determine the structural and thermal behaviors of each wall, SolidWorks' simulation tool, which uses finite element analysis, was used to analyze models based on the applied conditions [10]. The printing process starts with the printing of the formwork using a specific type of 3D printer, which is usually either the gantry frame type or a robotic arm system [1]. Once the formwork has been printed, the appropriate cavities are filled with concrete or insulation [1]. To model these walls in SolidWorks, each wall's formwork and cavities were initially sketched and modeled in SolidWorks. The overall shape of each wall was determined and created based on online images provided by different companies, the literature, and other resources. As the exact dimensions of each company's wall were unavailable, each wall was sketched using an estimation of certain dimensions based on these photos. It should be noted that the exterior and interior faces of every wall were connected via a continuous piece of concrete to close the sides of each wall. This allowed us to remain consistent for every wall modeled and to see how the overall shape of each wall affected its behavior. Additionally, there were several constants maintained for each wall so as to compare and contrast the walls more accurately. For example, each wall was taken to have an overall length of 1.8034 m (71 in.), an overall height of 2.438 m (96 in.), and a depth of 0.3048 m (12 in.). Additionally, each layer of concrete was taken to have a thickness of 0.0381 m (1.5 in.). Furthermore, each edge of the wall was filleted to avoid local "singularities", which are defined in SolidWorks as unbounded stress values caused by sharp edges that occur during finite element analysis (but do not appear in the real world) [11,12]. To verify that there are no local singularities in each wall, SolidWorks' built-in hotspot detection function can be used (using its default settings). This function runs throughout the entire model to ensure that there are no singularities present throughout the model during the structural study of each wall [13].

Once the concrete portions of each wall had been modeled, the material properties were assigned to each model to accurately represent the behavior of a concrete wall. For this purpose, the properties of concrete were researched, and then these properties were assigned to a new material created in SolidWorks. First, a compressive strength of 27,579,000 Pa (4000 psi) for the concrete was found in the software database [14]. It should be noted that this value is towards the upper limit of the range defined by the companies in the industry for the compressive strength of their printed materials (between 2000 psi to 3500 psi) [15]. Then, the elastic modulus was found, according to Section 19.2.2 of the ACI 318 Code (http://aghababaie.usc.ac.ir/files/1506505203365.pdf and https://mattia.ir/wp-content/uploads/2020/10/ACI-318R-19.pdf; accessed on 28 December 2022), using the equation 57,000 × (compressive strength in psi)$^{1/2}$, which yielded a value of approximately 25,000,000,000 Pa (about 3,600,000 psi) [16]. Additionally, Poisson's ratio was taken to be 0.2 [14]. Then, the shear modulus was found using the equation elastic modulus/(2 × (1 + Poisson's ratio)), which yielded a value of 10,420,000,000 Pa (1,511,293 psi). Additionally, the density and the tensile strength were found to be 2402.77 kg/m^3 (150 lb/ft^3) and 3,447,000 Pa (500 psi), respectively [14]. Finally, the thermal conductivity and the specific heat were found to be 3 W/m·K (1.734 BTU/h·ft·°F) [17] and 750 J/kg·K (0.179 BTU/lb·°F), respectively [14]. As for the failure criterion, Mohr–Coulomb stress was used, since this is more accurate for brittle materials (such as concrete) [18]. All of these values were assigned to a material that we defined and called "concrete" in the SolidWorks software (which can be found in Figure 1), and we then assigned the properties of this "concrete" to the concrete that was previously defined and modeled (as discussed above). In this study, concrete was defined as a linear elastic material. Though, in tension, this is true until failure, in compression, concrete is linear and elastic up to the point where it nonlinearly reaches its ultimate compressive strength [19]. It was noted that the stress–strain curve of concrete is linear elastic up to approximately half of its compressive strength [20]. Therefore, for this study, the linear-elastic portion of concrete's stress–strain curve was used to evaluate the stiffness of different walls while still uncracked. Thus, the compressive strength was taken to be 2000 psi, and the tensile strength was taken to be 500 psi, which were considered as the failure points for this study. This showed the stiffness of uncracked sections and displayed the weak points in each wall, as well as the overall behavior of each wall before cracking occurred.

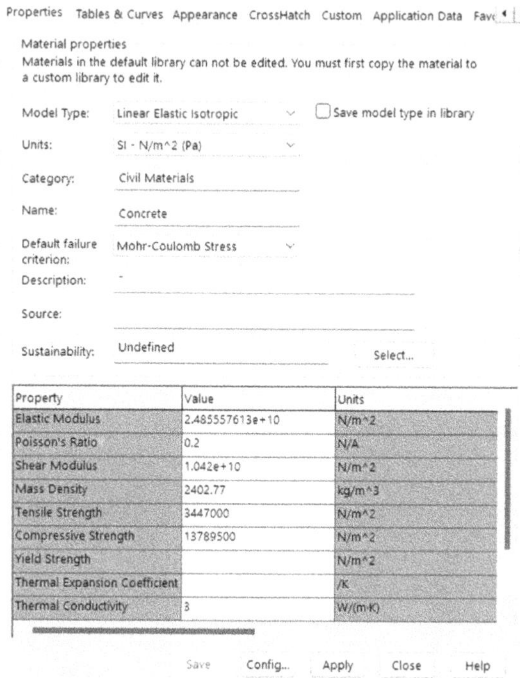

Figure 1. Creation of Concrete Material Properties.

Once the "concrete" material had been created and assigned to the model, the structural study (also called simulation in SolidWorks) could be set up. The first step in this process was to create fixities in the model. For this study, it was determined that the top and bottom faces (sections) of the wall would be fixed (could not move in any direction). These fixities could produce results most similar to the real world, because the bottom of the wall was connected to the footing embedded in the ground (thus being unable to move in any direction), and the top of the wall was connected to the concrete slab of the roof (e.g., for a single-story building), which was also assumed to be unable to move in any direction. Though this top portion may be able to move slightly in a building, it is more conservative to assume that it cannot move in any direction, as it results in larger wall end moments due to such fixity. For this reason, both the top and bottom of the wall were assumed to be fixed. Next, a pressure of 717.25 Pa (14.98 lb/ft^2) was applied to the exterior surface of the wall in the direction towards the interior of the wall. The value of this pressure was calculated using the wind pressure equation from ASCE 7-16, assuming a wind speed of 90 mph, C exposure category, and II risk category [21]. This calculation can be found in Table 1. It should also be noted that each abbreviation in Table 1 represents a specific parameter: Kz is the coefficient based on velocity and pressure at height z of the structure, Kzt is the topographic factor of the region, Kd is the factor based on the direction of the wind, Ke is the factor based on the ground's elevation, V is the wind velocity used for the study, and qh is the calculated pressure of the wind perpendicular to the wall. These parameters were chosen for the walls due to their use in the real-world applications of companies using this technology to print the walls of one-story homes using these wall shapes [22]. After the results were recorded using this pressure, a separate pressure of 717.25 Pa (14.98 lb/ft^2) was applied in place of the previous pressure to the exterior surface of the wall in the direction away from the interior of the wall.

Table 1. Calculation of Pressure Using ASCE 7-16.

Variable	Value	ASCE 7-16 Section
Kz	0.85	26.10.1
Kzt	1	26.8.2
Kd	0.85	26.6
Ke	1	26.9
V (mph)	90	
Exposure Category	C	26.7.3
Risk Category	II	1.5–1
qh (psf)	14.98	26.10–1
qh (Pa)	717.25	26.10–1

As mentioned above, the main goal of this study was to determine the flexural ability of different walls based on their shapes. In order to do this, a pressure was applied to the exterior face of each wall, as discussed and calculated above. However, it is recognized that there are other factors that determine whether or not a wall will fail. For example, factors such as extreme wind loads in hurricanes, earthquake loads, and support settlement (due to low soil-bearing capacity) could be limiting factors for the safety of a wall. Though these considerations are recognized for the actual design of buildings, the main goal of this study was not to analyze these behaviors. Instead, a study of the behavior of each wall, with the same applied wind pressure inducing tensile and compressive stresses in the wall models based on the wall's shape alone, was the main objective. Then, since this was a comparative study, the results for each wall were compared and contrasted to those of the other walls studied. In this way, weak points in specific walls (due to their shape), relative to other walls, could be identified.

Following the applied loads, the object was meshed in SolidWorks, allowing us to perform finite element analysis. In general, the default mesh options were used in SolidWorks to create the mesh. However, certain characteristics were checked to ensure that the mesh was of high quality. This means that the tetrahedron that comprises the mesh of the model possess nodes at the corners of the tetrahedron and the midpoints of the lines connecting the corners of the tetrahedron [23]. This high-quality mesh allows for more accurate results during the analysis part of the study [23]. Additionally, after the mesh was created, the aspect ratio of the mesh was checked. To ensure a mesh of a high quality, 90% of the elements or more had to possess an aspect ratio of less than 3 based on the longest and shortest connecting nodes on each tetrahedron comprising the mesh [23]. As a result, a smaller aspect ratio results in a better mesh because each tetrahedral is more uniformly shaped [23]. If the mesh generated with the default mesh settings in SolidWorks creates a mesh with less than 90% of the elements having an aspect ratio of less than 3, a denser (also referred to as a finer) mesh will be created until more than 90% of the elements have an aspect ratio of less than 3. This essentially means that the tetrahedron comprising the mesh will be smaller in size, resulting in more tetrahedrons comprising the mesh.

Finally, the process includes the performance of strength analysis using the Mohr–Coulomb failure criterion, which provides more accurate results for brittle materials such as concrete [18] and allows for a factor of safety plot to be generated for the model for several scenarios [18]:

$$\text{For principal stresses in tension,} \quad FOS = \frac{\sigma_{tensile\ limit}}{\sigma_1} \tag{1}$$

$$\text{For principal stresses in compression,} \quad FOS = \frac{\sigma_{compressive\ limit}}{|\sigma_3|} \tag{2}$$

$$\text{For } \sigma_1 \text{ in tension and } \sigma_3 \text{ in compression, } FOS = \left(\frac{\sigma_1}{\sigma_{tensile\ limit}} + \frac{|\sigma_3|}{\sigma_{compressive\ limit}} \right)^{-1} \quad (3)$$

Additionally, SolidWorks helps to define the principal stresses [24], where portions of a structure can be broken down into smaller cubes, each experiencing normal and shear stresses [24]. To calculate the maximum normal stresses these cubes experience, SolidWorks rotates these cubes so that the shear stresses are negligible [24]. At this point, the only stresses that remain on the rotated portion are the normal stresses [24]. In general, each face of the cube has a corresponding face on the opposite end of the cube with the same stress [24]. As a result, there are a total of three different stresses acting on the faces of this rotated cube, the largest called the first principal stress, the second-largest stress called the second principal stress, and the smallest called the third principal stress [24], as shown in the visualization in Figure 2.

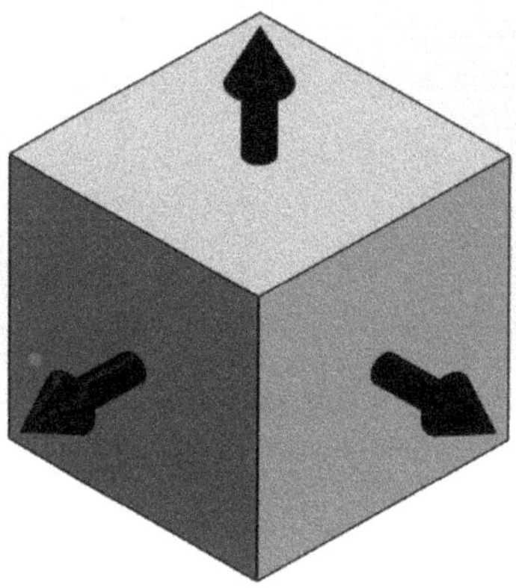

Figure 2. Three Principal Stresses on a Cube Element.

Essentially, the factor of safety values show how close each portion of the model is to failure, where the lower the factor of safety is, the closer that portion of the model is to failure. To accurately display the strength of the most vulnerable portion of each wall, the middle portion of each wall (also known as the midplane) must be the focus of the stress evaluation. Initially, several results, due to each load case, will be shown and discussed, including an overall plot of the factor of safety for each side of each wall. Then, a plot for three points along the midplane of the wall will be shown: a plot for the exterior of the wall, a plot for the interior of the wall, and a plot for the portion of the exterior concrete meeting the insulation of the wall. Together, these data will help to determine the flexural ability of each wall and where potential failure points could occur in each wall.

Additionally, with 3D concrete printing being such a new technology, certain assumptions must be made due to a lack of information available. For example, some companies use rebar in cavities of the printed shell parts of their walls and surround this rebar with cast concrete [25]. However, to remain consistent across walls made by different companies, no rebar was modeled in this study, because some companies use different spacings and other companies do not use rebar at all [26]. It should be noted that the effect of rebar prior to cracking is negligible, and since this analysis considers uncracked sections, neglect of the rebars for such modeling is reasonable. This enables the overall goal of this study to be met, i.e., showing the effect that the shape of the wall has on the uncracked strength of the wall.

Additionally, when beginning to print, companies print their overall framework (as a mold) and leave the cavities open (to be filled later) [1]. Then, once the framework has been created, companies fill in several of the cavities inside the wall with concrete and/or insulation [1]. For this study, the concrete that is later poured into cavities and the existing extruded concrete from the nozzle (acting as formwork) are assumed to act as one monolithic concrete component. Furthermore, insulation is ignored in the structural analysis part of the study. More specifically, the insulation cavity (where insulation is placed in the wall) is kept as an open void, since insulation provides little to no structural stiffness or, thus, resistance. Additionally, each wall studied is assumed to possess the same amount of material strength, so that the only variable between walls is their shape. Finally, since the main goal of this study is to determine the effect that the shape of the wall, alone, has on the structural performance, the quality of bonding between subsequent layers of extruded concrete placed on top of each other is not considered in this study, meaning the assumption is that there are no cold joints or gaps between layers. Additionally, it was assumed in this study that the cast concrete core of each wall does not separate (i.e., de-bond) from the printed formwork under any of the loading conditions considered.

Finally, a total of five different wall configurations were analyzed in this study. The overall shape of each wall was based on the shapes found in the industry through the research conducted (especially from [22,25–30]). In all, the five walls consisted of a symmetric two-cavity wall (as seen in Figure 3), an internal small-arched two-cavity wall (as seen in Figure 4), a zigzag cavity wall (as seen in Figure 5), a trapezoidal cavity wall (as seen in Figure 6), and an internal large-arched multicavity wall (as seen in Figure 7).

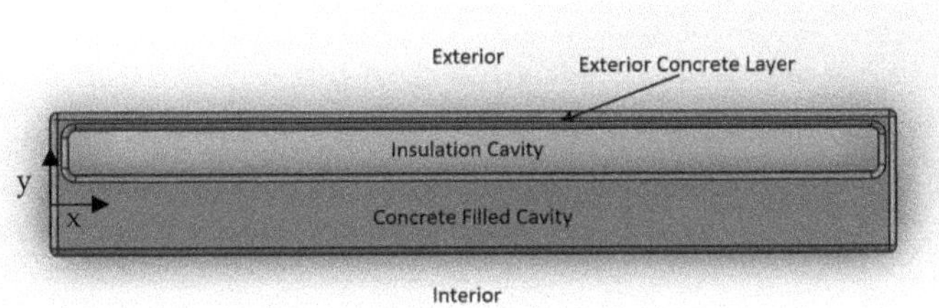

Figure 3. Symmetric Two-Cavity Wall Printed Formwork Filled with Concrete, Labeled. Modeled in reference to [26].

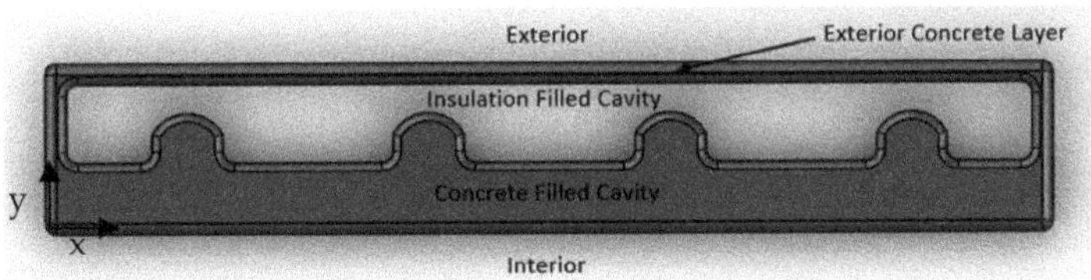

Figure 4. Internal Small-Arched Two-Cavity Wall Printed Formwork Filled with Concrete, Labeled. Modeled in reference to [25].

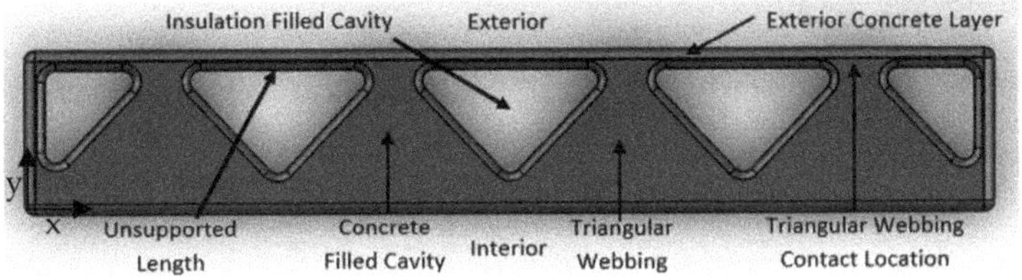

Figure 5. Zigzag Cavity Wall Printed Formwork Filled with Concrete, Labeled. Modeled in reference to [27,28].

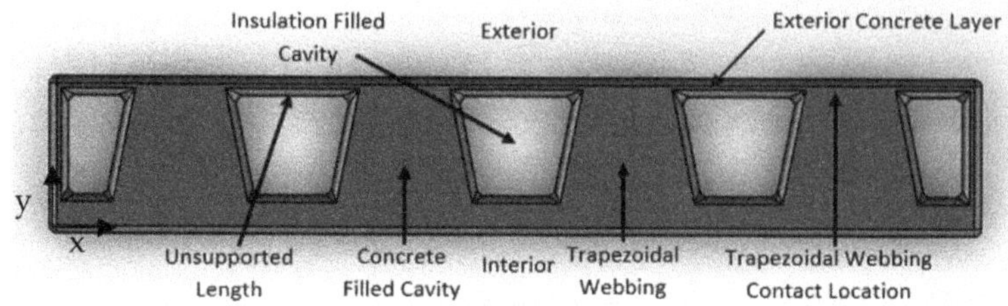

Figure 6. Trapezoidal Cavity Wall Printed Formwork Model with Fillets and Labels. Modeled in reference to [29].

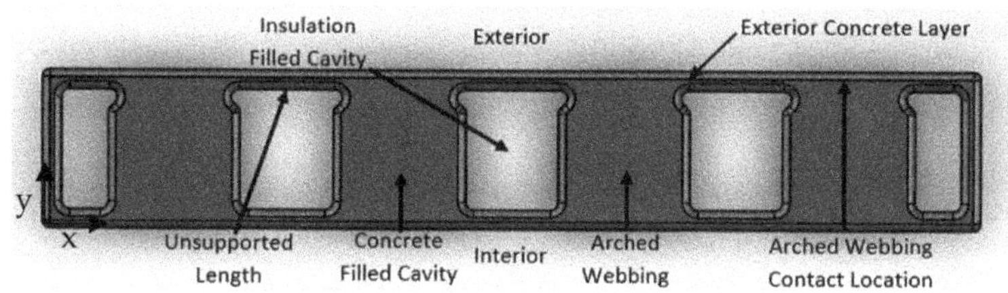

Figure 7. Internal Large-Arched Multicavity Wall Printed Formwork Model with Concrete Labeled. Modeled in reference to [22,30].

3.2. Thermal Study Setup in SolidWorks

At this point, the structural wall model (the part of the wall made of concrete) from the structural study still had an empty cavity present where the insulation was set to be placed. As a result, the insulation was modeled to fit within the remaining cavities for each wall. For the modeling of the insulation, the sketch dimensions were based on the remaining open cavities for each wall. Again, the height of the insulation was taken to be 8 ft so as to extend from the bottom to the top of each wall. Once the insulation had been modeled, an assembly was created to fit the insulation into the remaining empty cavity of the wall. Once completed, the concrete and insulation for this study were both modeled.

Next, the properties of the insulation material (rigid polyurethane foam) were determined. Since insulation was not used for the structural study, only the density and the thermal properties of the insulation were defined. The density, thermal conductivity, and specific heat were defined based on the literature [31], being 30 kg/m^3, 0.025 W/m·K, and 1500 J/kg·K, respectively. These properties were then assigned to an insulation material that we created and gave the name "insulation" in SolidWorks, which can be seen in Figure 8.

To set up the thermal study, several components needed to be defined. First, the component interactions between members of the assembly (e.g., the insulation and concrete) needed to be defined as either bonded or insulated. Insulated component interaction stops the flow of heat from one component to another (where they are in contact) [32]. However, for this study, this is not desirable, because the insulation has its own insulating behavior that serves to stop the flow of heat (and this does not increase on the contact surface). On the other hand, bonded contact interaction allows the concrete and insulation to act as one component with different thermal properties [32]. For this study, this is preferable, as it allows the flow of heat through the wall to be carried out as it would be in the real world. It should also be noted that the fillet for the edge where the insulation cavity meets the concrete at the top and bottom of each wall must be removed. The removal of this fillet aims to ensure that the insulation and concrete remain in full contact for the entire time during the thermal study, which would not be possible if these edges were filleted.

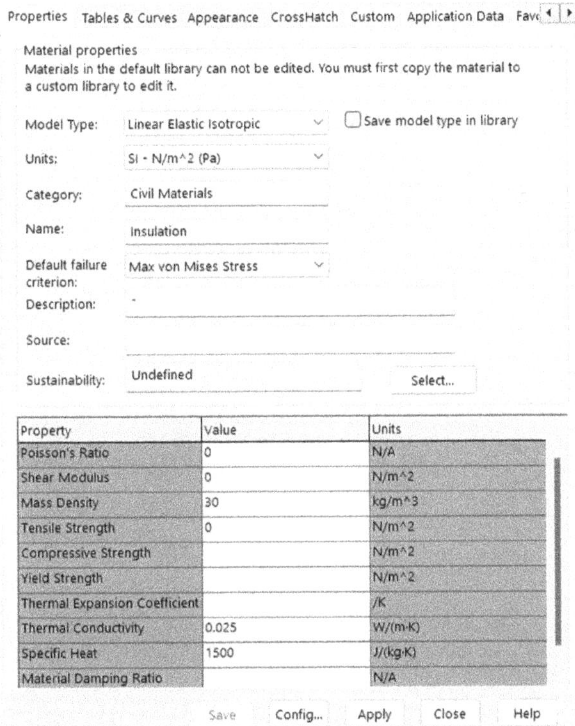

Figure 8. Creation of Insulation Material Properties.

Once the component interaction has been defined, the difference in temperature between the exterior and interior wall surfaces can be generated. To determine the thermal performance of each wall, a warm temperature was applied to the interior side of the wall and a cold temperature was applied to the exterior side of the wall. More specifically, a temperature of 295 K (approximately room temperature) was applied to the interior side of the wall to replicate an inside wall exposed to indoor temperatures. On the other hand, the exterior side of the wall was subjected to a convection thermal load. It was subjected to a heat transfer coefficient of 5 $W/m^2{\cdot}K$ (approximately that of standing air) and a bulk ambient temperature of 273.15 K to replicate a wall on the outside of a house that is exposed to cold air [33]. Finally, all of the meshing techniques and checks previously used for the structural study were also replicated for this study.

Proceeding with this approach, several thermal results were obtained to display the thermal performance of each wall. More specifically, an overall plot of the temperatures for each portion of the wall is shown to display the locations where more or less thermal energy is transferred from the interior to the exterior. Additionally, the amount of heat power lost on the exterior surface of the wall is also demonstrated. This shows the amount of heat that each wall loses due to the change in temperature [34].

4. Solid Concrete Wall Structural Study

As an example, the behavior of a solid concrete wall with the same conditions as those used for the other walls in this study was modeled in SolidWorks. More specifically, the top and bottom faces of the wall were taken to be fixed, and a pressure of 717.25 Pa was applied to the wall. Then, the result of this study was compared to a hand calculation of the factor of safety value (using the Mohr–Coulomb failure criterion) of this same wall. The SolidWorks result can be seen in Figure 9. In this figure, the central portion of the back face of the wall is in tension, which can be seen with yellow shading. Upon closer examination, the factor of safety value for this portion of the wall was 304.8. The loading conditions and shear and moment diagrams for this wall can be seen in Figure 10. To complete the hand calculation, the top and bottom faces of the wall were fixed (as they were in SolidWorks). Then, equilibrium conditions were applied to determine the respective moment and force values of the wall. Following this, the shear and moment diagrams were drawn. The principal stress at the center of the wall was then calculated using the flexural formula. This principal stress value was then divided by the tensile limit (as performed in SolidWorks) to determine the factor of safety value for the central portion of the back face of the wall. The value for this portion of the wall resulted in a factor of safety of 300. Thus, the two values are comparable.

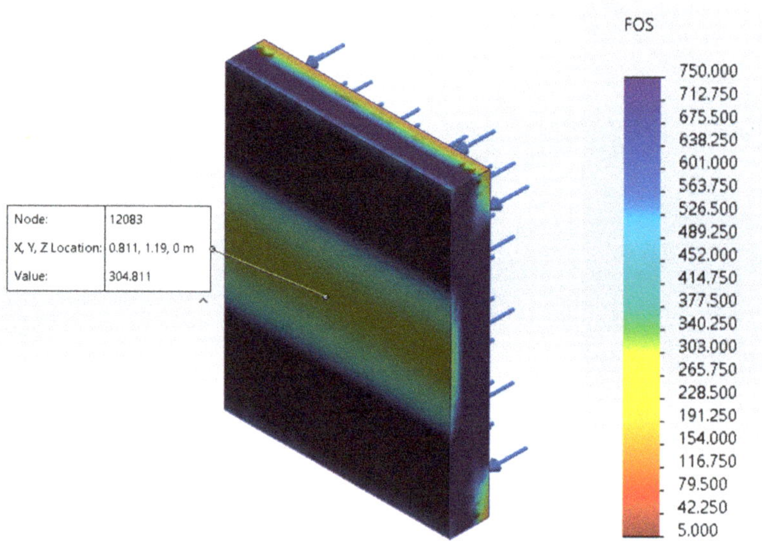

Figure 9. Factor of Safety Plot of a Solid Concrete Wall.

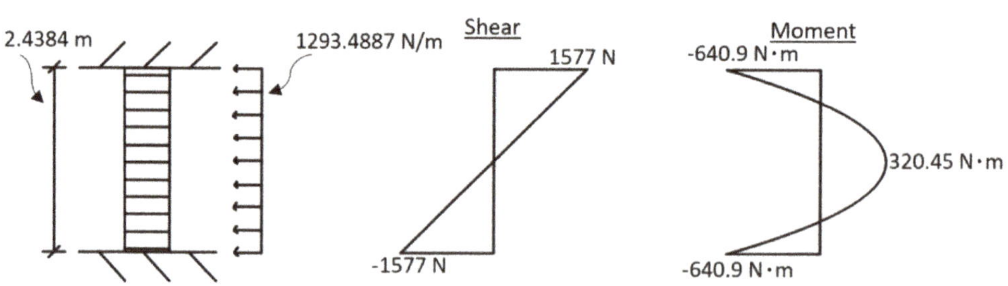

Figure 10. Loading and Moment and Shear Diagrams for Wind Loading Conditions.

At the midpoint:

$$Moment = \frac{w \times L^2}{12} = \frac{(1293.4887)(2.4384)^2}{12} = 640.902 \text{ N·m}$$

$$Moment\ of\ Inertia = \frac{1}{12} \times b \times h^3 = \frac{1}{12} \times 1.8034 \times 0.3048^3 = 0.0042556 \text{ m}^4$$

Stress calculation at the midpoint of the wall:

$$\sigma = \frac{M \times y}{I} = \frac{(320.451) \times (0.1524)}{0.0042556} = 11,475.85 \text{ Pa}$$

Factor of safety calculation at the midpoint of the wall:

$$FOS = \frac{tensile\ limit}{\sigma} = \frac{3,447,000}{11,475.85} = 300.4$$

Additionally, to display the effect that a seismic load could have on different portions of a wall in a structure, the seismic loading conditions were determined and converted to a distributed load for a 100 ft by 25 ft structure with 1 ft thick concrete walls and a flat concrete slab above these 8 ft walls. The seismic loadings for two locations were determined (New York, NY, USA and San Francisco, CA, USA) using the equivalent lateral force method in ASCE 7-16 for a building in risk category II and soil class C. First, the seismic coefficients for each were found using the ASCE 7 Hazard Tool Online, as can be seen in Table 2. Then, the appropriate coefficients were found to determine the amount of seismic loading present according to ASCE 7-16. It should be noted that the abbreviations in Table 2 stand for different parameters: SMS is the spectral response acceleration for short periods, SM1 is the spectral response acceleration for 1 s periods and a damping of 5%, SDS is the spectral response acceleration for the short periods used for design, SD1 is the spectral response acceleration at 1 s used for design, TL is the transition period for long periods, and SDC is the Seismic Design Category of a building in this location. First, the seismic loading for New York was determined. In this way, the equivalent lateral force was determined to be 24.375 K. This force was then converted to a distributed load along a 1.8034 m wide section of the wall (as seen in this study). Ultimately, this distributed seismic load resulted in a factor of safety value of 36.9 for the interior center of the wall. The calculations for this step can be found below, and a drawing of the loading and a shear and moment diagram can be found in Figure 11.

Table 2. Seismic Coefficients for New York City, New York [35].

Coefficient	Value
SMS	0.375
SM1	0.089
SDS	0.25
SD1	0.06
TL (s)	6
SDC	B

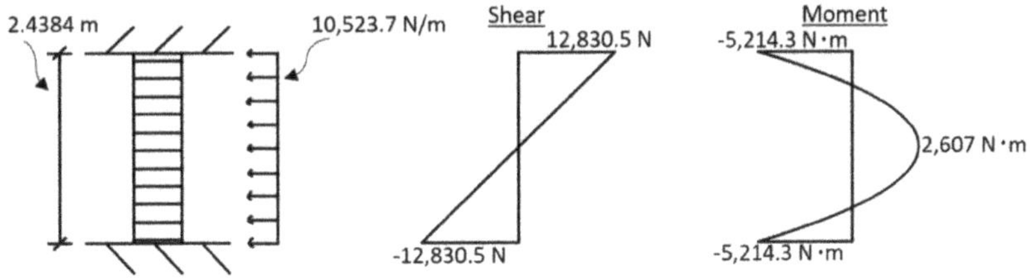

Figure 11. Loading and Moment and Shear Diagrams for Conditions in New York City, New York.

For concrete structures from ASCE 7-16 [21]:

$$R = 5$$

$$I_e = 1.0$$

$$T_a = 0.1 \times (number\ of\ stories) = 0.1 \times (1) = 0.1\ \text{s}$$

Weight of structure:

$$W_{wall} = \left[2 \times (8' \times 1' \times 25') + 2 \times (8' \times 1'8\ 100')\right] \times 150\ \text{psf} = 300\ \text{kips}$$

$$W_{roof} = \left(\frac{6''}{12} \times 25' \times 100'\right) \times 150\ \text{pcf} = 187.5\ \text{kips}$$

$$W_{total} = 300\ \text{kips} + 187.5\ \text{kips} = 487.5\ \text{kips}$$

Loading calculation [21]:

$$C_{Smax} = \frac{S_{D1}}{T_a \times \left(\frac{R}{I_e}\right)} = \frac{0.06}{0.1 \times \left(\frac{5}{1.0}\right)} = 0.12$$

$$C_{Smin} = 0.044 \times S_{DS} \times I_e = 0.044 \times 0.25 \times 1.0 = 0.011 \geq 0.01$$

$$C_S = \frac{S_{DS}}{\left(\frac{R}{I_e}\right)} = \frac{0.25}{\frac{5}{1.0}} = 0.05$$

$$V = C_S \times W_{total} = 0.05 \times 487.5\ \text{kips} = 24.375\ \text{kips}$$

Converting the load to newtons:

$$V = 24.375\ \text{kips} \times \frac{4448.22\ \text{N}}{1\ \text{kip}} = 108,425.3625\ \text{N}$$

Converting the loading to a distributed load for a 1.8034-meter-wide section of the wall:

$$V = \frac{108,425.3625\ \text{N} \times 1.8034\ \text{m}}{7.62\ \text{m}} = 25,660.7\ \text{N}$$

$$w = \frac{V}{Wall\ Height} = \frac{25,660.7\ \text{N}}{2.4384\ \text{m}} = 10,523.7\ \text{N/m}$$

At the midpoint:

$$Moment = \frac{w \times L^2}{12} = \frac{(10,523.7)(2.4384)^2}{12} = 5214.3\ \text{N·m}$$

$$Moment\ of\ Inertia = \frac{1}{12} \times b \times h^3 = \frac{1}{12} \times 1.8034 \times 0.3048^3 = 0.0042556\ \text{m}^4$$

Stress calculation at the midpoint of the wall:

$$\sigma = \frac{M \times y}{I} = \frac{(2607.16) \times (0.1524)}{0.0042556} = 93,366.6\ \text{Pa}$$

Factor of safety calculation at the midpoint of the wall:

$$FOS = \frac{tensile\ limit}{\sigma} = \frac{3,447,000}{93,366.6} = 36.9$$

Once completed, the same process was repeated to determine the loading for San Francisco, California. The seismic coefficients can be found in Table 3. In this case, the equivalent lateral load was found to be 117 K. This was then converted to a distributed load along a 1.8034 m wide section of the wall. Ultimately, this resulted in a factor of safety value of 7.7. The calculations for this loading and factor of safety determination can be found below. Additionally, Figure 12 shows the drawing of the loading and the shear and moment diagrams.

Table 3. Seismic Coefficients for San Francisco, California [35].

Coefficient	Value
SMS	1.8
SM1	0.84
SDS	1.2
SD1	0.56
TL (s)	12
SDC	D

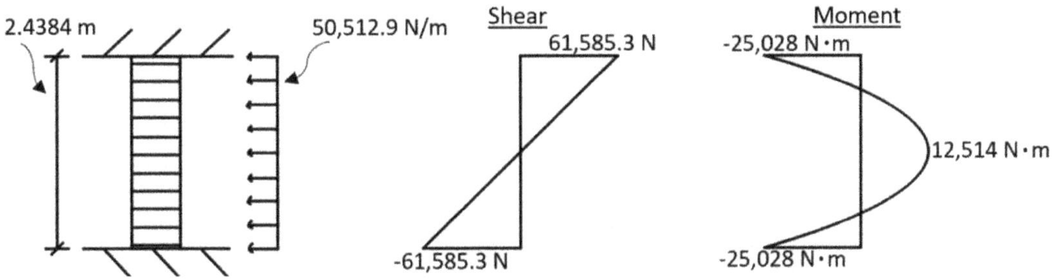

Figure 12. Loading and Moment and Shear Diagrams for Conditions in San Francisco, California.

Loading calculation [21]:

$$C_{Smax} = \frac{S_{D1}}{T_a \times \left(\frac{R}{I_e}\right)} = \frac{0.56}{0.1 \times \left(\frac{5}{1.0}\right)} = 1.12$$

$$C_{Smin} = 0.044 \times S_{DS} \times I_e = 0.044v1.2 \times 1.0 = 0.0528 \geq 0.01$$

$$C_S = \frac{S_{DS}}{\left(\frac{R}{I_e}\right)} = \frac{1.2}{\frac{5}{1.0}} = 0.24$$

$$V = C_S \times W_{total} = 0.05 \times 487.5 \text{ kips} = 117 \text{ kips}$$

Converting the load to newtons:

$$V = 117 \text{ kips} \times \frac{4448.22 \text{ N}}{1 \text{ kip}} = 520,439.4 \text{ N}$$

Converting the loading to a distributed load for a 1.8034-meter-wide section of the wall:

$$V = \frac{108,425.3625 \text{ N} \times 1.8034 \text{ m}}{7.62 \text{ m}} = 123,170.7 \text{ N}$$

$$w = \frac{V}{Wall\ Height} = \frac{123,170.7 \text{ N}}{2.4384 \text{ m}} = 50,512.9 \text{ N/m}$$

At the midpoint:

$$Moment = \frac{w \times L^2}{12} = \frac{(50,512.9)(2.4384)^2}{12} = 25,028.3 \text{ N·m}$$

$$Moment\ of\ Inertia = \frac{1}{12} \times b \times h^3 = \frac{1}{12} \times 1.8034 \times 0.3048^3 = 0.0042556 \text{ m}^4$$

Stress calculation at the midpoint of the wall:

$$\sigma = \frac{M \times y}{I} = \frac{(12,514.1) \times (0.1524)}{0.0042556} = 448,151.8 \text{ Pa}$$

Factor of safety calculation at the midpoint of the wall:

$$FOS = \frac{tensile\ limit}{\sigma} = \frac{3,447,000}{448,151.8} = 7.7$$

Though the value for seismic loading produces lower factor of safety values, the loading used in this study was 717.25 Pa (14.98 psf) (from the wind loading). The overall goal of this study was to compare and contrast different walls based on their shape. As a result, the application of any pressure (such as 717.25 Pa) would still enable us to accomplish this goal. More specifically, a 717.25 Pa (14.98 psf) applied pressure would still produce factor of safety values relative to the other walls and highlight where potential weak points in each wall type could develop. As a result, the comparative study could be carried out using this pressure.

5. Zigzag Cavity Wall

5.1. Zigzag Cavity Wall Structural Study

5.1.1. Pressure Applied on the Exterior Face towards the Interior

The zigzag cavity wall was the third wall to be modeled and studied [27,28]. Here, there is a printed zigzag portion running through the middle of the wall, which creates cavities throughout for insulation and concrete. The sketch used in this study for a zigzag cavity wall can be found in Figure 13, together with its dimensions.

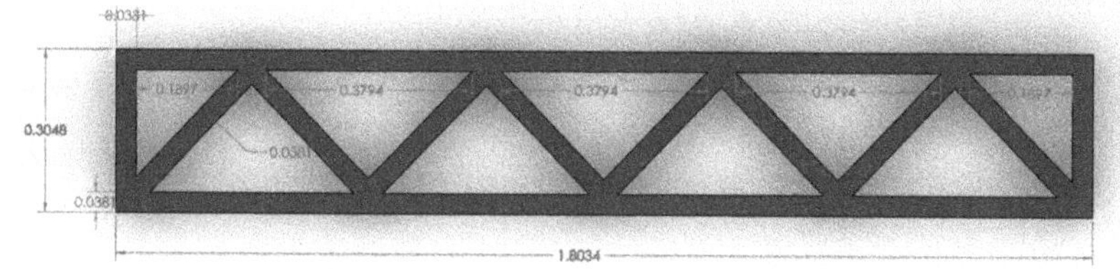

Figure 13. Zigzag Cavity Wall Printed Formwork Model. Modeled in reference to [27,28].

After the creation of the formwork in SolidWorks, all of the sharp edges present in the wall were filleted. A 0.015 m fillet was used to reduce these sharp edges and prevent the formation of local singularities during the study. Additionally, the wall was extruded to a height of 2.438 m (8 ft), which was the same as that of the other walls studied.

Once the formwork was modeled, the cavities could be filled with concrete. It is not entirely clear which cavities companies fill with concrete and insulation. Thus, to remain consistent with other walls, the cavities closer to the interior of the wall were filled with concrete in this study, and the exterior cavities were filled with insulation. Once the filled concrete cavity was modeled in SolidWorks (as seen in Figures 14 and 15), the concrete material properties were applied to the modeled portions of the wall. At this point, the structural study was set up.

Figure 14. Zigzag Cavity Wall Printed Formwork with Interior Cavity Filled with Concrete.

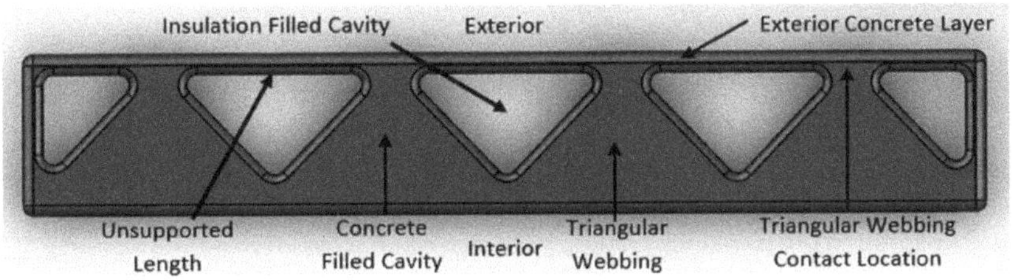

Figure 15. Zigzag Cavity Wall Printed Formwork Filled with Concrete, Labeled.

Again, the top and bottom of the wall were fixed, meaning it could not move in any direction. Next, 717.25 Pa (14.98 psf) of pressure was applied to the exterior face of the wall in the direction towards the interior of the wall. Subsequently, the meshing of the wall was completed and checked in SolidWorks. More specifically, it was ensured that the aspect ratios of at least 90% of the elements were less than 3. Additionally, the mesh was checked for hotspots using the built-in tool in SolidWorks. The meshing met both of these criteria; thus, the factor of safety results were then checked and gathered for the wall.

The first set of information gathered was related to the isometric factor of safety plot of the entire wall. As seen in Figure 16, the wall possesses several locations that have lower factor of safety values. More specifically, the isometric view on the left of Figure 16 shows that the wall acts in a manner similar to a vertically oriented fixed–fixed wall. In more detail, since there is significant webbing connecting the interior and exterior faces of the wall, the wall appears to behave more like a solid fixed–fixed wall, that is, a wall fixed on the top and bottom. More specifically, if a pressure is applied to its face (as for this wall), and it is fixed at its top and bottom faces, then tension will be present next to its supports and in the middle of its back face. In the case of this wall, there are lower factor of safety values for the center of the interior face of the wall. The wall also has low factor of safety values at the top and bottom of its exterior face. As previously discussed, these lower factor of safety values indicate that these portions of the wall are in tension, since the tensile limit of concrete is much lower than the compressive limit of concrete. For a better understanding, Figure 17 displays two images of the deflected shape of this wall from the perspectives of the side and midplane.

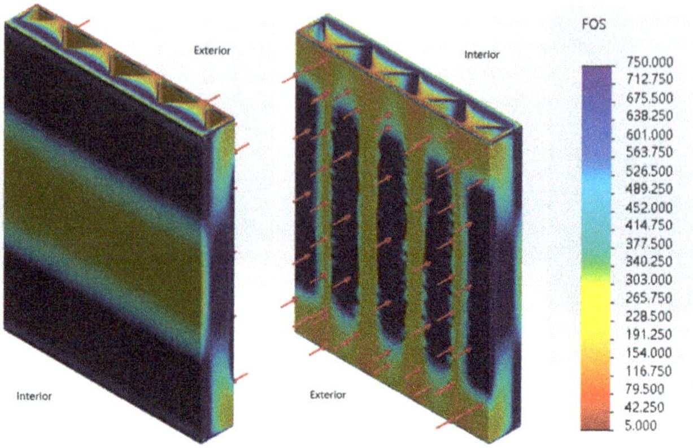

Figure 16. Zigzag Cavity Wall Structural Study Isometric Results After Applying Pressure to the Interior.

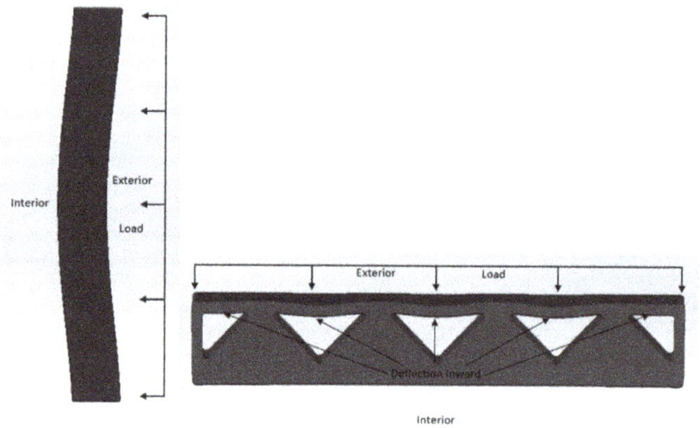

Figure 17. Zigzag Cavity Wall's Side and Midplane (Not to Scale).

In addition, it is clear from this isometric view that the concrete-filled triangular webbing connecting the interior and exterior faces transfers stress from the exterior to the interior of the wall. This is clear from the yellow color present at the tips of the triangles that slowly transitions to a blue color. This is a key characteristic of this wall, as the previous two walls did not contain webbing in the middle of the wall that allowed stress to be transferred from its exterior to interior faces.

Figure 18 displays the results of the study for the exterior concrete layer of the zigzag cavity wall. From these results, we can see that there are five unsupported lengths (where the insulation cavity meets the exterior concrete layer) that also act in a similar manner to fixed–fixed walls, one of which is labeled "unsupported length" in Figure 15. More specifically, each time the triangular webbing makes contact with the exterior face, it provides rigidity and support to the wall at this location. This triangular webbing contact location is displayed in Figure 15. Thus, the unsupported lengths of the exterior concrete layer that span between these contact locations act in a similar way to a fixed–fixed wall. The results in Figure 18 show this behavior. More specifically, the triangular webbing contact locations experience tension (and, thus, lower factors of safety) on the exterior face. This behavior can be seen in the left picture in Figure 18. On the interior side, the unsupported lengths experience tension at their center. This behavior can be seen in the middle picture in Figure 18. As previously discussed, this behavior is consistent with a fixed–fixed wall.

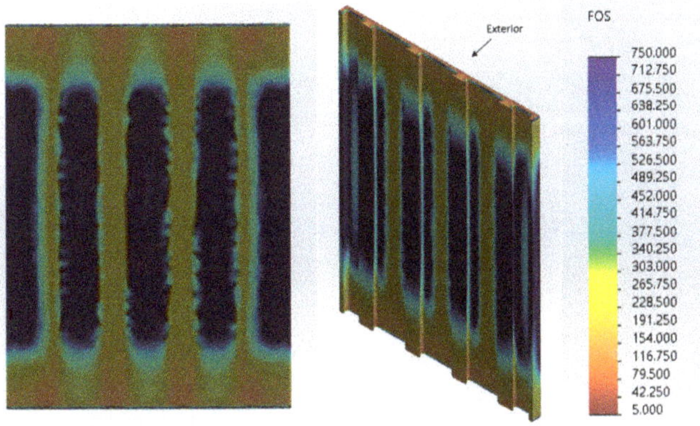

Figure 18. Zigzag Cavity Wall Exterior Concrete Layer Results After Applying Pressure to the Interior. Left: Exterior Side of Exterior Concrete Layer; Middle: Interior Side of Exterior Concrete Layer; Right: Color Scale of Results.

The following graphs in the figures below graphically display the factor of safety values for several locations on the midplane of each wall. The first of these graphs in Figure 19 shows the factor of safety values along the exterior edge of the zigzag cavity wall. This graph shows that there are relatively low factor of safety values (around 200) at the triangular webbing contact locations, which indicates that tension is present. On the other hand, this graph displays high factor of safety values at the unsupported portions of the wall (around 750). This indicates that this portion of the wall is in compression.

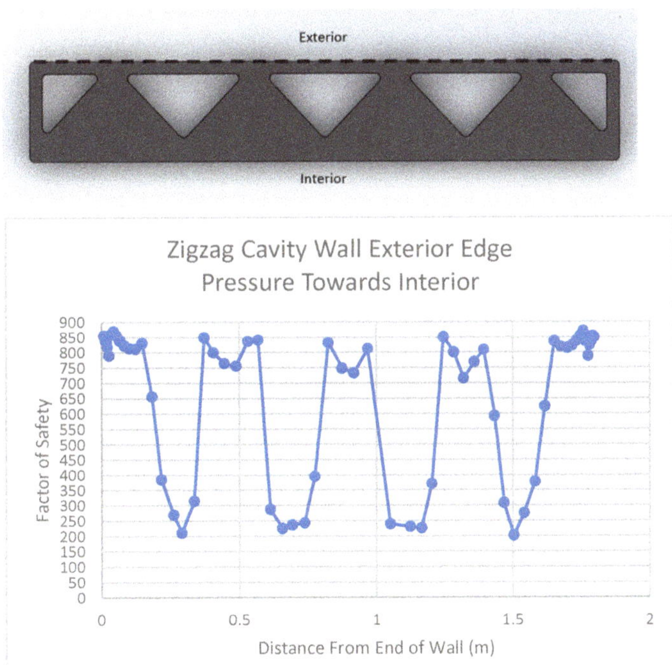

Figure 19. Results for Exterior Edge of Zigzag Cavity Wall After Applying Pressure to the Interior.

The graph displayed in Figure 20 shows the factor of safety values for the zigzag wall where the insulation cavity meets the exterior concrete layer. In this case, there are lower factor of safety values at the center of each unsupported length. For this reason, there are five drops in the factor of safety plot that reach a value as low as approximately 200.

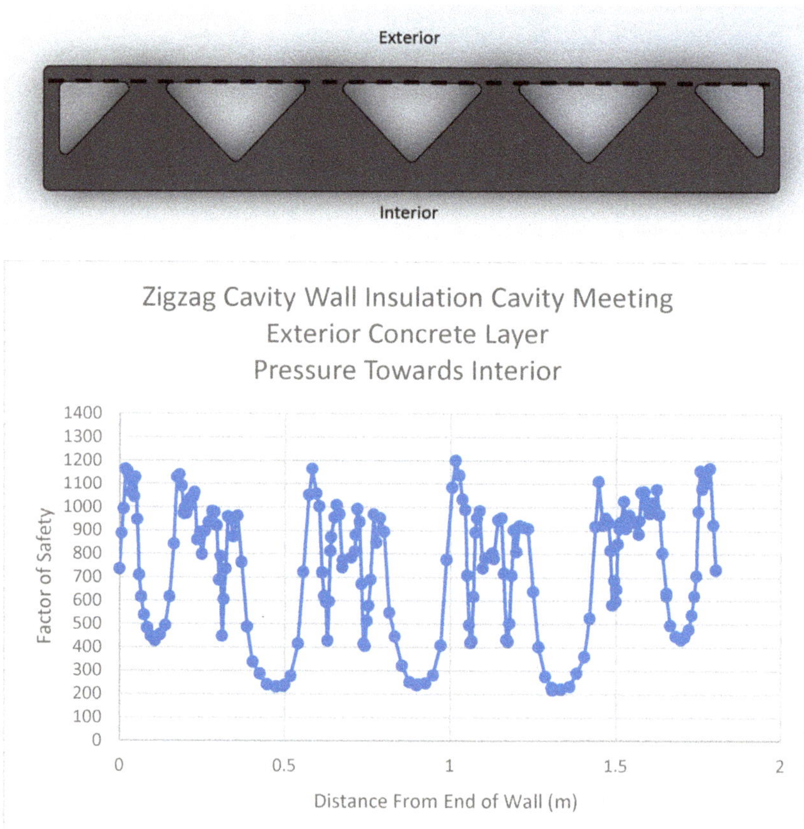

Figure 20. Results for Insulation Cavity Meeting the Exterior Concrete Layer of the Zigzag Cavity Wall After Applying Pressure to the Interior.

The factor of safety graph in Figure 21 shows the values for the interior side of the wall. The values all remain in a relatively small range (between 250 and 270). This is due to the fact that the interior side of the wall in the midplane is in tension, because it acts in a similar way to a vertically oriented fixed–fixed wall, as previously discussed. Nevertheless, there still appears to be a drop in the factor of safety (larger tension stress) where there is a continuous connection between the exterior and interior faces of the wall, thus showing that there is a transfer of stress due to the triangular webbing that connects these two faces.

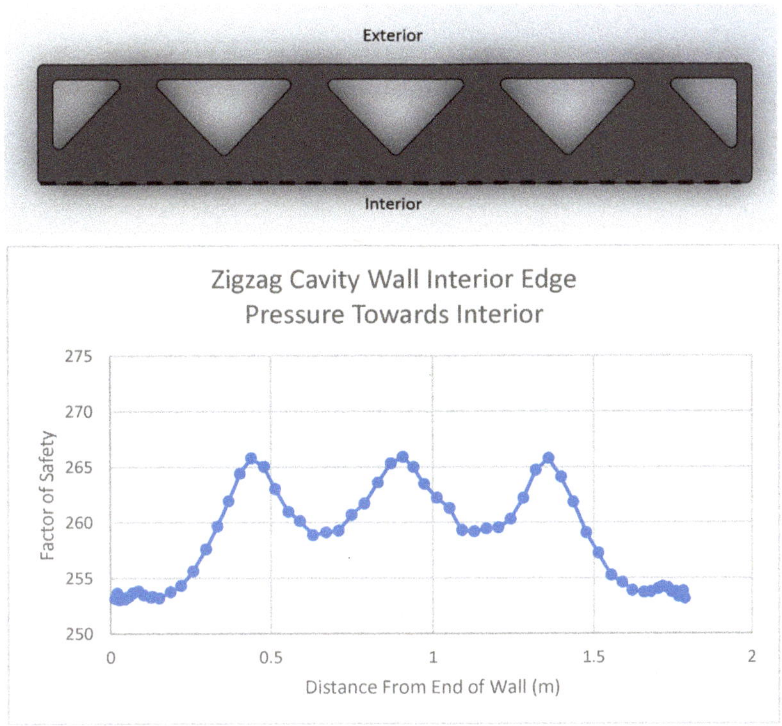

Figure 21. Results for Interior Edge of Zigzag Cavity Wall After Applying Pressure to the Interior.

5.1.2. Pressure Applied on the Exterior Face away from the Interior

Next, the study was repeated for a pressure acting on the exterior face but away from the interior of the wall. This study resulted in factor of safety values similar to those obtained when pressure was applied to the interior of the wall. However, the interior and exterior faces of the wall exhibited behaviors different from before. When pressure was applied to the interior of the wall, the interior face of the wall experienced tension at its center. However, as pressure was applied in the opposite direction at this point, the interior face of the wall then experienced compression at its center. Similarly, the exterior face of the wall experienced tension at its center, whereas when the pressure was flipped, it experienced compression at its center. Thus, when the pressure's direction is flipped, the tension and compression regions in the wall are also flipped. The isometric view of the factor of safety plot for this scenario can be seen in Figure 22. In addition to Figure 22, Figure 23 displays the deflected shape of this wall in its side profile and midplane.

As mentioned in the previous paragraph, flipping the pressure's direction also flips the tension and compression regions in the wall. This is no different for the exterior concrete layer, as can be seen in Figure 24. In this case, the unsupported lengths now experience tension on the exterior face of the exterior concrete layer. On the other hand, the triangular webbing contact locations now experience tension on the interior face of the exterior concrete layer. This is consonant with the previous statement that flipping the direction of the pressure will put regions previously in tension into compression.

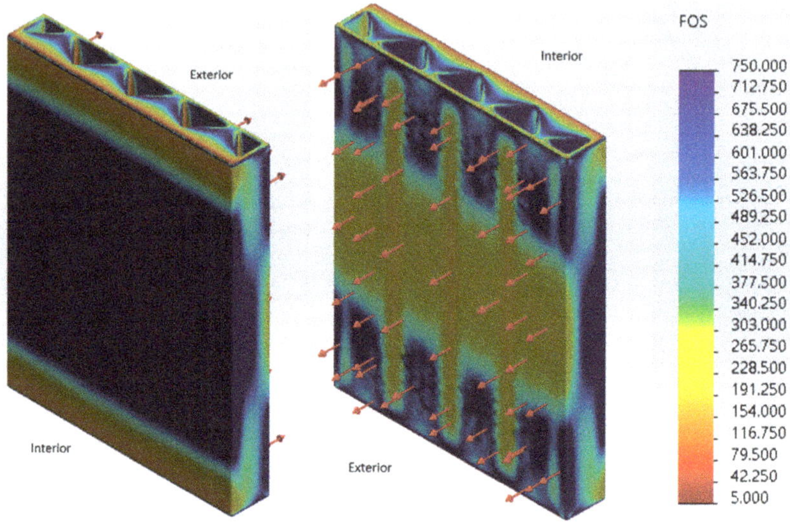

Figure 22. Zigzag Cavity Wall Structural Study Isometric Results for Pressure Applied Away from the Interior.

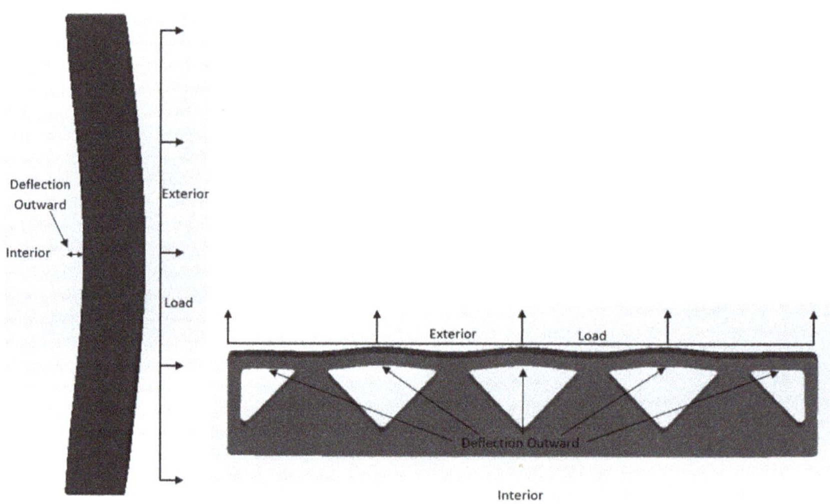

Figure 23. Deflected Shape of Zigzag Cavity Wall's Side and Midplane (Not to Scale).

Next, the factor of safety graphs were generated to display the midplane behavior of the wall. First, the exterior edge of the wall is shown in Figure 25. This graph clearly shows that the unsupported lengths of the wall possess lower factor of safety values and are thus in tension. On the other hand, the triangular webbing contact locations possess higher factor of safety values, indicating that the wall is further from failure at this point.

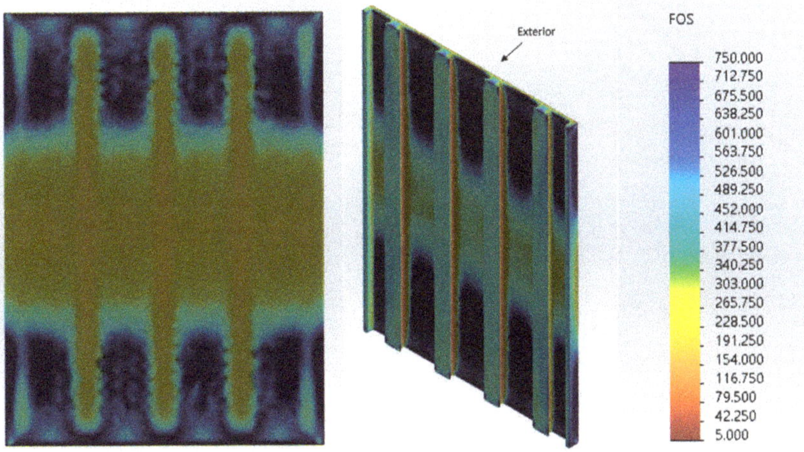

Figure 24. Zigzag Cavity Wall Exterior Concrete Layer Results for Pressure Applied Away from the Interior. Left: Exterior Side of Exterior Concrete Layer; Middle: Interior Side of Exterior Concrete Layer; Right: Color Scale of Results.

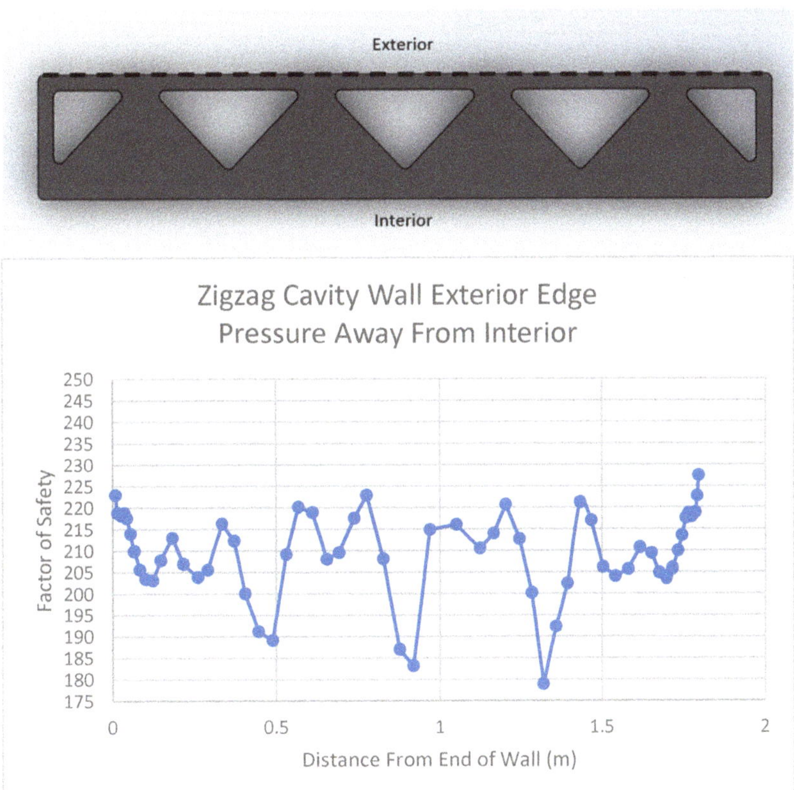

Figure 25. Zigzag Cavity Wall Exterior Edge Results for Pressure Applied Away from the Interior.

The next graph in Figure 26 shows the portion of the wall where the insulation cavity meets the exterior concrete layer. This graph clearly shows that in this portion of the wall, the unsupported lengths have higher factor of safety values (over 325). On the other hand, the triangular webbing contact locations have relatively lower factor of safety values of approximately 100, which indicates that this portion of the wall is in tension.

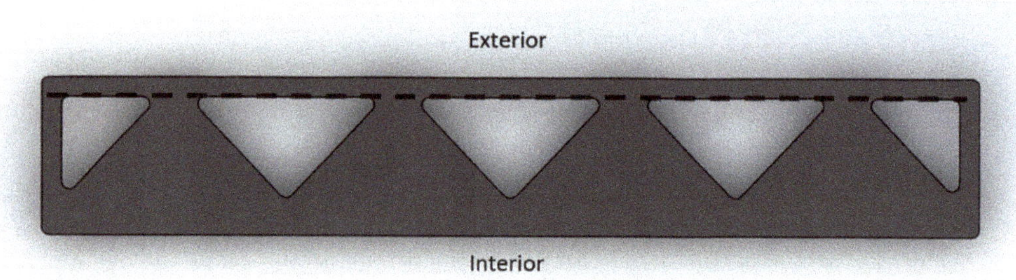

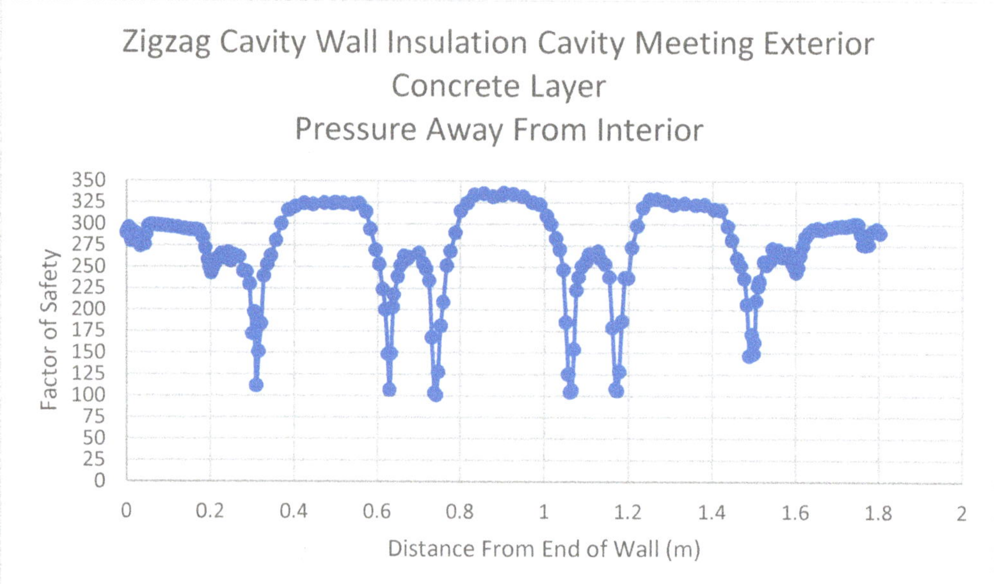

Figure 26. Results for Insulation Cavity Meeting the Exterior Concrete Layer of the Zigzag Cavity Wall for Pressure Applied Away from the Interior.

The graph in Figure 27 displays the factor of safety values for the interior edge of the zigzag wall. The factor of safety values for this portion of the wall are relatively high (reaching upwards of 1000). This is due to the fact that the interior side of the wall in its midplane is in compression for this direction of pressure. This behavior can clearly be seen in Figure 22, which shows an isometric view of the results. It should again be noted that there are lower factor of safety values where the concrete is continuous throughout the wall. This indicates that the webbing is able to transfer stress from the exterior face to the interior face.

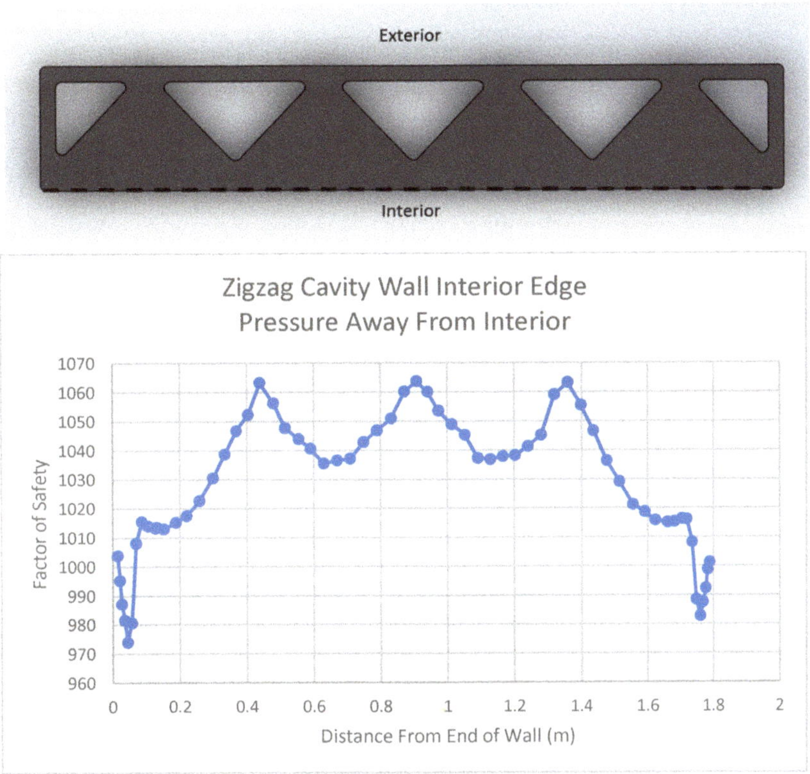

Figure 27. Zigzag Cavity Wall Interior Edge Results for Pressure Applied Away from the Interior.

5.2. *Zigzag Cavity Wall Thermal Study*

After the results for the structural study were obtained, the thermal study of the zigzag cavity wall was conducted. First, the concrete portion of the wall (without fillets) was filled with a modeled insulation. This insulation was then assigned the insulation properties discussed in the setup section. The assembly and the labeled cross section of the assembly can be seen in Figure 28.

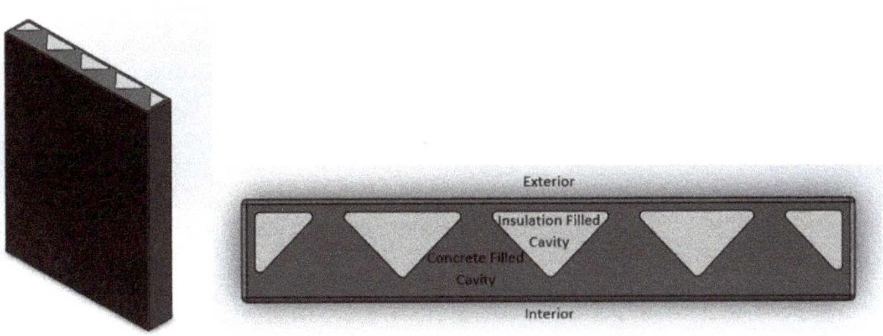

Figure 28. Zigzag Cavity Wall Concrete and Insulation Assembly.

Once the wall assembly was modeled, the insulation and concrete portions of the wall were connected as a bonded connection in SolidWorks. To represent a wall exposed to cold outdoor temperatures, a convection thermal load was applied to the exterior of the wall. This convection thermal load was given a heat transfer coefficient of 5 $W/m^2 \cdot K$ with a bulk ambient temperature of 273.15 K [33]. Additionally, a 295 K temperature was assigned to the interior side of the wall. The wall was then meshed in SolidWorks. Again, it was ensured that 90% or more of the elements in the mesh had an aspect ratio of less than 3. Subsequently, the results were gathered.

Figure 29 shows an isometric view of the temperature distribution within the wall examined in this study. From the figure, it is clear there is less heat loss where there is insulation. This can be seen from the teal color shown in the middle image in Figure 29. On the other hand, there are clear thermal bridging locations present throughout the wall. For example, in every location where the concrete is continuous (from the exterior to the interior face), there appears to be thermal bridging. For example, on the exterior face of the wall, there is clear yellow shading on the exterior face (where the concrete is continuous) that represents thermal bridging locations. These thermal bridging locations can be seen more clearly in Figure 30, which shows an overhead view of the wall. Again, there is less heat loss where the insulation comes into contact with the exterior concrete layer (shown in blue).

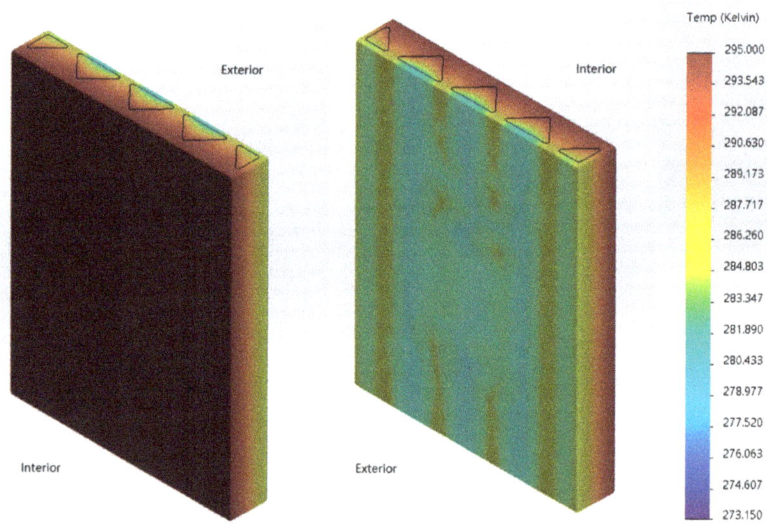

Figure 29. Zigzag Cavity Wall Isometric Temperature Plot.

Additionally, the heat power of the exterior face of the wall was calculated using SolidWorks. When calculated, it was found that the wall lost a total of 211.83 W of power. This heat loss is a direct result of the warm temperature traveling along the continuous concrete portions of the wall to the exterior face of the wall. Again, concrete possesses a higher thermal conductivity value than the insulation. Thus, when continuous, the concrete results in thermal bridging. The results of the heat power calculation in SolidWorks can be seen in Figure 31.

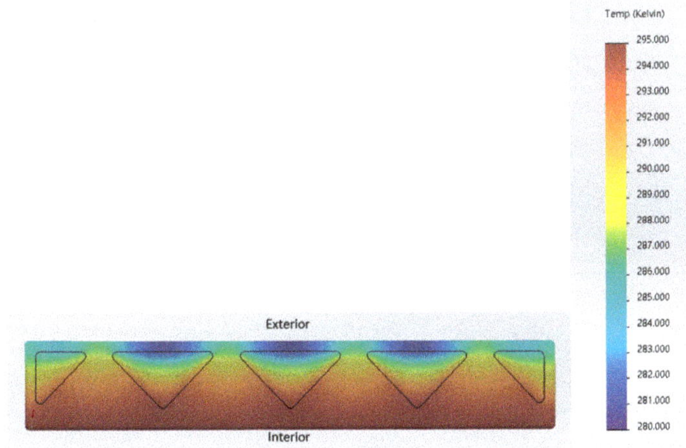

Figure 30. Zigzag Cavity Wall Overhead Temperature Plot.

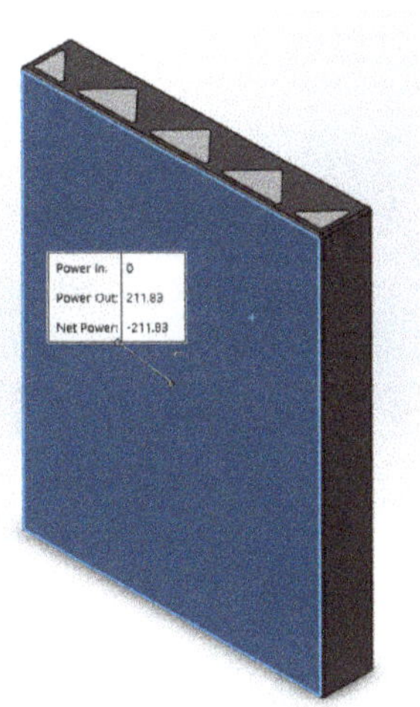

Figure 31. Zigzag Cavity Wall Net Power.

6. Relative Stiffness Determination for Different Wall Shapes

To further develop the structural aspect discussion, the moment of inertia values for each wall's cross section were calculated using SolidWorks and then compared to the moment of inertia value of a solid wall with the same dimensions. It should be noted that the same solid wall used for the example calculation in a previous section was also used for this calculation. Additionally, the moment of inertia of each wall was taken about the x-axis, which is the axis parallel to the width of each wall. To clarify, the width is the 1.8034 m dimension in this study. In addition, the moment of inertia was calculated for walls with no filleted edges and for the filled concrete cavities and the insulation cavities left empty. Thus, the shapes that were used can be found in Figures 3–7, with the fillets removed for this calculation. In all, these values provide a very useful indication of the relative stiffness of each wall studied. Additionally, Table 4 shows the areas of printed concrete, poured concrete, and insulation for the studied walls.

Table 4. Cross Sectional Area of Each Wall Studied.

Wall Type	Printed Concrete (m^2)	Poured Concrete (m^2)	Filled Insulation (m^2)
Symmetric Two-Cavity Wall	0.22	0.16	0.16
Internal Small-Arched Multicavity Wall	0.24	0.10	0.20
Zigzag Cavity Wall	0.24	0.15	0.15
Trapezoidal Cavity Wall	0.27	0.14	0.14
Internal Large-Arched Multicavity Wall	0.25	0.13	0.17

Figure 32 shows the moment of inertia values relative to the moment of inertia value of a solid wall with the same dimensions. From this figure, it is clear which walls produce the largest moment of inertia values. The symmetric two-cavity wall and the internal small-arched two-cavity wall produced relatively low moment of inertia values. On the other hand, the zigzag cavity wall, the trapezoidal cavity wall, and the internal large-arched multicavity wall all produced higher moment of inertia values. From this finding and the flexural formula ($\sigma = My/I$), it is clear that the amount of stress generated in these walls is inversely proportional to the moment of inertia of the wall. This means that the larger the moment of inertia is, the less stress is generated in the wall. Accordingly, it is clear that a larger moment of inertia value is beneficial for reducing the amount of stress present within the wall's cross section. Additionally, the walls that contain webbing connecting the interior and exterior faces of the wall (the zigzag cavity wall, the trapezoidal cavity wall, and the internal large-arched multicavity wall) are seen to possess relatively larger moment of inertia values. Thus, it can be inferred that these walls generate less stress (due to an applied moment) as a result. Consequently, the walls with webbing (that have higher moment of inertia values) possess more stiffness relative to the walls without webbing.

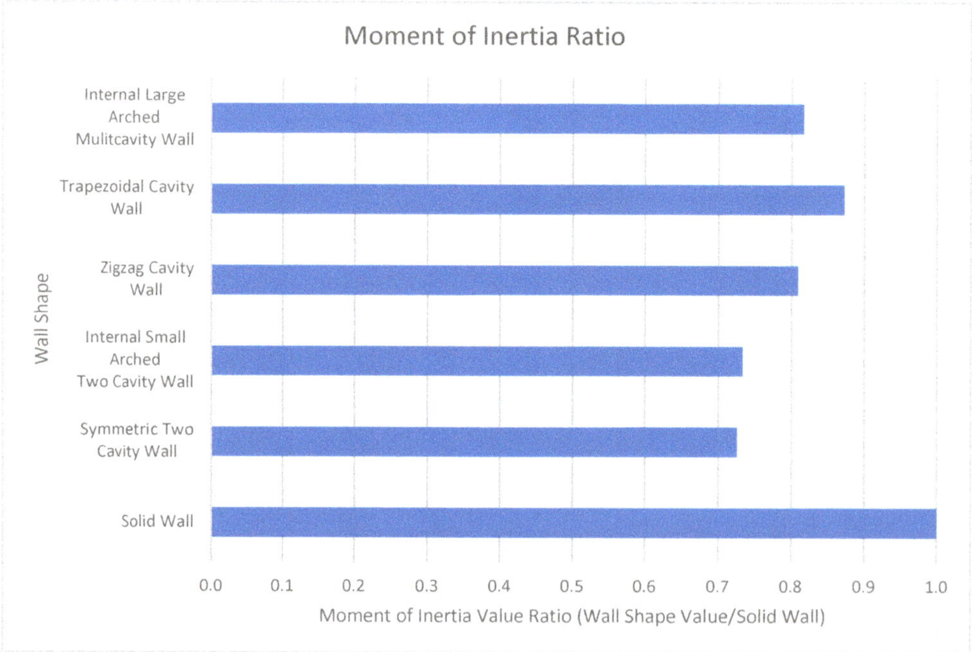

Figure 32. Moment of Inertia Ratio Chart.

7. Discussion of Results

The structural results of each wall show several revealing trends that should be noted. The walls used in this study and their results can be found in Figures 3–7. First, the concrete connecting the exterior and interior faces of these walls allowed for the entire thickness of the wall to act in resisting the stress applied to the wall. It is seen that in the first two walls studied (the symmetric two-cavity wall and the internal small-arched two-cavity wall), there is no concrete webbing on the interior and exterior faces of the walls. As a result, the thinner exterior concrete layer was responsible for resisting the majority of the applied load. This was not the case for the last three walls that were studied (zigzag cavity wall, trapezoidal cavity wall, and internal large-arched multicavity wall). In these walls, there was significant webbing present to connect the interior and exterior faces of the wall. As a result, pressure could be transferred from the exterior face (where the pressure was applied) to the interior face of the wall, thus showing that the entire thickness of the wall was utilized to resist the applied pressure. The factor of safety values for each wall studied (when pressure was applied to the interior of the wall) can be seen in Figures 33–35.

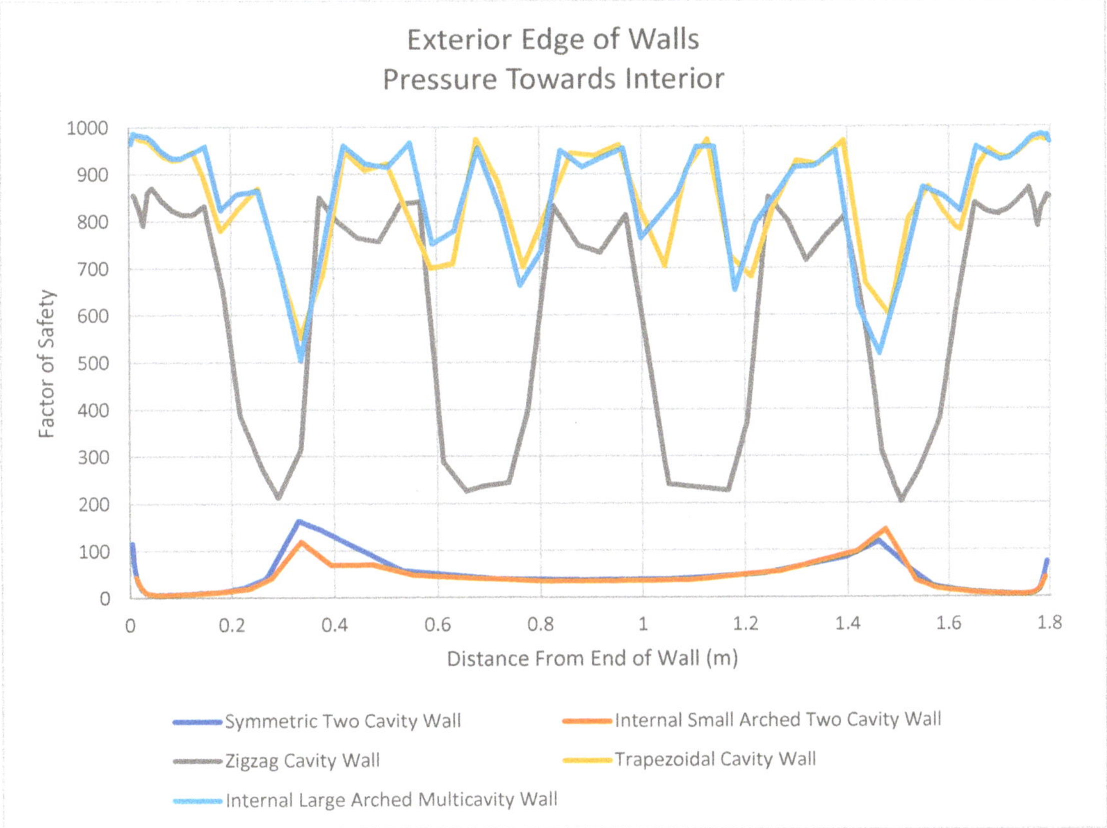

Figure 33. Factor of Safety Values for Exterior Edge of Walls in the Midplane When Pressure Is Directed Towards the Interior.

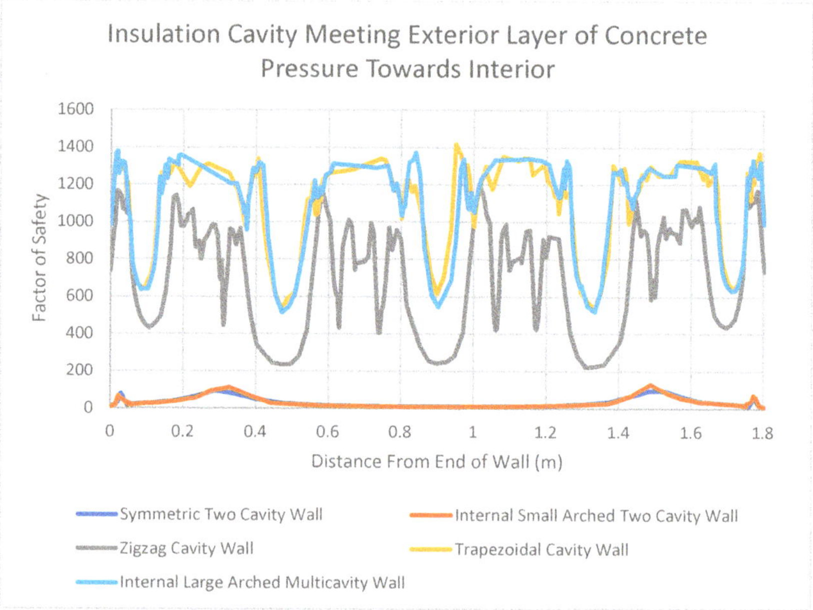

Figure 34. Factor of Safety Values for Insulation Cavity Meeting Exterior Layer of Concrete in the Midplane When Pressure Is Directed Towards the Interior.

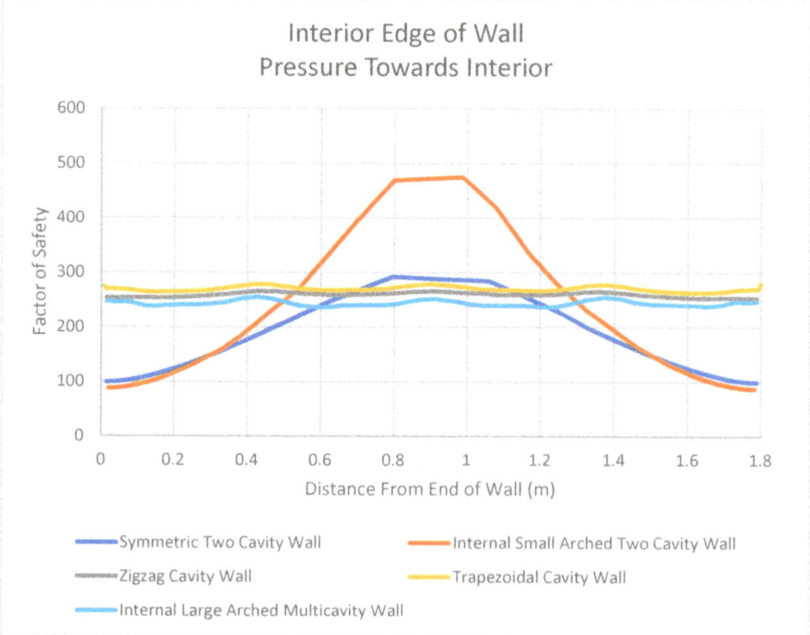

Figure 35. Factor of Safety Values for Interior Edge of Walls in the Midplane When Pressure Is Directed Towards the Interior.

Additionally, it was seen that the first two walls studied (the symmetric two-cavity wall and the internal small-arched two-cavity wall) had much lower factor of safety values where the insulation cavity came into contact with the exterior concrete layer. When pressure was applied to the interior of the wall, the middle portion of this exterior concrete layer had factor of safety values as low as approximately 10 for these two walls, which can be seen in the graph in Figure 34. The remaining three walls that were studied (zigzag cavity wall, trapezoidal cavity wall, and internal large-arched multicavity wall) had higher factor of safety values at these locations. More specifically, the lowest factor of safety value occurred at the center of the unsupported length of each wall's exterior concrete layer (when pressure was applied to the interior). The factor of safety value reached as low as 200 for the zigzag cavity wall at this location, which can also be seen in Figure 34. Though much larger in value than 10, this factor of safety value of 200 is logical. It was mentioned that the unsupported lengths of these walls acted in a manner analogous to a fixed–fixed wall. When calculating the stress for a fixed–fixed wall that is transversely loaded (as is the case in this study), the flexural formula can be utilized ($\sigma = M \times y/I$).

When the unsupported lengths of the wall are reduced due to webbing acting as supports (as seen in the zigzag cavity wall, as compared with the symmetric two-cavity wall and internal small-arched two-cavity wall), the cross section of the wall does not change. Thus, the y and I variables in the flexural formula do not change. However, the moment value in the flexural formula does change. It should be noted that the moment value for this type of loading can be calculated using the equation $w \times L^2/12$, where w is the distributed load and L is the unsupported length. For the first two walls, the unsupported length of the wall is approximately 1.7272 m. When the webbing in the zigzag cavity wall is introduced, the unsupported length is reduced to approximately 0.39 m, while the pressure does not change. Therefore, plugging 1.7272 m into the moment equation above and comparing it to the moment value for 0.39 m yields a moment value that is approximately 20 times larger. Carrying this relationship over to the flexural formula results in a stress that is approximately 20 times larger for the first two walls than for the zigzag cavity wall. When comparing the factor of safety values for the symmetric two-cavity wall and the zigzag cavity wall examined in this study (10 to 200), the exact same relationship between the walls is found. As a result, the walls with webbing tend to limit the amount of stress developed in the exterior concrete layer of the wall and tend to perform better than walls with no webbing at all.

Similar behaviors can be seen when the direction of the pressure is applied away from the interior of the wall. For example, Figures 36–38 display the results of each wall for three locations along the midplane due to pressure applied away from the interior of the wall. In this figure, one can see relationships similar to those seen when pressure was directed towards the interior of the wall. More specifically, the walls without webbing present larger unsupported lengths. As a result, the factor of safety values were lower for these walls than the walls with webbing. Additionally, the walls with webbing were more able to transfer stress from the exterior face to the interior face of the wall. This can also be seen in Figure 38, where the factor of safety values drop at the interior locations where webbing is present.

However, this was not the case when comparing the walls in the thermal study. As seen in the temperature distribution plots for each wall, the walls that contain webbing between the interior and exterior face of the wall contain more thermal bridging locations than those without webbing. For example, the first two walls studied (the symmetric two-cavity wall and internal small-arched two-cavity wall) had only two thermal bridging locations. On the other hand, the last three walls studied (the zigzag cavity wall, trapezoidal cavity wall, and internal large-arched multicavity wall) had several thermal bridging locations. This behavior was quantified using the net power function in SolidWorks. This function provides the ability to quantify the amount of heat lost to the exterior surface of each wall. From these calculations, it was clear that the walls without webbing (the symmetric two-cavity wall and internal small-arched two-cavity wall) lost between 55 W and 60 W. On the other hand, the walls with webbing lost over 200 W. Figure 39 displays the lost power that was calculated for each wall throughout the study.

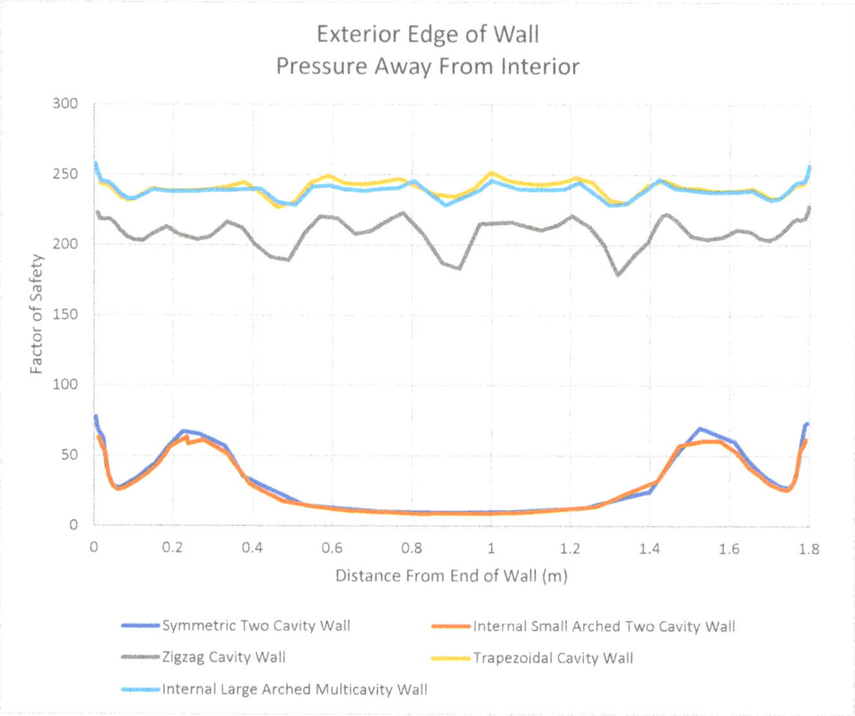

Figure 36. Factor of Safety Values for Exterior Edge of Walls in the Midplane When Pressure Is Directed Away from the Interior.

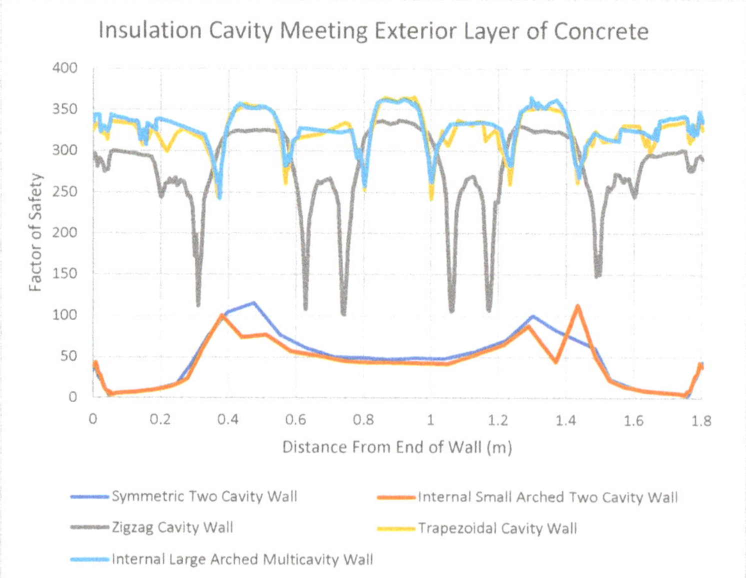

Figure 37. Factor of Safety Values for Insulation Cavity Meeting Exterior Layer of Concrete in the Midplane When Pressure Is Directed Away from the Interior.

Figure 38. Factor of Safety Values for Interior Edge of Walls in the Midplane When Pressure Is Directed Away from the Interior.

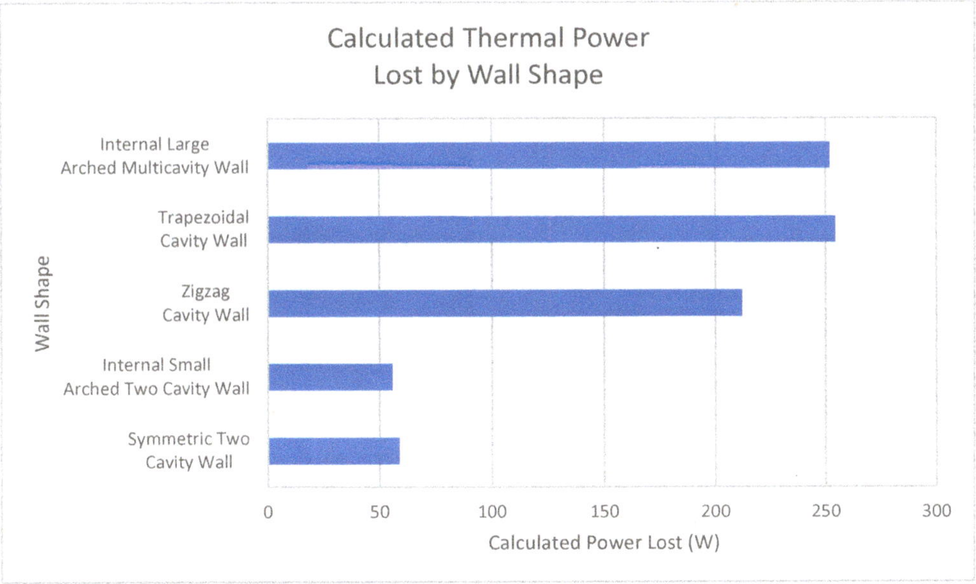

Figure 39. Calculated Thermal Power Lost According to Wall Shape.

8. Conclusions

The structural and thermal studies showed that there is no single best wall configuration. Certain characteristics of each wall allow it to perform better in certain categories. For example, a wall with webbing possesses better flexural ability than a wall without webbing. However, a wall without webbing loses less heat due to thermal bridging. In conclusion, each wall studied has its benefits and drawbacks. For certain locations and applications, a wall with better flexural ability may be key. However, for other cases, a wall that performs well thermally may be more important. For this reason, each wall has a proper use and can be utilized by practicing companies. It may even become common for companies to use different wall types based on the locations of their buildings.

From the results of the structural study, it appears that webbing is required in order for stress to be transferred from the exterior concrete layer to the interior of the wall. This can be seen in the difference between the symmetric two-cavity wall and the zigzag cavity wall in Figure 35. The webbing allows the stress to be transferred to the interior face of the wall for the zigzag cavity wall (as seen from the drops in the factor of safety values), whereas this is not the case for the symmetric two-cavity wall. Therefore, it can be concluded that reinforcements placed on the interior sides of the walls with webbing would be effective in increasing the strength of the walls. On the other hand, reinforcements placed on the interior sides of the wall (where the filled concrete is present, as shown in Figures 3 and 4) would not significantly increase the strength of walls without webbing (such as the symmetric two-cavity wall and the internal small-arched two-cavity wall).

While this study provides a brief review of the performance of different wall shapes, there is more work to be done. First, this study did not take into account the individual layers stacked on top of each other (and the bonding between them). Additionally, this study used the properties of typical concrete. Follow-up studies using the material properties of printed concrete and developing optimized shapes based on the results would be highly beneficial and complementary.

Author Contributions: Conceptualization, N.D.B. and A.M.M.; methodology, N.D.B. and A.M.M.; software, N.D.B.; validation, A.M.M.; formal analysis, N.D.B. and A.M.M.; investigation, N.D.B. and A.M.M.; resources, N.D.B. and A.M.M.; data curation, N.D.B. and A.M.M.; writing—original draft preparation, N.D.B.; writing—review and editing, A.M.M.; visualization, N.D.B.; supervision, A.M.M.; project administration, A.M.M. All authors have read and agreed to the published version of the manuscript.

Funding: This research received no external funding.

Data Availability Statement: All of the data are provided within the manuscript.

Conflicts of Interest: The authors declare no conflict of interest.

References

1. Elfatah, A.S.A. 3D Printing in Architecture, Engineering and Construction. *Eng. Res. J.* **2019**, *162*, 119–137. [CrossRef]
2. Zhang, J.; Wang, J.; Dong, S.; Yu, X.; Han, B. A review of the current progress and application of 3D printed concrete. *Compos. Part A Appl. Sci. Manuf.* **2019**, *125*, 105533. [CrossRef]
3. García-Alvarado, R.; Moroni-Orellana, G.; Banda-Pérez, P. Architectural evaluation of 3D-printed buildings. *Buildings* **2021**, *11*, 254. [CrossRef]
4. Bos, F.; Wolfs, R.; Ahmed, Z.; Salet, T. Additive manufacturing of concrete in construction: Potentials and challenges of 3D concrete printing. *Virtual Phys. Prototyp.* **2016**, *11*, 209–225. [CrossRef]
5. Verian, K.P.; Carli, M.D.; Bright, R.P.; Maandi, E. Research Development in 3DCP: Cured-on-Demand with Adhesion Enhancement Delivery System. ResearchGate, Laticrete International. 2018. Available online: https://www.researchgate.net/profile/Kho-Verian/publication/328757411_Research_Development_in_3DCP_Cured-on-Demand_with_Adhesion_Enhancement_Delivery_System/links/5be23cb292851c6b27ac6370/Research-Development-in-3DCP-Cured-on-Demand-with-Adhesion-Enhancement-Delivery-System.pdf (accessed on 2 January 2023).

6. Buswell, R.A.; Leal de Silva, W.R.; Jones, S.Z.; Dirrenberger, J. 3D printing using concrete extrusion: A roadmap for Research. *Cem. Concr. Res.* **2018**, *112*, 37–49. [CrossRef]
7. Dey, D.; Panda, B. An experimental study of thermal performance of 3D printed concrete slabs. *Mater. Lett.* **2023**, *330*, 133273. [CrossRef]
8. Pessoa, S.; Guimarães, A.S.; Lucas, S.S.; Simões, N. 3D printing in the construction industry—A systematic review of the thermal performance in buildings. *Renew. Sustain. Energy Rev.* **2021**, *141*, 110794. [CrossRef]
9. Wu, H.; Hu, R.; Yang, D.; Ma, Z. Micro-macro characterizations of mortar containing construction waste fines as replacement of cement and sand: A comparative study. *Constr. Build. Mater.* **2023**, *383*, 131328. [CrossRef]
10. SolidWorks. Solidworks Simulation. 2022. Available online: https://www.solidworks.com/product/solidworks-simulation (accessed on 28 December 2022).
11. Frei, W. Should I Fillet the Geometry in My Electromagnetic Heating Analysis? COMSOL. 2017. Available online: https://www.comsol.com/blogs/should-i-fillet-the-geometry-in-my-electromagnetic-heating-analysis/#:~:text=When%20building%20finite%20element%20models,a%20fillet%20avoids%20this%20singularity (accessed on 28 December 2022).
12. Bi, Z. *Finite Element Analysis Applications*; Academic Press: Cambridge, MA, USA, 2019.
13. SolidWorks. Stress Hot Spot Property Manager. Stress Hot Spot PropertyManager—2021—SOLIDWORKS Help, SolidWorks. 2021. Available online: https://help.solidworks.com/2021/english/SolidWorks/cworks/hidd_stress_hot_spot.htm (accessed on 28 December 2022).
14. The Engineering Toolbox. Concrete Properties. Engineering ToolBox. 2008. Available online: https://www.engineeringtoolbox.com/concrete-properties-d_1223.html (accessed on 28 December 2022).
15. ICON. Vulcan. ICON. 2022. Available online: https://www.iconbuild.com/vulcan (accessed on 30 March 2022).
16. *ACI 318-19*; Building Code Requirements for Structural Concrete. American Concrete Institute: Farmington Hills, MI, USA, 2019.
17. Talebi, H.R.; Kayan, B.A.; Asadi, I.; Hassan, Z. Investigation of Thermal Properties of Normal Weight Concrete for Different Strength Classes. *J. Environ. Treat. Tech.* **2020**, *8*, 908–914.
18. SolidWorks. Mohr-Coulomb Stress Criterion. Mohr-Coulomb Stress Criterion—2018—SOLIDWORKS Help, SolidWorks. 2018. Available online: https://help.solidworks.com/2018/english/solidworks/cworks/r_mohr-coulomb_stress_criterion.htm (accessed on 28 December 2022).
19. Elkady, A. #21 Abaqus Tutorial: Defining Concrete Damage Plasticity Model + Failure and Element Deletion. YouTube, YouTube. 2021. Available online: https://www.youtube.com/watch?v=wy84XGamn3g (accessed on 4 January 2023).
20. The Constructor. Stress-Strain Curve for Concrete. The Constructor. 2018. Available online: https://theconstructor.org/concrete/stress-strain-curve-concrete/6159/ (accessed on 18 January 2023).
21. *ASCE/SEI 7-16*; Minimum Design Loads and Associated Criteria for Buildings and Other Structures. American Society of Civil Engineers: Reston, VA, USA, 2017.
22. SQ4D. 3D Printed Houses by SQ4D. SQ4D. 2022. Available online: https://www.sq4d.com/ (accessed on 12 April 2022).
23. Petrock, S. Learn Solidworks Simulation in Less than 30 Min. Here's All You Need to Know to Get Started. YouTube, YouTube. 2021. Available online: https://www.youtube.com/watch?v=Gthsy6OOjo0 (accessed on 28 December 2022).
24. SolidWorks. Principal Stresses Definition. Principal Stresses Definition—2017—SOLIDWORKS Help. 2017. Available online: https://help.solidworks.com/2017/english/SolidWorks/cworks/c_principal_stresses_definition.htm?verRedirect=1 (accessed on 18 January 2023).
25. Risinger, M. Is This the Future of Construction? 3D Concrete Printed Homes. YouTube. 2021. Available online: https://www.youtube.com/watch?v=_MsOXrprYXs (accessed on 30 March 2022).
26. Gross, J. The Most Innovative 3D Printed House in the World [Mense-Korte]. YouTube. 2021. Available online: https://www.youtube.com/watch?v=qWBA-6NgIJg (accessed on 30 March 2022).
27. Gross, J. Secrets of 3D Printed Concrete Exposed | Bold Black Buffalo 3D Shares All. YouTube. 2021. Available online: https://www.youtube.com/watch?v=R7lAnIWy5lg (accessed on 30 March 2022).
28. WinSun. 3D Printing Architecture Profile. Yingchuang Building Technique Shanghai Co., Ltd. (Winsun). Available online: http://www.winsun3d.com/En/Product/pro_inner/id/1 (accessed on 21 September 2022).
29. Apis Cor. 3D Houses. Apis Cor. 2022. Available online: https://www.apis-cor.com/showroom (accessed on 30 March 2022).
30. CyBe. R & DRone Laboratory. CyBe. 2022. Available online: https://cybe.eu/cases/rdrone-lab/ (accessed on 30 March 2022).
31. BING. Thermal Insulation Materials Made of Rigid Polyurethane Foam. Federation of European Rigid Polyurethane Foam Associations. 2006. Available online: https://highperformanceinsulation.eu/wp-content/uploads/2016/08/Thermal_insulation_materials_made_of_rigid_polyurethane_foam.pdf (accessed on 28 December 2022).
32. SolidWorks. Types of Contact. Types of Contact—2020—SOLIDWORKS Help, SolidWorks. 2020. Available online: https://help.solidworks.com/2020/english/SolidWorks/cworks/ID_HELP_CONTACT_PAIR.htm?id=c5b1c459da654e7b9be880b23724a100#Pg0 (accessed on 28 December 2022).
33. Kosky, P.G.; Balmer, R.; Keat, W.; Wise, G. *Exploring Engineering: An Introduction to Engineering and Design*; Academic Press: Waltham, MA, USA, 2013.

34. SolidWorks. Heat Power/Energy Property Manager. Heat Power/Energy PropertyManager—2019—SOLIDWORKS Help, SolidWorks. 2019. Available online: https://help.solidworks.com/2019/english/SolidWorks/cworks/IDH_HELP_HEAT_POWER.htm?verRedirect=1 (accessed on 28 December 2022).
35. American Society of Civil Engineers. ASCE 7 Hazard Tool. ASCE 7 Hazard Tool. 2021. Available online: https://asce7hazardtool.online/ (accessed on 14 February 2023).

designs

MDPI

Article

Effects of Raster Angle on the Elasticity of 3D-Printed Polylactic Acid and Polyethylene Terephthalate Glycol

Mohammed Aqeel Albadrani

Department of Mechanical Engineering, College of Engineering, Qassim University, Unaizah 56452, Saudi Arabia; moa.albadrani@qu.edu.sa

Abstract: Because of its numerous advantages, 3D printing is widely employed for a variety of purposes. The mechanical characteristics of 3D-printed items are quite important. 3D-printed polylactic acid (PLA) is a common thermoplastic polymer due to its excellent characteristics and affordable cost. Because of its enhanced characteristics, polyethylene terephthalate glycol (PETG) has recently received a lot of attention. Despite PETG's potential appeal in the 3D-printing field, little research has been conducted to explore its qualities, such as the impacts of raster angle on elasticity, which could lead to the development of more accurate guidelines for inspection and assessment. In this regard, this study examines the mechanical characteristics of polylactic acid (PLA) and polyethylene terephthalate glycol (PETG) 3D-printing specimens with different raster angles. Test specimens with raster angles of 15° and 30° were printed, and the stress–strain responses were recorded and compared with the simulated profiles generated using ANSYS software. The results showed that the raster angle significantly affected the mechanical properties of both types of materials. The simulated profile matched well with the experimental profile only in the case of PLA printed with a raster angle of 15°. These findings imply that extra effort should be made to ensure that the raster angle is tailored to yield the optimal mechanical properties of 3D-printed products.

Keywords: FEA; additive manufacturing; 3D printing; polylactic acid; raster angle

Citation: Albadrani, M.A. Effects of Raster Angle on the Elasticity of 3D-Printed Polylactic Acid and Polyethylene Terephthalate Glycol. *Designs* **2023**, *7*, 112. https://doi.org/10.3390/designs7050112

Academic Editor: Obeidi Muhannad

Received: 8 August 2023
Revised: 21 September 2023
Accepted: 27 September 2023
Published: 29 September 2023

1. Introduction

The process of additive manufacturing (AM), commonly known as 3D printing, has grown in popularity over the last decade due to its adaptability in generating a wide range of items with complicated geometries and unique mechanical qualities at a cheap cost. The development in the availability of 3D printers and the affordability of printing materials has fueled the rise in the use of 3D printing in recent years [1]. It is a technology that allows for the fabrication of parts with complicated geometry by creating them layer by layer on an optimized platform using computer-aided manufacturing and design. This technology is capable of printing almost any material (e.g., biological materials, polymers, ceramics, alloys and metals, etc.), allowing for a diverse range of products in engineering applications such as aerospace, civil engineering, biomedical, and automotive sectors [1–3]. One critical aspect that influences the performance of 3D-printed objects is their mechanical properties [4]. However, there is still a limited understanding of how different 3D-printing parameters affect these mechanical properties [5–7].

One of the reasons for the complexity in studying the mechanical properties of 3D-printed objects is the presence of complex microstructures. Due to the layer-by-layer nature of the additive manufacturing process, these objects often exhibit intrinsic anisotropic mechanical behavior. Anisotropy means that the mechanical properties can vary depending on the direction in which the object is tested. Additionally, numerous process factors, such as printing speed, temperature, and material composition, further influence the resulting mechanical properties. To overcome these challenges and gain insights into the structural performance of 3D-printed objects, computational modeling techniques are commonly employed [8]. In this regard, a finite element analysis (FEA) is used as a tool; it represents a powerful computational modeling approach used to simulate the mechanical behavior of complex structures, including those produced through additive manufacturing. This tool can be used in 3D-printing parameters, for which the raster angle is regarded as one of the 3D-printing parameters; it has been identified as a critical variable affecting the mechanical properties of printed objects [9–11]. The raster angle refers to the orientation of the printed layers with respect to the direction of printing. By changing the raster angle, the distribution of stresses and strains within the printed object can be altered, leading to variations in its mechanical behavior.

In the context of examining the mechanical properties of 3D-printed objects, two commonly used filament materials are polylactic acid (PLA) and polyethylene terephthalate glycol (PETG). PLA is a thermoplastic material known for its high tensile strength and stiffness [12,13]. PETG is another prominent filament material used in 3D printing [14]. Understanding the mechanical response of these materials after 3D printing at various raster angles is crucial for assessing their usability in different applications [15,16]. A previous study has demonstrated that the raster direction, thickness of layers, and temperature of the nozzle are the primary process parameters determining mechanical qualities [17,18]. It was also discovered that, additional to the previously discussed printing variables, the color of the filament, infill type, building orientation, and printing direction all have a significant impact on the tensile properties of PLA parts [19,20]. In this regard, therefore, it is noted that there are few recent studies that provided a study of the mechanical properties, especially for 3D-printed objects using PLA or PETG. Hanon, M.M. et al. [21] investigated the influence of the raster angle on the mechanical properties of 3D-printed objects using PLA and PETG. They focused on several key features of the 3D-printed standard tensile test samples, including stiffness, strength, and ultimate tensile strength. They aimed to understand how the raster angle affected these properties for both materials. Additionally, they performed FEA simulations to compare the simulated response with the experimental results [21]. Cocovi-Solberg, D.J. et al. [22] presented a measuring and analysis of the mechanical properties, including stiffness, strength, and ultimate tensile strength by conducting experimental tests on the 3D-printed samples. These tests were conducted at different raster angles to investigate the influence of orientation on the mechanical behavior of the printed objects. The data obtained from the experiments provided insights into the material's response and helped establish correlations between the raster angle and mechanical properties. Several suggested models are proposed to replicate the geometry and material properties. In this regard, Abbot, D. et al. [23] developed finite element models that replicated the geometry and material properties of the printed samples. These models were used to simulate the mechanical response of objects under various loading conditions. By varying the raster angle in the simulation, this study studied and analyzed how the mechanical properties changed. Comparing the simulated responses with the experimental results allowed for the validation and refinement of the computational models. Due to the demonstration of the effectiveness of the recently developed integration approaches, numerical scenarios involving elastic fracturing using extended finite element approaches are presented and proven to be effective [24].

Another study gave valuable information on the influence of the raster angle on the mechanical properties of 3D-printed objects using PLA and PETG. The results indicated that the raster angle played a significant role in determining the stiffness, strength, and ultimate tensile strength (UTS) of the printed samples. The FEA simulations demonstrated good agreement with the experimental data, reinforcing the accuracy and reliability of the computational models [16]. Galeja, M. et al. [25] contributed to the study of the effect of different 3D-printing parameters, especially the raster angle, on the mechanical properties of 3D-printed objects. For example, by selecting an appropriate raster angle, it may be possible to enhance the strength and stiffness of printed objects in desired directions, making them more suitable for applications where mechanical performance is critical. Due to the importance of studying the different raster angle, especially due to its effect on

the mechanical properties of three dimensional printing, Skhandesh, S. et al. [26] studied the effects of raster angle on the mechanical properties, which were investigated in seven orientations θ: 0°, 15°, 30°, 45°, 60°, 75°, and 90°. This study used commercially available SCFR-ABS and a consumer-grade FFF 3D printer; specimens of fracture toughness, flexure, and tensile were manufactured. This study concluded that the change in θ from 0° to 15° reduced tensile strength and flexural strength. Another study examined and contrasted the mechanical characteristics of PETG with those of PLA and ABS, two well-known FDM materials. To examine the impact of five distinct raster angle orientations on mechanical qualities, a total of 75 tensile tests were conducted. The findings of this inquiry will be used to make suggestions for using PETG material in 3D printing that is suitable for the intended design and application. This could lead to more precise reference data for prospective manufacturing technology applications [27]. E. Soleyman. et al. [28] proposed a new shape memory polymer (SMP) for the first, which exhibits high controlled self-coiling and stress-shaped memory behaviors in 3D printable PETG thermoplastic constructions. The outcomes demonstrated that the first printed layer contains the majority of the PETG's pre-strain caused by printing. The wall impact experiment showed that raising the wall size weakens the shape transformation associated with the infill pattern. In addition, another study was presented to precisely resolve the underlying fourth-order complex differential equations (PDE) in linear elastic fracture mechanism (LEFM) issues, and a reliable Bezier-based multi-step technique was developed. The study's findings revealed that GnPs with the highest aspect ratio are best for enhancing the plate's elastic characteristics and may be preventing the spread of edge cracks [29].

Starting with recent research information, which is frequently marred by ambiguity concerning the influence of specific process parameters on the mechanical characteristics [19,30], additionally, most recent research has presented poor results on the effects of raster angle on the elasticity of 3D-printed polylactic acid and polyethylene terephthalate glycol. So, the field remains open for more research studies on this point because additive manufacturing has revolutionized various industries, offering new possibilities for rapid prototyping and functional applications. Additionally, studying the mechanical properties of 3D-printed objects significantly impacts their functional behavior and overall performance. However, the influence of different 3D-printing parameters on these properties, including the raster angle, is still not fully understood. In this regard, this study presents computational modeling using finite element analysis, combined with experimental testing, provides an effective approach to evaluate the structural performance of 3D-printed objects. In addition, it presents an examination of the effect of the raster angle on the mechanical properties, such as stiffness, strength, and ultimate tensile strength, that contribute to the understanding of material behavior and facilitate the optimization of printing parameters for specific applications.

2. Methodology

Using the universal test machine (UTM) as in Figure 1a, according to the requirements of the American Society for Materials and Testing standard, the tensile specimens were developed, processed, and printed in a flat dogbone shape. The shapes were examined in accordance with the Test Procedures for Tensile Properties of Plastics (ASTM D638) [31], a standard for plastic specimens. To investigate the influence of the raster angle on the mechanical properties of 3D-printed objects, standard tensile test samples were fabricated using a 3D printer. The generic type of filament fiber in 3D printing (PLA) is used in this study. The materials chosen for this study were polylactic acid (PLA) and polyethylene terephthalate glycol (PETG). The printing angles selected were 15° and 30°.

The dimensions of the test samples were determined based on the requirements of the standard tensile test. These dimensions, including length, width, and thickness. It is important to ensure that the samples meet the specifications of the standard test to ensure accurate and comparable results.

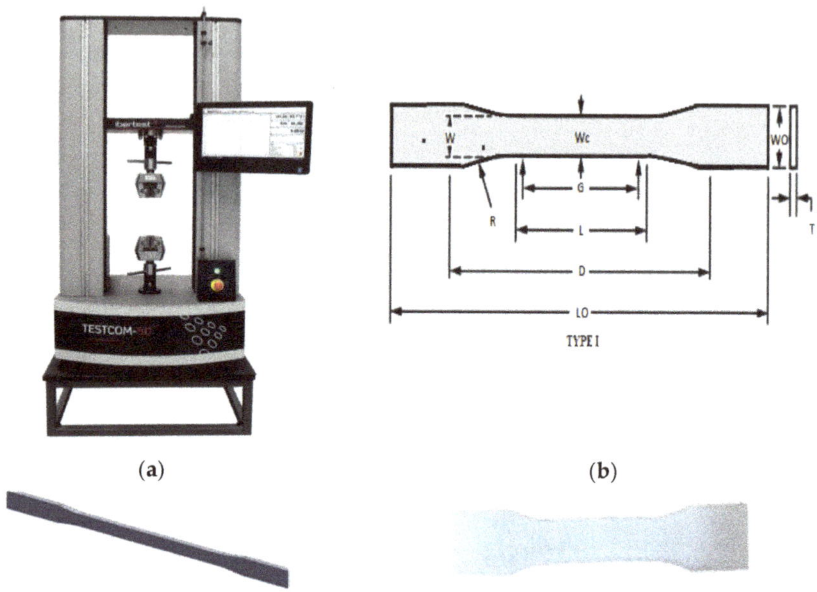

(**a**) (**b**)

(**c**) (**d**)

Figure 1. (**a**) Universal test machine, (**b**) dimensions and shape of test sample, (**c**) the SolidWorks dogbone sample, and (**d**) the dogbone sample in CURA.

To evaluate the mechanical properties of the printed samples, the ANSYS program was employed. ANSYS is widely used finite element analysis (FEA) software that allows for the simulation of mechanical behavior in complex structures. The use of the ANSYS program in this study allowed for a comprehensive analysis of the mechanical properties of the 3D-printed samples. It provided a means to predict and visualize the behavior of the samples under different loading conditions, enabling a deeper understanding of the influence of the raster angle on their performance. The program uses numerical methods to solve the governing equations and predict the response of the structure to external loads. Using the ANSYS program, simulation curves for the tensile test samples were generated. These simulation curves represented the expected mechanical behavior of the printed samples under different loading conditions. By varying the raster angle in the simulations, the effect of this parameter on the mechanical properties of the objects was studied. Using software for finite element analysis, the simulation was carried out. The same steps are used in all finite element analyses. The software used for FEA (finite element analysis) is called ANSYS Workbench Explicit. The CAD (computer-aided design) model, which is prepared for analysis and broken-down during splicing (differentiating into smaller pieces), is an important aspect of FEM. The calibration test simulation is carried out in order to compare the findings of the simulation and experiment in order to validate the simulation results.

The simulation curves provided valuable insights into how the mechanical properties, such as stiffness, strength, and ultimate tensile strength, varied with the printing angles of 15° and 30°. By comparing the simulation results with the experimental data obtained from testing the fabricated samples, the accuracy of the computational models can be validated.

Dynamic explicit/implicit in Ansys with a fixed support on the bottom of the sample as a boundary condition and another boundary condition with a displacement support on the top of the PLA and PETG samples were used.

In this study, dogbone tensile test samples were used to evaluate the mechanical properties of the 3D-printed objects. The dogbone geometry, specified by the ASTM D638 standard [31], was employed for its well-established design and suitability for measuring the tensile characteristics of both reinforced and unreinforced polymers [32]. When tested under specific preparation, humidity, temperature, and testing machine speed conditions, typical dumbbell-shaped test specimens used in ASTM D638 can be used to determine the tensile strength of both reinforced and unreinforced plastic. The procedures using ASTM D638 are as follows:-

- Each specimen's length and thickness should be measured to the nearest 0.025 mm (0.001 in.);
- Put the sample in the test machine's grips while being careful to line up their long axes with an imaginary line connecting the grips' connection points to the machine;
- Turn on the machine, noting the specimen's load-extension curve, the load and extension at the yield point (if one exists), and the pressure and extension at the point of rupture;
- Screw on the extension indication;
- Sett the testing speed to the appropriate rate and start the machine;
- Record the specimen's load-extension curve and extension at the yield point.

The dogbone geometry features shoulder sections at both ends of the sample, connected by a narrower gauge section. The shoulders are wider than the gauge section, resulting in a stress concentration in the middle when the sample is subjected to tensile forces. This stress concentration amplifies the likelihood of sample rupture, making it a favorable configuration for determining the ultimate tensile strength of the material.

During the tensile test, the sample is clamped at the shoulders, and an axial force is applied to the gauge section until it fractures. When the sample ruptures in the middle of the gauge section, it indicates that the material has reached its ultimate tensile strength. However, if the failure occurs at one of the ends or in the grip region, it suggests that improper loading or preexisting flaws in the material may be responsible for the failure.

By utilizing the dogbone geometry, this testing methodology ensures the highest probability of failure due to the ultimate tensile load. ASTM D638 type I is the most widely adopted standard for measuring the tensile properties of both reinforced and unreinforced polymers. It provides guidelines for sample dimensions, specimen preparation, testing conditions, and result calculations [32].

Figures 1 and 2 depict the dimensions and shapes of the 3D-printed test specimens used in the study. The specimens are 4 mm thick with an 85 mm gauge length. Figure 1 depicts more geometries and measurements. Figures 1 and 2 illustrate the specific geometry of the dogbone samples, including the overall length, shoulder width, gauge length, and thickness. Adhering to the ASTM D638 standard and using a dogbone design enables the reliable and consistent testing of the mechanical properties of the 3D-printed objects. Printing parameters can be tabulated as a wall, and the top and bottom thicknesses are 2 and 0.8 mm, respectively (See Table 1). The tabulated dimensional parameters are 13 mm for W—width and 57 mm for L—length of the narrow section. Overall, width and length are 19 and 165 mm, respectively, according to Table 2.

Table 1. Printing parameters.

Property	Value
Infill density	60%
Layer height	0.21 mm
Wall thickness	2 mm
Top/bottom thickness	0.84 mm
Printing temperature	230 °C
Build plate temperature	60 °C
Speed value	60 mm/s

Table 2. Printing dimensional parameters.

Name of Dimensions (See Drawings)	Value (mm)
W—width of narrow section	13
L—length of narrow section	57
WO—width overall	19
LO—length overall	165
G—gauge length	50
D—distance between grips	115
R—radius of fillet	76
T—thickness	3.5

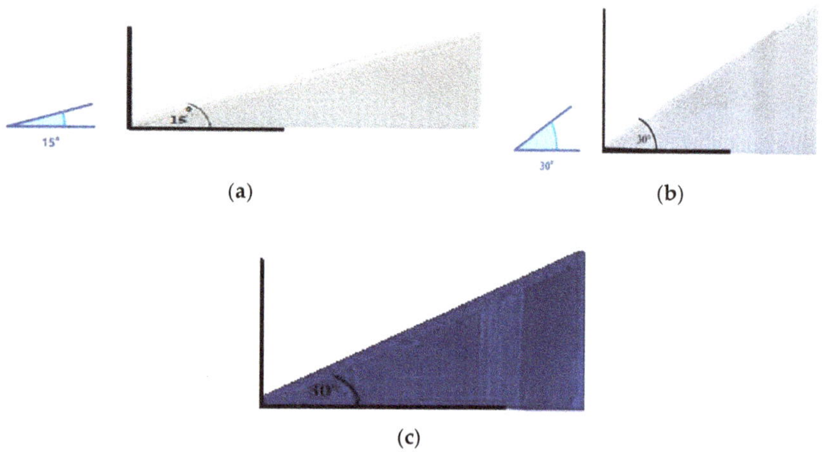

Figure 2. (**a**) Dogbone sample 15 degree for PLA/PETG, (**b**) dogbone sample 30 degree for PETG, and (**c**) dogbone sample 30 degree for PLA.

By employing the dogbone tensile test samples, this study aimed to evaluate the mechanical behavior of the 3D-printed objects under tensile loading. The standardized geometry and testing procedure allow for meaningful comparisons and provide valuable insights into the material's tensile strength, stiffness, and other mechanical properties. The elastic modulus calculated in this study uses the slope approach.

3. Results and Discussion

3.1. Experimental Results of PLA

In this study, the mechanical properties of PLA filaments were thoroughly investigated, with a focus on the impact of different printing angles on these properties. Stress-relieving heat treatment was used to establish a uniform temperature distribution during the printing process. Initially, it was found that PLA filaments exhibited high strength but limited ductility, as evidenced by the data presented in Table 3. To address this limitation, the effects of printing angles of 15° and 30° on the ductility of PLA can be explored. The data presented in Table 3 are the best sample measured. As shown in this table, the densities (g/cm^3) were 1.24, 1.24, 1.28, and 1.28; the elasticities (MPa) were 191.23, 111.66, 56.17, and 52.9; and the maximum tensile strengths (MPa) were 25.53, 5.73, 4.484, and 8.39, for PLA 15°, PLA 30°, PETG 15°, and PETG 30°, respectively.

Upon conducting the tensile tests, several key observations were made. The sample printed at a 15° angle exhibited reduced stiffness compared to the original material. This indicates that printing at this angle led to a more pliable structure, which is advantageous in applications requiring flexibility. Additionally, the 15° print angle resulted in the improved strength and higher elasticity of the PLA material. The enhanced strength implies a

greater ability to withstand applied loads, while the increased elasticity suggests improved deformability under stress.

Table 3. Mechanical properties for PLA and PETG samples.

Material Properties	PLA 15°	PLA 30°	PETG 15°	PETG 30°
Density (g/cm^3)	1.24	1.24	1.28	1.28
Elasticity (MPa)	191.23	111.66	56.17	52.90
Max. tensile strength (MPa)	25.53	5.73	4.484	8.39

In contrast, the PLA samples printed at a 30° angle exhibited a distinct mechanical behavior. Notably, the material displayed higher elasticity and flexibility, which led to failure occurring before reaching the yield point. This is illustrated by the linear stress–strain profile observed throughout the entire range in Figure 3. The 30° print angle enhanced the PLA's ability to deform without permanent damage, indicating its suitability for applications where flexibility and resilience are desired.

The results emphasize the significant influence of print angle on the mechanical properties of PLA. By altering the angle, it is possible to tailor the material's stiffness, elasticity, strength, and ductility to meet specific application requirements. These findings provide valuable insights for optimizing the 3D-printing process and achieving desired mechanical properties in PLA-based objects.

It is important to note that the study's outcomes are focused specifically on PLA filaments and their response to different printing angles. While PLA is a widely used material, other factors such as filament composition, printing parameters, and post-processing techniques may also impact the final mechanical properties of 3D-printed objects. Further research is needed to explore these aspects and expand our understanding of how different materials and printing conditions interact to influence the mechanical behavior of printed objects.

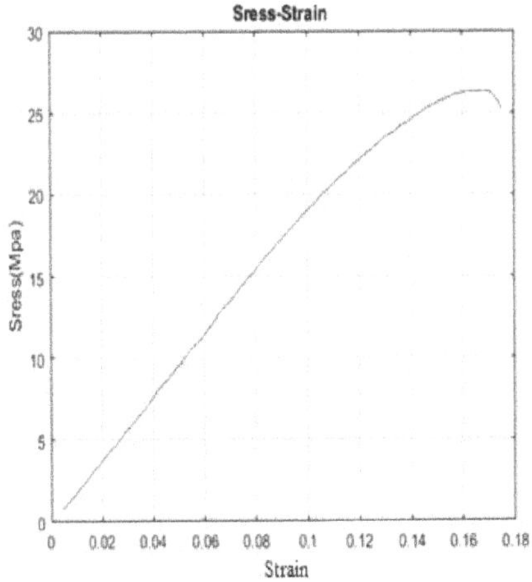

(a)

(b)

Figure 3. *Cont.*

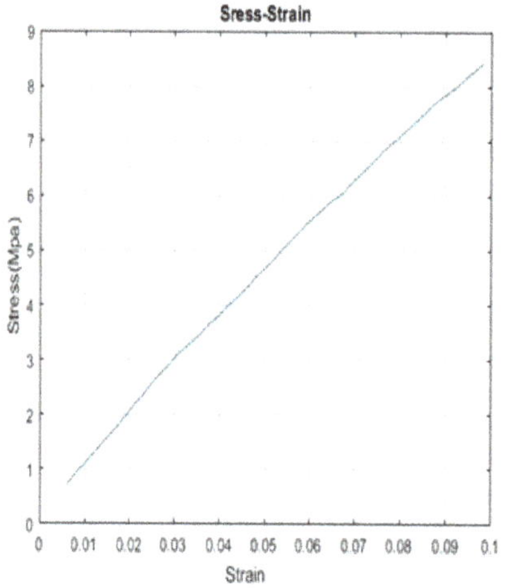

(**c**)

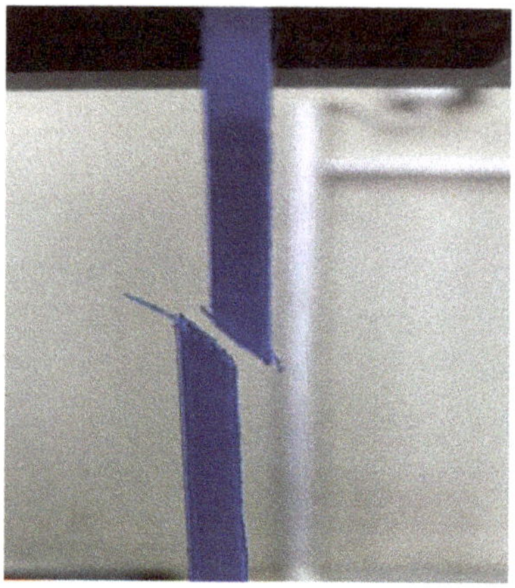

(**d**)

Figure 3. Mechanical properties of the 3D-printed specimen: (**a**) stress–strain profile for PLA 15°, (**b**) photograph of a failed 3D-printed PLA 15° specimen sample, (**c**) stress–strain profile for PLA 30°, and (**d**) photograph of a failed 3D-printed PLA 30° specimen.

Overall, the findings underscore the potential of utilizing various printing angles as a means of tailoring the mechanical properties of 3D-printed PLA objects. This knowledge can guide the design and fabrication of PLA-based components in diverse fields, including biomedicine, aerospace, and consumer goods, where specific mechanical characteristics are crucial for optimal performance.

3.2. Simulation Results of PLA

The ANSYS program has been used to simulate the stress–strain response of the dogbone specimens and compare the experimental results with the simulation curves. Regarding the interface ANSYS tensile test for meshing, and element size of 0.0009 was used to obtain adequate meshing; for boundary conditions, explicit dynamics were used by employing fixed support and the steps of using subroutines in ANSYS. This analysis aimed to investigate the relationship between the printing angle and the mechanical behavior of the samples.

Figure 4 presents the experimental curves for each sample, which closely aligned with the simulation curves associated with their respective printing angles. This close match between the experimental and simulated responses allowed one to obtain a clear understanding of how different printing angles affected the mechanical properties of the samples. Notably, a significant difference was observed when the printing angle was varied.

For the PLA samples printed at a 15° angle, notable improvements in mechanical properties were observed. These improvements included enhanced ductility, flexibility, and strength [28]. Moreover, the experimental results of these samples closely resembled the simulation curve, indicating good agreement between the experimental and simulated responses, with the exception of sample 2. This suggests that the mechanical properties of PLA printed at a 15° angle were predictable and consistent with the predictions from the simulation.

In contrast, the PLA samples printed at a 30° angle exhibited a reduction in certain mechanical properties, particularly in terms of ductility. These samples were more brittle and showed a noticeable decrease in strength and elasticity. However, it is important to note that the experimental results for these samples were not entirely valid as only two of them matched the simulation curve [29]. This discrepancy between the experimental and simulated results suggests a deviation from the expected behavior, potentially due to other influencing factors or variations in the printing process.

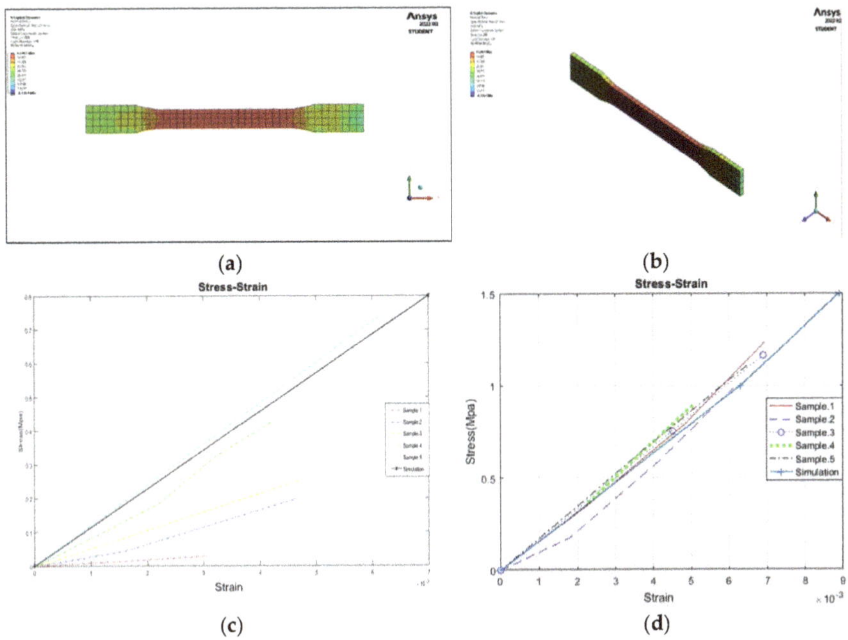

Figure 4. (**a**) Simulation for PLA, (**b**) simulation for PLA (3D), (**c**) stress–strain results chart of PLA−15° experimental/simulation, and (**d**) stress–strain results chart of PLA−30° experimental/simulation.

The findings derived from the comparison between the experimental and simulated results further highlight the significance of the printing angle in influencing the mechanical properties of PLA samples. The PLA printed at a 15° angle demonstrated improved performance, whereas the 30° printing angle led to a reduction in certain mechanical properties. These results provide valuable insights into how adjusting the printing angle can affect the final mechanical behavior of 3D-printed PLA objects, underscoring the importance of considering this parameter when aiming to achieve desired material properties.

It is important to acknowledge that the experimental and simulated results presented in this study focused specifically on PLA samples and their response to different printing angles. Other factors such as filament composition, printing parameters, and post-processing techniques could also impact the final mechanical properties of 3D-printed objects. Therefore, it is crucial to conduct further research to explore these factors and gain a comprehensive understanding of their influence on the mechanical behavior of printed objects.

The findings from this study provide valuable insights into the relationship between the printing angle and the mechanical properties in PLA samples. By adjusting the printing angle, it is possible to tailor the mechanical behavior of 3D-printed PLA objects to meet specific application requirements. These insights contribute to the optimization of the 3D-printing process and the production of PLA-based components with desired mechanical characteristics in various fields, including biomedicine, aerospace, and consumer goods.

3.3. Experimental Results of PETG

The mechanical properties of the PETG filament materials were compared with PLA in this study, and it was found that PETG exhibited greater strength [30]. The experimental results for the two different printing angles, 15° and 30°, were analyzed to understand their impact on the mechanical properties of PETG. Figure 5 illustrates the findings, showing improvements in the mechanical properties of the samples printed at a 30° angle.

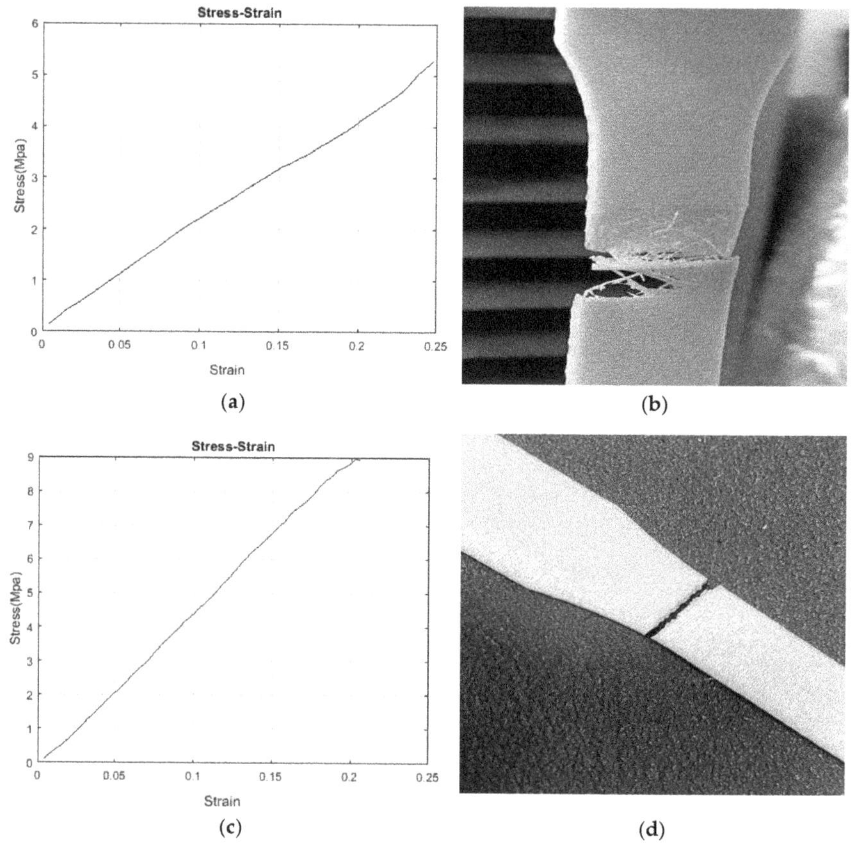

Figure 5. Mechanical properties of the 3D-printed specimen: (**a**) stress–strain profile for PETG 15°, (**b**) photograph of a failed 3D-printed PETG 15° specimen sample, (**c**) stress–strain profile for PETG 30°, and (**d**) photograph of a failed 3D-printed PETG 30° specimen.

The PETG samples printed at a 30° angle demonstrated higher strength than those printed at a 15° angle. Additionally, there was a slight increase in stiffness and a reduction in ductility and flexibility for the 30° samples. This indicates that the printing angle significantly influenced the mechanical behavior of PETG, affecting its strength, stiffness, ductility, and flexibility. The samples printed at a 30° angle exhibited improved the overall mechanical performance.

On the other hand, the experimental results for the PETG samples printed at a 15° angle indicated slightly increased ductility and elasticity compared to the 30° samples. This suggests that a lower printing angle can lead to enhanced ductility and elasticity in PETG. However, it is important to note that the improvements in these properties were not as significant as the gains observed in strength and stiffness for the samples printed at a 30° angle.

These findings provide valuable insights into how adjusting the printing angle can optimize the mechanical properties of 3D-printed PETG objects for different applications. By selecting an appropriate printing angle, it is possible to enhance the strength and stiffness of PETG, which may be desirable in applications where structural integrity is crucial. However, if improved ductility and flexibility are the main requirements, a lower printing angle such as 15° may be more suitable.

It is important to consider that other factors, such as filament composition, printing parameters, and post-processing techniques, can also influence the mechanical properties of 3D-printed PETG objects. Therefore, further research is needed to investigate the combined effects of these factors along with the printing angle to fully understand their impact on the mechanical behavior of PETG.

The insights gained from this study provide guidance for optimizing the 3D-printing process and producing PETG-based components with tailored mechanical properties. This knowledge can be valuable in various industries, including automotive, aerospace, and consumer goods, where PETG's strength and other mechanical characteristics play a crucial role in the performance and durability of printed objects.

3.4. Simulation Results of PETG

The sample printed at a 15° angle demonstrated a slight decrease in stiffness and ductility compared to the original material. However, it is important to acknowledge that the experimental results for this particular sample were not entirely reliable as only one sample closely aligned with the simulation curve, while the remaining samples exhibited significant deviations, as depicted in Figure 6.

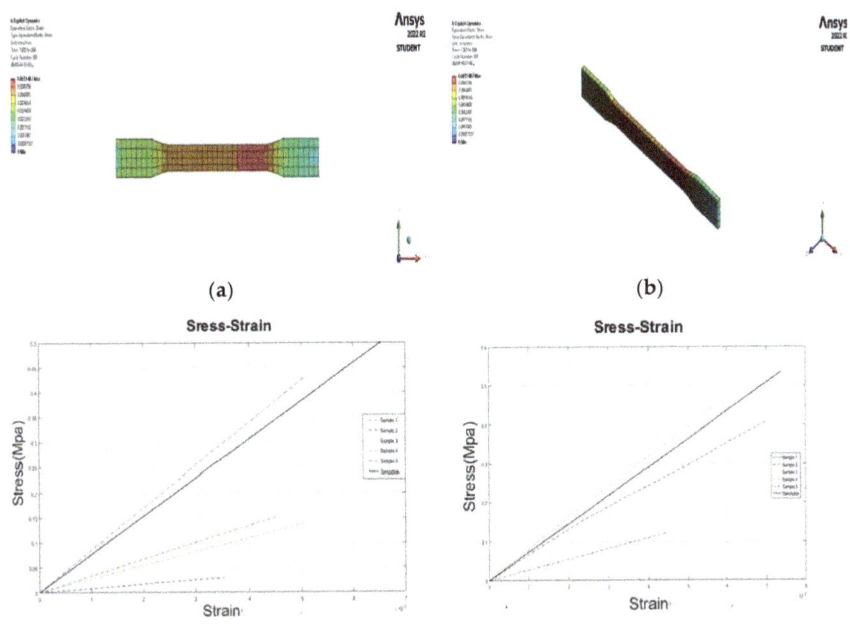

Figure 6. (**a**) Simulation for PETG, (**b**) simulation for PETG (3D), (**c**) stress–strain results chart of PETG −15° experimental/simulation, and (**d**) stress–strain results chart of PETG −30° experimental/simulation.

Despite the limitations in the experimental results, it is still possible to observe some trends in the mechanical properties of the sample printed at a 15° angle. The findings indicate that this printing angle resulted in improvements in strength and stiffness compared to the original material. This suggests that altering the printing angle can have a discernible impact on enhancing the mechanical properties of the printed samples.

Additionally, there were reductions in ductility and flexibility observed for the sample printed at a 15° angle. These changes in mechanical behavior indicate that the printing angle can influence the material's response to applied loads and its ability to deform without fracturing. By modifying the printing angle, it is possible to achieve different combinations of mechanical properties that align with the requirements of specific applications.

However, it is crucial to acknowledge the limitations of the experimental results as the majority of the samples printed at a 15° angle deviated significantly from the simulated profile. This discrepancy between the experimental and simulated responses suggests potential factors influencing the mechanical behavior that were not accounted for or variations in the printing process.

To gain a comprehensive understanding of the impact of the 15° printing angle on the mechanical behavior of the printed objects, further investigation and analysis are warranted. It is necessary to conduct a more extensive experimental study with a larger sample size to obtain reliable and statistically significant results. Moreover, additional factors such as printing parameters, material composition, and post-processing techniques should be considered to elucidate their contributions to the observed mechanical properties.

By conducting thorough investigations and addressing the limitations, authors can refine their understanding of the influence of the 15° printing angle and potentially optimize the printing parameters to achieve desired mechanical properties in future 3D-printed objects. This knowledge can pave the way for the development of tailored materials and structures with improved mechanical performance for various applications. The results in Figures 4 and 6, which indicate the simulation of PETG and PLA, were based on the difference in the raster angle.

4. Conclusions

The aim of this research is to complement previous research in the field of 3D printers; therefore, this research has presented the effects of raster angle on the elasticity of 3D-printed polylactic acid and polyethylene terephthalate glycol. These printing angles were 15° and 30°. Accordingly, this study has provided valuable insights into the significant impact of the raster angle on the mechanical properties of 3D-printed PLA and PETG materials. By manipulating the printing angles to 15° and 30° for PLA and PETG, respectively, notable improvements were observed in the mechanical characteristics of these materials using the simulated profiles generated using ANSYS software. This study concluded that

- The specimens are 60% solid, which is the infill for the rest of the infilling; this is the primary reason for anisotropy.
- For PLA, the samples printed at a 15° angle exhibited enhancements in strength, elasticity, and flexibility, while experiencing a slight reduction in stiffness compared to the original material. These findings suggest that the 15° printing angle optimized the mechanical properties of PLA, making it more suitable for applications where increased strength and flexibility are desired.
- In the case of PETG, the samples printed at a 30° angle showed improvements in strength, elasticity, and flexibility. However, there was a reduction in ductility compared to the original material. These findings suggest that the 30° printing angle optimized the mechanical properties of PETG, making it more suitable for applications where increased strength and flexibility are desirable.

Generally, this study emphasizes the criticality of carefully optimizing 3D-printing parameters, particularly the raster angle, to achieve the desired mechanical properties of printed PLA and PETG objects. These findings highlight the complexity of the 3D-printing process and underscore the need for further research to fully comprehend and control the mechanical behavior of printed materials. Such insights are essential for guiding the design and fabrication of 3D-printed objects across diverse fields, including biomedicine, aerospace, and automotive engineering, where tailored mechanical properties are of utmost importance.

Continued investigations into the effects of printing parameters, material compositions, and post-processing techniques will contribute to a deeper understanding of the relationships between 3D-printing parameters and mechanical behavior. This knowledge can further facilitate the development of advanced 3D-printing strategies, enabling the production of customized objects with optimized mechanical properties to meet the specific requirements of various applications.

This study examines the mechanical characteristics of polylactic acid (PLA) and polyethylene terephthalate glycol (PETG) 3D-printing specimens with 15° and 30° raster angles only. This allows the field to study the mechanical characteristics of polylactic acid (PLA) and polyethylene terephthalate glycol (PETG) 3D-printing specimens at other different angles.

Supplementary Materials: The following supporting information can be downloaded at: https://www.mdpi.com/article/10.3390/designs7050112/s1, Figure S1: The steps of using subroutines in ANSYS.

Funding: This research received no external funding.

Institutional Review Board Statement: Not applicable.

Data Availability Statement: Data is contained within the article or Supplementary Materials. The datasets used during the current study are available from the corresponding author on reasonable request.

Acknowledgments: The author would like to thank the Deanship of Scientific Research, Qassim University for funding the publication fee of this project.

Conflicts of Interest: The author declares no conflict of interest. The funders had no role in the design of the study; in the collection, analyses, or interpretation of data; in the writing of the manuscript; or in the decision to publish the results.

References

1. Albadrani, M.A. Failure Prediction in 3D Printed Kevlar/Glass Fiber-Reinforced Nylon Structures with a Hole and Different Fiber Orientations. *Polymers* **2022**, *14*, 4464. [CrossRef] [PubMed]
2. Mohanavel, V.; Ashraff Ali, K.S.; Ranganathan, K.; Allen Jeffrey, J.; Ravikumar, M.M.; Rajkumar, S. The roles and applications of additive manufacturing in the aerospace and automobile sector. *Mater. Today Proc.* **2021**, *47*, 405–409. [CrossRef]
3. Rahmatabadi, D.; Aberoumand, M.; Soltanmohammadi, K.; Soleyman, E.; Ghasemi, I.; Baniassadi, M.; Abrinia, K.; Zolfagharian, A.; Bodaghi, M.; Baghani, M. A New Strategy for Achieving Shape Memory Effects in 4D Printed Two-Layer Composite Structures. *Polymers* **2022**, *14*, 5446. [CrossRef] [PubMed]
4. Rahmatabadi, D.; Soltanmohammadi, K.; Aberoumand, M.; Soleyman, E.; Ghasemi, I.; Baniassadi, M.; Abrinia, K.; Bodaghi, M.; Baghani, M. Development of Pure Poly Vinyl Chloride (PVC) with Excellent 3D Printability and Macro- and Micro-Structural Properties. *Macromol. Mater. Eng.* **2023**, *308*, 2200568. [CrossRef]
5. Abeykoon, C.; Sri-Amphorn, P.; Fernando, A. Optimization of fused deposition modeling parameters for improved PLA and ABS 3D printed structures. *Int. J. Lightweight Mater. Manuf.* **2020**, *3*, 284–297. [CrossRef]
6. Schwartz, J.J.; Hamel, J.; Ekstrom, T.; Ndagang, L.; Boydston, A.J. Not all PLA filaments are created equal: An experimental investigation. *Rapid Prototyp. J.* **2020**, *26*, 1263–1276. [CrossRef]
7. Lee, D.; Wu, G.-Y. Parameters Affecting the Mechanical Properties of Three-Dimensional (3D) Printed Carbon Fiber-Reinforced Polylactide Composites. *Polymers* **2020**, *12*, 2456. [CrossRef]
8. Rahmatabadi, D.; Soltanmohammadi, K.; Pahlavani, M.; Aberoumand, M.; Soleyman, E.; Ghasemi, I.; Baniassadi, M.; Abrinia, K.; Bodaghi, M.; Baghani, M. Shape memory performance assessment of FDM 3D printed PLA-TPU composites by Box-Behnken response surface methodology. *Int. J. Adv. Manuf. Technol.* **2023**, *127*, 935–950. [CrossRef]
9. Ezeh, O.H.; Susmel, L. Fatigue strength of additively manufactured polylactide (PLA): Effect of raster angle and non-zero mean stresses. *Int. J. Fatigue* **2019**, *126*, 319–326. [CrossRef]
10. Rajpurohit, S.R.; Dave, H.K. Flexural strength of fused filament fabricated (FFF) PLA parts on an open-source 3D printer. *Adv. Manuf.* **2018**, *6*, 430–441. [CrossRef]
11. Gonabadi, H.; Chen, Y.; Yadav, A.; Bull, S. Investigation of the effect of raster angle, build orientation, and infill density on the elastic response of 3D printed parts using finite element microstructural modeling and homogenization techniques. *Int. J. Adv. Manuf. Technol.* **2021**, *118*, 1485–1510. [CrossRef]
12. Lyu, Y.; Chen, Y.; Lin, Z.; Zhang, J.; Shi, X. Manipulating phase structure of biodegradable PLA/PBAT system: Effects on dynamic rheological responses and 3D printing. *Compos. Sci. Technol.* **2020**, *200*, 108399. [CrossRef]
13. Yonezawa, A.; Yamada, A. Deterioration of the mechanical properties of FFF 3d-printed PLA structures. *Inventions* **2020**, *6*, 1. [CrossRef]

14. Ajay Kumar, M.; Khan, M.S.; Mishra, S.B. Effect of machine parameters on strength and hardness of FDM printed carbon fiber reinforced PETG thermoplastics. *Mater. Today Proc.* **2020**, *27*, 975–983. [CrossRef]
15. Yao, T.; Deng, Z.; Zhang, K.; Li, S. A method to predict the ultimate tensile strength of 3D printing polylactic acid (PLA) materials with different printing orientations. *Compos. Part. B Eng.* **2019**, *163*, 393–402. [CrossRef]
16. Hsueh, M.H.; Lai, C.J.; Chung, C.F.; Wang, S.H.; Huang, W.C.; Pan, C.Y.; Zeng, Y.S.; Hsieh, C.H. Effect of Printing Parameters on the Tensile Properties of 3D-Printed Polylactic Acid (PLA) Based on Fused Deposition Modeling. *Polymers* **2021**, *13*, 2387. [CrossRef] [PubMed]
17. Linul, E.; Marsavina, L.; Stoia, D.I. Mode I and II fracture toughness investigation of Laser-Sintered Polyamide. *Theor. Appl. Fract. Mech.* **2020**, *106*, 102497. [CrossRef]
18. Stoia, D.I.; Marsavina, L.; Linul, E. Mode I fracture toughness of polyamide and alumide samples obtained by Selective Laser Sintering additive process. *Polymers* **2020**, *12*, 640. [CrossRef]
19. Kiendl, J.; Gao, C. Controlling toughness and strength of FDM 3D-printed PLA components through the raster layup. *Compos. Part B* **2020**, *180*, 107562. [CrossRef]
20. Yao, T.; Ye, J.; Deng, Z.; Zhang, K.; Ma, Y.; Ouyang, H. Tensile failure strength and separation angle of FDM 3D printing PLA material: Experimental and theoretical analyses. *Compos. Part B* **2020**, *188*, 107894. [CrossRef]
21. Hanon, M.M.; Marczis, R.; Zsidai, L. Anisotropy evaluation of different raster directions, spatial orientations, and fill percentage of 3D printed PETG tensile test specimens. In *Key Engineering Materials*; Trans Tech Publ.: Schwyz, Switzerland, 2019.
22. Cocovi-Solberg, D.J.; Worsfold, P.J.; Miró, M. Opportunities for 3D printed millifluidic platforms incorporating on-line sample handling and separation. *TrAC Trends Anal. Chem.* **2018**, *108*, 13–22. [CrossRef]
23. Abbot, D.W.; Kallon, D.V.; Anghel, C.; Dube, P. Finite element analysis of 3D printed model via compression tests. *Procedia Manuf.* **2019**, *35*, 164–173. [CrossRef]
24. Mousavi, S.E.; Sukumar, N. Generalized Gaussian quadrature rules for discontinuities and crack singularities in the extended finite element method. *Comput. Methods Appl. Mech. Eng.* **2010**, *199*, 3237–3249. [CrossRef]
25. Galeja, M.; Hejna, A.; Kosmela, P.; Kulawik, A. Static and dynamic mechanical properties of 3D printed ABS as a function of raster angle. *Materials* **2020**, *13*, 297. [CrossRef] [PubMed]
26. Skhandesh, S.; Ozgur, K. Effect of raster angle on mechanical properties of 3D printed short carbon fiber reinforced acrylonitrile butadiene styrene. *Compos. Commun.* **2022**, *32*, 101163.
27. Sepahi, M.; Abusalma, H.; Jovanovic, V.; Eisazadeh, H. Mechanical Properties of 3D-Printed Parts Made of Polyethylene Terephthalate Glycol. *J. Mater. Eng. Perform.* **2021**, *30*, 6851–6861. [CrossRef]
28. Soleyman, E.; Aberoumand, M.; Rahmatabadi, D.; Soltanmohammadi, K.; Ghasemi, I.; Baniassadi, M.; Abrinia, K.; Baghani, M. Assessment of controllable shape transformation, potential applications, and tensile shape memory properties of 3D printed PETG. *J. Mater. Res. Technol.* **2022**, *18*, 4201–4215. [CrossRef]
29. Kabir, H.; Aghdam, M.M. A generalized 2D Bézier-based solution for stress analysis of notched epoxy resin plates reinforced with graphene nanoplatelets. *Thin-Walled Struct.* **2021**, *169*, 108484. [CrossRef]
30. Heidari-Rarani, M.; Ezati, N.; Sadeghi, P.; Badrossamay, M.R. Optimization of FDM process parameters for tensile properties of polylactic acid specimens using Taguchi design of experiment method. *J. Thermoplast. Compos. Mater.* **2020**, *35*, 2435–2452. [CrossRef]
31. Lachica, L. *Making Dogbone Tensile Test Samples*; TestResources: Shakopee, MN, USA, 2013.
32. *ASTM D638*; Standard Test Method for Tensile Properties of Plastics. ASTM: West Conshohocken, PA, USA, 2022.

designs MDPI

Article

Rheological Behaviour of ABS/Metal Composites with Improved Thermal Conductivity for Additive Manufacturing

Vicente F. Moritz [1,2,*], Harald Prévost [3], Janaína S. Crespo [4], Carlos A. Ferreira [2] and Declan M. Devine [1,*]

1 PRISM Research Institute, Technological University of the Shannon, Dublin Rd., N37 HD68 Athlone, Ireland
2 Postgraduate Programme in Mining, Metallurgical and Materials Engineering (PPGE3M), Federal University of Rio Grande do Sul (UFRGS), Bento Gonçalves Av. 9500, Porto Alegre 91509-900, Brazil; ferreira.carlos@ufrgs.br
3 ENSIACET, Institute National Polytechnique de Toulouse, Emile Monso Rd., 31030 Toulouse, France; harald.prevost@gmail.com
4 Postgraduate Programme in Materials Science and Engineering (PPGMAT), University of Caxias do Sul (UCS), Francisco Getúlio Vargas St. 1130, Caxias do Sul 95070-560, Brazil; jscrespo@ucs.br
* Correspondence: vicente.moritz@ufrgs.br (V.F.M.); declan.devine@tus.ie (D.M.D.)

Abstract: Metal-reinforced polymer composites are suitable materials for applications requiring special thermal, electrical or magnetic properties. Three-dimensional printing technologies enable these materials to be quickly shaped in any design directly and without the need for expensive moulds. However, processing data correlating specific information on how the metal particles influence the rheological behaviour of such composites is lacking, which has a direct effect on the processability of these composites through melt processing additive manufacturing. This study reports the compounding and characterisation of ABS composites filled with aluminium and copper particulates. Experimental results demonstrated that the tensile modulus increased with the incorporation of metal particles; however, there was also an intense embrittling effect. Mechanical testing and rheological analysis indicated poor affinity between the fillers and matrix, and the volume fraction proved to be a crucial factor for complex viscosity, storage modulus and thermal conductivity. However, a promising set of properties was achieved, paving the way for polymer–metal composites with optimised processability, microstructure and properties in melt processing additive manufacturing.

Keywords: polymer–matrix composites (PMCs); particle reinforcement; rheological properties; thermal properties; 3D printing

Citation: Moritz, V.F.; Prévost, H.; Crespo, J.S.; Ferreira, C.A.; Devine, D.M. Rheological Behaviour of ABS/Metal Composites with Improved Thermal Conductivity for Additive Manufacturing. *Designs* **2023**, *7*, 133. https://doi.org/10.3390/designs7060133

Academic Editor: Tiago Pinto Ribeiro

Received: 19 October 2023
Revised: 14 November 2023
Accepted: 16 November 2023
Published: 17 November 2023

1. Introduction

Polymer–metal composites are of particular interest for applications in which either thermal or electrical conductivity is required since polymers are not intrinsically conductive [1–6]. For instance, concern has arisen due to warpage and distortion caused by thermal expansion on polymer-based 3D-printed parts. Finite element analysis (FEA) simulations have been employed to model the mechanical and thermal behaviours of parts built through fused deposition modelling (FDM) [1,3,7]. Therefore, materials with improved thermal properties would address the final product's potential distortion in the fabrication of large-dimension structures such as electromagnetic interference shields and antennas [1]. A combination of heat dissipation and electrical insulation is usually desired in encapsulated electronic devices, preventing a potentially jeopardising overheating situation while enabling the retention of the electrical functionality [4,8]. Moreover, inserts for injection hybrid moulds (which are manufactured through 3D printing techniques) have also proven to benefit from inorganic fillers, exhibiting improved heat dissipation and greater longevity and yielding parts with properties comparable to those produced with conventional injection moulding [9–11]. In this context, metal-reinforced polymer–matrix composites have become an alternative material to overcome these issues as the addition of fillers, either particles or fibres simply with conductivity greater than that of polymers, aimed at improving mechanical characteristics combined with light weight, can also impart distinct thermal, electrical, optical or magnetic properties to the material [5,7,12,13].

The effects of a number of conductive fillers on the properties of different polymer materials have been investigated, with particular emphasis on copper and aluminium powders, and various experimental approaches and numerical models have been employed to assess the thermal conductivity of such composites [14–20]. For instance, Rahmati and Dickens (2007) [21] and Pontes et al. (2008) [22] carried out studies on aluminium-filled 3D-printed epoxy inserts for hybrid moulds, evaluating the thermal and mechanical properties and reporting promising outcomes relative to enhancing the thermal performance of the moulds. Ranjan et al. (2023) [23] 3D-printed ABS parts to fabricate composites with aluminium metal spray reinforcement, and then applied machine learning methodologies to optimise the process variables and the flexural properties. Akrout et al. (2023) [24] investigated the structural, thermal, micromechanical and tribological properties of ABS/Cu composites for bearing applications. Compared to neat ABS, the composites showed a higher hardness and elastic modulus. Significant improvements in ABS's mechanical and thermal properties suggest that a suitable selection of material composition and other parameters may lead to a FDM filament of great potential, yielding high-performance, functional prototypes for a wide range of applications manufactured with the FDM process [7,25].

From the point of view of the material, the acrylonitrile–butadiene–styrene copolymer (ABS) is known for its excellent mechanical and impact strengths, dimensional stability, good chemical resistance, low water absorption and high filling capacity, arising as a popular thermoplastic material of choice for FDM applications as it offers satisfactory processability and a relatively low melting temperature [5,26–29]. The manufacturing of ABS composites through additive manufacturing (AM) techniques, such as FDM, has been demonstrated [30–34], and melt flow index (MFI) analysis has been a popular testing choice to evaluate the flow behaviour trends of ABS composites developed for 3D printing [35–41]. Isa et al. (2015) [35] and Sa'ude et al. (2016) [36] studied the behaviour of ABS/copper composites, and Kumar et al. (2019) [38] carried out similar work on ABS/aluminium composites, with all of them targeting the development of materials for extrusion-based AM processes; nonetheless, their results indicated that the filler has distinct effects on MFI depending on its content, suggesting that there would be no direct dependence between the loading content and MFI. However, only knowing a material's MFI is insufficient to process it, since this type of measurement represents the flow of a material under the pressure applied during its testing, and it does not encompass considerations due to shear conditions a polymer chain will be subjected to as it undergoes the processing stage [42–44]. Moreover, MFI lacks accuracy and reproducibility as it is a single-point viscosity value relative to the shear rate and applied pressure at a fixed temperature, prone to inherent measurement errors as well as experimental errors [44–46].

On the other hand, considering that the flowability of a polymer is inversely proportional to the dynamic viscosity, it is well known that the melt flow behaviour (MFB) is an important parameter for 3D printing as the printed parts' precision and interlayer adhesion may be affected by both the shear thinning effect (which determines the pressure needed to push the material through the nozzle) and the temperature parameters (which will ultimately govern the mechanical properties and integrity of the printed parts) [43,47–50]. Dynamic oscillatory rheometry has been employed to determine the appropriate processing conditions for high-performance thermoplastics in the form of AM feedstock, as it provides a more comprehensive rheological profile of such materials, such as including other linear viscoelastic properties of interest, e.g., melt storage modulus (G') and complex viscosity (η^*), and it has been successfully utilised to characterise ABS composites [51–57].

However, there is a gap of knowledge regarding the dynamic rheological properties of ABS/metal composites [58]. Within this frame of reference, this work aims to provide a better understanding of the effect of incorporating metallic particles on the rheological and viscoelastic properties of ABS, which is essential to determine the feasibility of these polymer/metal compounds as potential candidates for thermally conductive thermoplastic composites' AM feedstock. In this study, metal particle-reinforced ABS composites were produced through extrusion compounding and injection moulding, with aluminium and copper particles incorporated into the polymer matrix. In addition, mechanical, rheological, viscoelastic, thermal and morphological characterisation techniques were employed to evaluate the effects of the metallic fillers' loadings on the composites' performance, as a suitable set of properties may be challenging to achieve, especially for highly loaded systems.

2. Experimental

2.1. Materials

Commercial-grade ABS was supplied by LG Chemical under trade name TR-557I as a transparent, impact-resistant material. Aluminium powder (250 mesh, >99.7%, air-atomised) and copper powder (325 mesh, >99.0%, water-atomised) were supplied by East Coast Fibreglass.

2.2. Metallic Powders Characterisation

Aluminium and copper powder were analysed in order to determine their surface and physicochemical properties, such as specific surface area, mean particle size and crystalline structure. Surface adsorption characteristics were studied in a Quantachrome Nova 1000e unit with nitrogen 5.0 as testing gas, and the specific surface area (A_{sp}) was calculated using the Brunauer–Emmett–Teller (BET) isothermal model. Laser scattering particle size distribution measurements were performed in a Horiba LA-950 granulometer. X-ray diffraction was carried out in a Siemens/Bruker D5000, with 2θ angle ranging from $10°$ to $80°$ at a $0.05°\cdot s^{-1}$ step, utilising a Cu-$K\alpha$ tube (λ = 1.54056 Å).

2.3. Compounding and Extrusion

Metal-reinforced composites were prepared by mixing and extruding ABS with the metallic powders in a co-rotating twin-screw Leistritz Macromatex II (Ø 27 mm, 36:1 L/D ratio), with a temperature profile of 200–230 °C from the throat to nozzle at 30 rpm. Prior to compounding, ABS and metallic powders were dried at 80 and 110 °C, respectively, for 2 h. Following extrusion, obtained composites were pelletised. Composite formulations were designed with two variable factors at two levels; the polymer/metal weight ratios are presented in Table 1, along with the theoretical volume fractions of polymer, f_{ABS}, and the metal filler, f_{met}.

Table 1. Metal powder contents for the composites prepared.

	ABS [wt%]	Al [wt%]	Cu [wt%]	f_{ABS}	f_{met}
ABS-30%Al	70	30	-	0.855	0.145
ABS-60%Al	40	60	-	0.627	0.373
ABS-30%Cu	70	-	30	0.951	0.049
ABS-60%Cu	40	-	60	0.848	0.152

wt% = 100 × (phase mass)/(total mass).

2.4. Injection Moulding and Processing Optimisation

An Arburg 370E Allrounder injection moulding machine (Ø 30 mm screw) was utilised in this study with a temperature profile of 195–220 °C from the throat to nozzle. Composite pellets and virgin ABS were pre-dried at 70 °C for four hours. Specimens (ASTM D638 type I tensile, unnotched ASTM D6110 Charpy impact and Ø 25 mm disc) were moulded. Allowing for the increased viscosity due to the filler contents, the injection process parameters needed to be modified in order to reach an optimal moulding condition for each formulation (Table 2), preventing issues such as short shots and flashing.

Table 2. Injection moulding parameters employed for process optimisation.

	ABS	ABS-30%Al	ABS-60%Al	ABS-30%Cu	ABS-60%Cu
Injection Pressure [bar]	805	1050	1890	900	1070
Holding Pressure [bar]	500	500	1600	500	500
Mould Temperature [°C]	50 ± 5	50 ± 5	50 ± 5	50 ± 5	50 ± 5
Cooling Time [s]	30	30	45	30	30

2.5. Mechanical Testing

Stress–strain tensile tests were performed with injection-moulded ASTM D638 type I specimens in a Zwick Roell (Zwick GmbH & Co. KG, Ulm, DE) universal testing machine with a 10 kN loadcell and testXpert III (version 1.5) software by the same manufacturer, based on ASDM D638-14. Test speed was 5 mm·min^{-1}, and gauge length was 50 mm. A total of 16 specimens were tested for each formulation.

Unnotched Charpy impact test was carried out based on ASTM D6110-10 in a Ceast Resil 6844 (Ceast Instron SpA, Turin, Italy) digital machine with a 4.0 J hammer. Specimens were mechanically notched so that the material remaining in the specimen under the notch was 10.16 ± 0.05 mm thick. Twelve specimens were tested for each formulation. Hardness tests were carried out on a CV Instruments Shore D Durometer (CV Instruments Ltd., Sheffield, UK) digital machine, with a test load of 5.0 kg. A total of 64 measurements were taken from each sample. For both tests, the values reported were calculated as the average of the recorded values.

2.6. Dynamic Rheology

Oscillatory dynamic rheology of the composites was studied using a Discovery HR30 rheometer (TA Instruments, New Castle, DE, USA), featuring a parallel-plate geometry fixture with Ø 25 mm and 1.0 mm gap. To ensure that the strain utilised was within the linear viscoelastic region, amplitude sweep tests were performed within a strain range from 0.1% to 100% at 1.0 Hz. Frequency sweep experiments, also known as small angle oscillatory shear (SAOS), were carried out within a frequency range from 100 to 0.01 Hz under a strain of 1%. In SAOS studies, the flow curves follow a power law equation, $\eta^* = K \times \omega^n$, that expresses the shear thinning parameter n, which provides a semi-quantitative measurement of the filler dispersion [52,53]. Temperature of 195 °C and nitrogen atmosphere were applied for both test modes. Three specimens of each formulation were analysed in each testing mode.

2.7. Dynamic Mechanical Thermal Analysis

Dynamic mechanical analysis (DMA) was used to investigate the storage modulus (E') and the glass transition temperature (T_g). A DMA Q800 (TA Instruments, New Castle, US) machine was used for the analysis using single cantilever mode with a free bending length of 17.5 mm. The specimens were 12 mm wide and 2 mm thick. A 1 µm oscillation amplitude, 1 Hz oscillation frequency and 5 °C·min^{-1} heating rate from room temperature to 160 °C were employed. Three specimens of each material were analysed. T_g was determined as the $tan(\delta)_{DMA}$ peak temperature.

2.8. Thermal Conductivity

The thermal conductivity (k) of specimens was measured using a H111A Heat Transfer Unit (P.A. Hilton Ltd., Andover, UK). A disc specimen (Ø 25 mm, thickness 2 mm) was fitted between two parallel-plate copper blocks. A voltage and current of 80 V and 0.109 A were applied, and six specimens were tested. Finally, the surface temperatures were estimated according to the manufacturer's instructions, and k (for the materials) was calculated based on Fourier's Unidirectional Heat Transfer Law.

2.9. Scanning Electronic Microscopy

The morphology of the impact fracture surfaces was observed through a field emission scanning electron microscope (FE-SEM) (Tescan Mira 3, Tescan UK/Oxford Instruments, Cambridge, UK), with a 10 kV energy beam using the backscattered electrons (BSE) mode. Prior to analysis, specimens were mounted on a metallic stub and gold-coated for 30 s at 0.13 mbar vacuum. Combined with the FE-SEM instrument, a surface chemical composition mapping was carried out through energy dispersive spectroscopy (EDS) equipped with a silicon drift detector (SDD) to evaluate the dispersion of the metal particles within the ABS matrix.

2.10. Statistical Analyses

Analysis of variance (ANOVA), a statistical method, was employed to assess whether the metallic powder type and content were significant for the composites' properties. All the values were analysed at a 99% confidence interval, and differences were considered significant through the *p*-value test when $p \leq \alpha = 0.01$. In order to determine differences between sample groups, Tukey's HSD pairwise post hoc test was applied [11,59].

3. Results and Discussion

3.1. Physicochemical Properties of Metallic Powders

The aluminium and copper powder's specific surface areas were estimated through the BET model [60] and were 2.14 and 3.41 $m^2 \cdot g^{-1}$, respectively, which are very low values and indicate an absence of porosity. The granulometric distribution data are shown in Table 3. The metallic powders presented mean particle sizes of 27.95 ± 9.59 µm (Al) and 23.55 ± 9.92 µm (Cu), which are consistent with the differences in specific surface area and material density. Both powders demonstrated a unimodal frequency distribution of particle sizes (Figure 1), with ca. 63% of their particles within the medium silt range (14.5–28.5 µm). Aluminium had 30.16% of its particles falling within the coarse silt range, while copper presented a more balanced distribution between fine silt (19.09%) and coarse silt (16.46%). It is also important to mention that the powders were synthesised through distinct processes. The aluminium powder was produced through gas atomisation, a process that usually yields fine, regular and spherical particles, while copper powder was manufactured through water atomisation, which produces particles with highly irregular morphology [61,62]. The powders' synthesis is therefore likely related to both the specific surface area and the particle size distribution.

Table 3. Surface adsorption characteristics and granulometry distribution of the metallic powders.

		Aluminium	Copper
	Specific surface area	2.14 $m^2 \cdot g^{-1}$	3.41 $m^2 \cdot g^{-1}$
	Mean size ± SD	27.95 ± 9.59 µm	23.55 ± 9.92 µm
	Mode size	27.88 µm	21.41 µm
	Diameter on 10%	17.17 µm	13.34 µm
	Diameter on 50%	26.69 µm	21.69 µm
Diameter Frequency	Very fine silt (3.5–7.0 µm)	-	0.49%
	Fine silt (7.0–14.5 µm)	6.21%	19.09%
	Medium silt (14.5–28.5 µm)	62.98%	63.27%
	Coarse silt (28.5–57.0 µm)	30.16%	16.46%
	Very fine sand (57.0–115.0 µm)	0.64%	0.69%

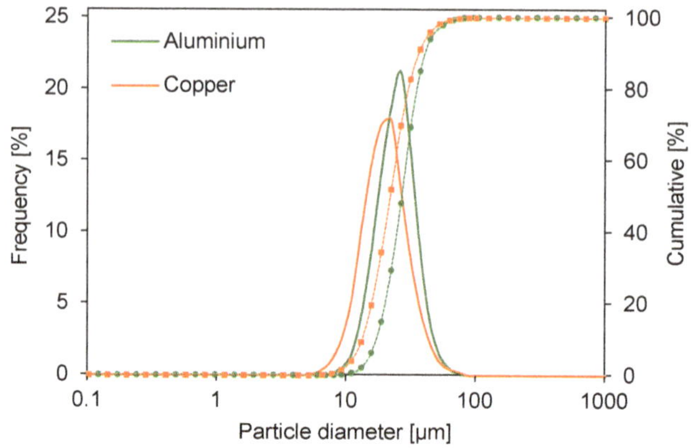

Figure 1. Frequency and cumulative distributions of particle sizes of aluminium and copper powders obtained with laser scattering granulometry. Frequency (solid line) and cumulative (dotted line) distributions of particle sizes obtained with laser-scattering granulometry.

The XRD patterns of both powders are presented in Figure 2. The powder diffraction data were processed with Rietveld refinement using the GSAS-II software [63]. Both aluminium and copper powders were found to match the diffraction patterns for pure metallic phases recorded on the software's database and as previously reported in the literature, confirming the absence of oxidation. Aluminium had peaks at 2θ values of 38.6°, 44.8°, 65.2° and 78.3° [63,64], and copper had peaks at 43.3°, 50.4° and 74.1° [63,65].

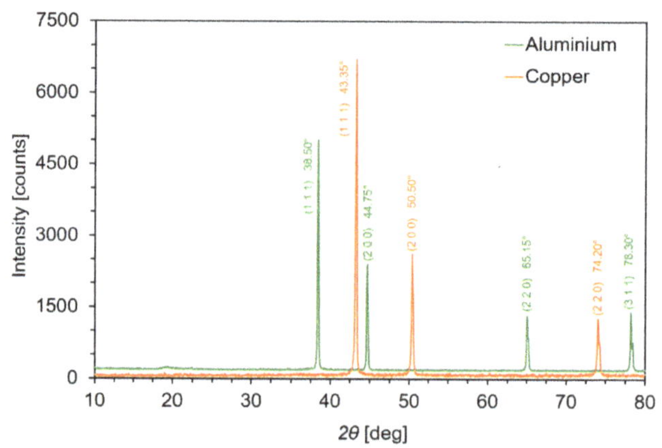

Figure 2. X-ray diffraction patterns of the metallic powders.

3.2. Mechanical Properties

Representative stress–strain tensile curves of the neat polymer and the produced composites are shown in Figure 3, and the obtained mechanical properties are presented in Table 4. In Figure 3, it is possible to observe that neat ABS presented the typical thermoplastic's stress–strain curve, with an elastic region, a prominent yield point and ductile break at greater strains [13]. Three out of four composites presented brittle behaviour. The tensile results indicate that the polymer/metal weight ratio has a significant effect on Young's modulus (E), tensile strength (σ_{max}), stress at break (σ_B) and strain at break (ε_B), and that there is a significant interaction between the weight ratio and type of filler (ANOVA, $p < 0.01$).

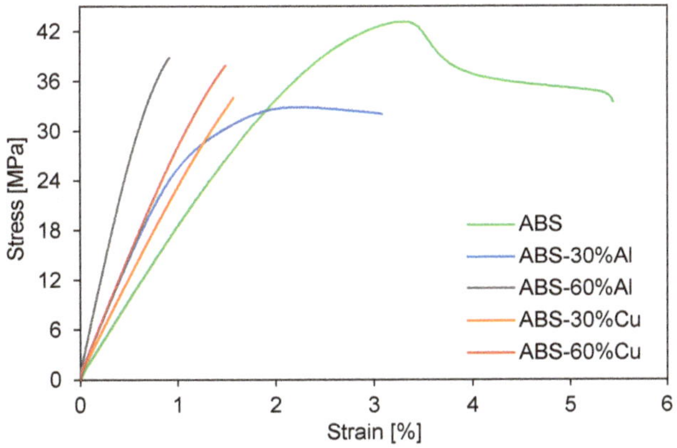

Figure 3. Representative stress–strain tensile curves of ABS and composites.

Table 4. Mechanical performance of injection-moulded ABS composites and respective standard deviations.

	ABS	ABS-30%Al	ABS-60%Al	ABS-30%Cu	ABS-60%Cu
Young's Modulus (E) [MPa]	1885.0 ±18.97	2837.5 ^{A} ±227.84	4784.4 ±47.04	2329.4 ±16.52	2853.1 ^{A} ±18.52
Yield Strength (σ_Y) [MPa]	43.4 ±0.36	32.4 ±0.54	- -	- -	- -
Strain at Yield (ε_Y) [%]	3.3 ±0.04	2.3 ±0.07	- -	- -	- -
Tensile Strength (σ_{max}) [MPa]	43.4 ±0.36	32.4 ^{B} ±0.67	39.4 ^{C} ±2.36	33.7 ^{B} ±1.71	38.1 ^{C} ±1.56
Stress at Break (σ_B) [MPa]	31.9 ^{D} ±3.64	31.3 ^{D} ±1.08	39.4 ^{E} ±2.36	33.7 ^{D} ±1.71	38.1 ^{E} ±1.56
Strain at Break (ε_B) [%]	5.2 ±1.44	2.9 ±0.58	1.0 ^{F} ±0.12	1.6 ^{F} ±0.11	1.5 ^{F} ±0.09
Impact Strength (IS) [kJ·m^{-2}]	13.71 ±0.304	2.70 ±0.090	2.03 ±0.072	3.12 ±0.092	4.30 ±0.081
Impact Resistance (IR) [kJ·m^{-1}]	139.33 ±3.088	27.40 ±0.912	20.67 ±0.731	31.72 ±0.931	43.68 ±0.825
Shore D Hardness	78.1 ±0.78	79.4 ±0.98	83.3 ^{G} ±1.27	83.2 ^{G} ±0.66	84.1 ±0.69

Values of average and standard deviations for σ_Y and ε_Y of ABS-30%Al refer to 12 specimens, as 4 specimens out of 16 presented a brittle fracture. All specimens of ABS-60%Al, ABS-30%Cu and ABS-60%Cu presented a brittle fracture during tensile testing; thus, no values of σ_Y and ε_Y are recorded. ^{A,B,C,D,E,F,G} Means that share a letter have no statistical differences.

The incorporation of metallic particles changed the mechanical response of ABS. The metallic particles significantly raised E for ABS-30%Al and ABS-60%Cu, whose values are close due to the similar filler volume fraction, in comparison to neat ABS (ca. 50%, $p < 0.01$), and for ABS-60%Al (over 250%, $p < 0.01$). ABS exhibited a mean yield point at $\sigma_Y = 43.3$ MPa and $\varepsilon_Y = 3.3\%$, while ABS-30%Al featured yielding at $\sigma_Y = 32.4$ MPa and $\varepsilon_Y = 2.3\%$ followed by a ductile break at $\sigma_B = 31.3$ MPa and $\varepsilon_B = 2.9\%$. All copper-loaded specimens failed before yielding during the tensile test, as well as all specimens of ABS-60%Al and 4 specimens out of 16 of ABS-30%Al. This reveals a major embrittling effect of the metallic filler on the matrix, representing a considerable loss in terms of toughness.

Overall, the mechanical behaviour observed under tensile loading follows a trend of increasing E and reducing ε_B as the filler content increases, as also reported by previous studies [1,9,41,57,66]. Moreover, E displays a linear dependence (R^2 = 0.9881) upon the volume fraction of the filler, regardless of the filler's nature (Figure 4). Young's modulus is not subjected to particle–matrix interfacial adhesion effects because debonding is not yet observed for small loads or displacements, i.e., the range within which the modulus is measured [66]. However, Masood and Song (2004, 2005) [9,10] reported a dependency of E and σ_{max} upon the size and content of metallic fillers: the larger the particle size, the greater the tensile modulus and strength.

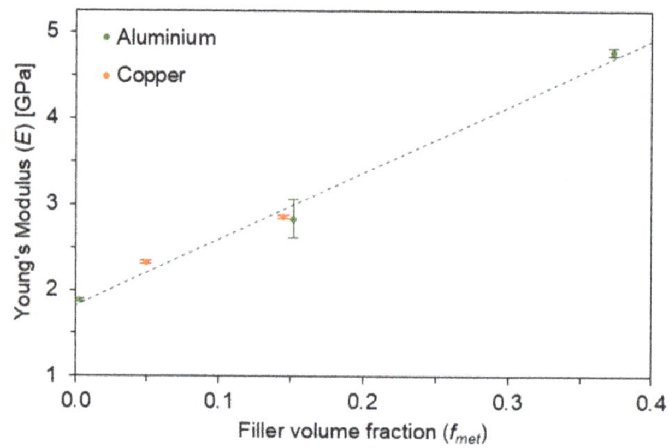

Figure 4. Young's modulus linear dependency upon reinforcement volume fraction.

Characteristics such as σ_{max} first decreased for a 30 wt% filler content and then, with a 60 wt% metal powder, increased to ca. 90% of that of neat ABS (Figure 5), while a significant increase of σ_B was observed for ABS-60%Al and ABS-60%Cu. Embrittlement, increase in E and decrease in ε_B are typically observed for ABS/aluminium composites [41,57]. In spite of the changes in injection moulding parameters due to the composites' increased viscosity, the injection pressure has minimal to low influence on the mechanical properties of injection-moulded ABS [67]. The most important parameter for E, σ_{max} and ε_B was identified as melt temperature, with impact levels weighting between 43 and 87%, whilst injection pressure made up between 2 to 22% for the same properties [67].

Although ABS-30%Al and ABS-60%Cu have similar filler volume fractions, σ_{max} and ε_B of such composites are significantly distinct ($p < 0.01$) and ABS-30%Al has ductile behaviour. This indicates that aluminium and copper promote distinct fracture mechanisms due to different levels of polymer–particle interface interactions. The polymer–metal interface is a crucial factor, and its integrity can be affected by the particle's morphology and stability, impairing the stress transfer between the matrix and the filler [13,66]. Since the formulations did not contain a coupling agent to promote enhanced interface interactions, there may be insufficient bonding between the metal particle and the polymer matrix [7].

ABS-30%Al and ABS-60%Cu have σ_{max} of 32.4 and 38.1 MPa, respectively, suggesting that the ABS–copper interface may be less weak than the ABS–aluminium one. As σ_{max} first drops for a 30 wt% filler content and then rises with 60 wt%, metal particles likely act as defects and have a weakening effect on ABS due to the stress concentration at a lower loading, later reinforcing the matrix as barriers to crack growth at higher concentrations [13,66]. Another reason for this may lie in the non-homogeneous dispersion of voids and particles, possibly leading agglomerations to cause stress concentrations [39].

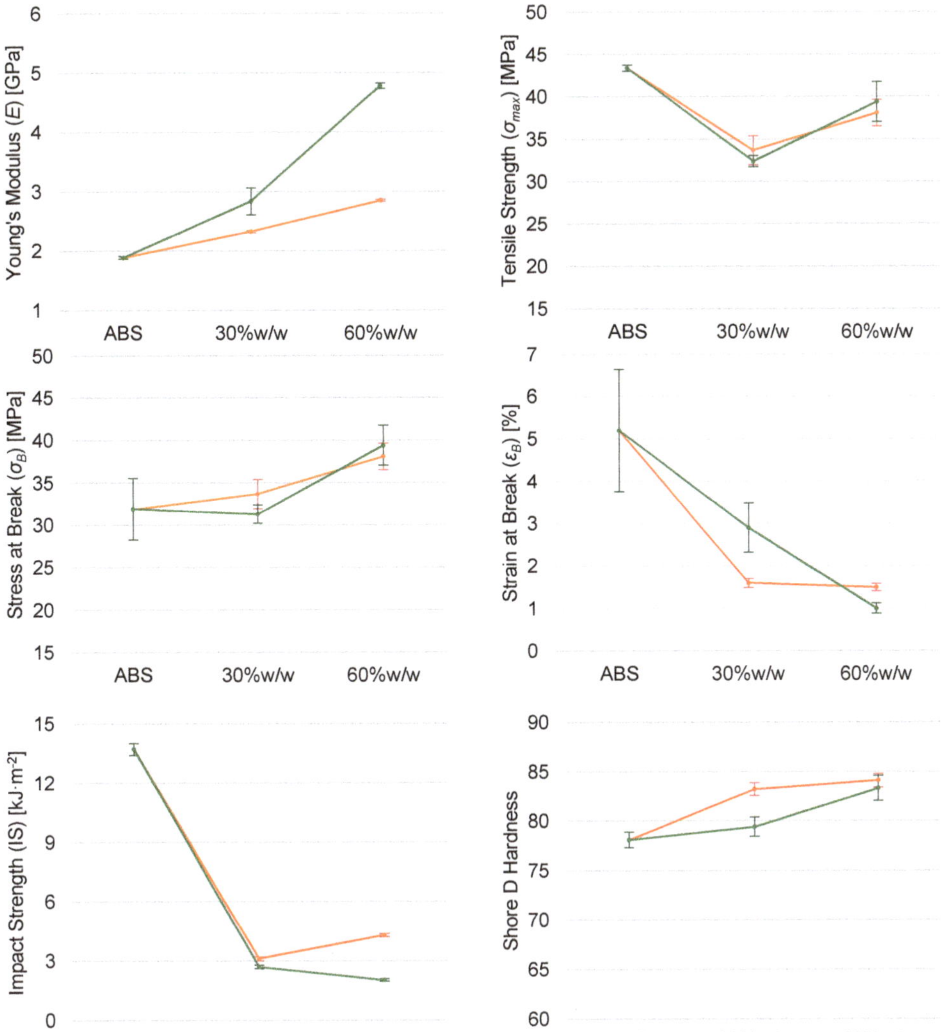

Figure 5. Interaction plots of selected mechanical properties of ABS/aluminium (green) and ABS/copper (orange) composites as functions of the metal powder content.

In ABS-30%Al, the poor interface bonding would cause ABS regions adjacent to the particle agglomerates to remain unaffected, withstanding a certain degree of plastic deformation. This is not observed for ABS-60%Al. The energy that can be absorbed by the ABS matrix is limited by the metal particles, leading to crack initiation and growth [68]. As the filler will not deform, it will rather debond; at the same time, the particles induce a concentration of tensile stress in the matrix, promoting multiple crazing sites [68].

The interaction between ABS and copper particles likely led the interface to stiffen neighbouring polymer segments from lower filler volume fractions, increasing the composite's brittleness. Increasing the filler concentration leads to the formation of clusters. Moreover, polymer–particle interfaces would act as breaking sites; therefore, as these agglomerations grow with the filler content, the surface-to-volume ratio drops, decreasing the amount of breaking sites, thus leading σ_B to rise [13].

The resulting brittleness observed in E is endorsed by a decrease in IS, with values for ABS-30%Al and ABS-60%Al as low as 20% and 15% of that of ABS, while ABS-30%Cu and ABS-60%Cu were as low as 23% and 31% relative to the pure material (Table 4 and Figure 5). The same trend was observed for IR. Such values of IS and IR suggest a poor adhesion between the matrix and reinforcement as a low-energy cleavage path leads to reduced energy absorption through the matrix [4,66,69]. In neat ABS, a fracture is driven by two simultaneous mechanisms, shear yielding and crazing, with shear-yielding bands acting as craze terminators and rigid particulates promoting enhanced shear yielding [68].

In general, the fracture toughness of ductile matrixes such as ABS is inversely proportional to the filler content due to a poor interfacial adhesion, particle agglomeration and non-homogeneous dispersion [66,70]. However, in spite of the reduction in toughness, the copper reinforcement particles likely promoted extra mechanisms for energy dissipation under a certain range of filler content, which is possibly associated with debonding and shear yielding, hence increasing the impact strength to some extent [66,68]. This would explain why increasing the copper content to 60 wt% enhanced the toughness behaviour as it raised IS by 38% relative to the 30 wt% copper content.

This effect was not observed for the aluminium composites as ABS-60%Al presented IS 25% lower than ABS-30%Al. In this case, the higher filler volume of aluminium reinforcement possibly led particles to agglomerate and act as stress concentration points, causing the impact properties to reduce, and supporting the hypothesis that the ABS–copper interface may be stronger than the ABS–aluminium one.

The cross-section fracture surfaces resulting from the Charpy test can be observed in Figure 6. ABS specimens presented clear evidence of deformation as the impact rupture took place, displaying the typical characteristics of a ductile rupture mechanism, including the presence of whitening due to the shear flow under stress [69]. These signals were absent on all the composites. Blisters can be observed on the composites cross-sections, likely caused by the moulding process, and could relate to the reduction in impact strength. Additionally, macroscopic aluminium agglomerations can be observed within the ABS matrix. Blisters observed on ABS-30%Cu were the largest overall. The amount and size of voids could also be related to the higher impact and tensile strength of ABS-60%Cu, as the non-homogeneous dispersion of blisters and particles might act as a point that creates stress concentrations. In Charpy impact testing, both the content and type of filler were found to be significant factors, and the interaction between the weight ratio and type of filler also had significant effects ($p < 0.01$). In addition, all the means were statistically different.

The values obtained for Shore D hardness showed an increasing trend alike to that of E, demonstrating that the stiffening effect due to the incorporation of the metal powders was also noticeable on the composites' surfaces. The highest hardness was recorded for ABS-60%Cu, a significant increase of 7.7% ($p < 0.01$) in comparison to neat ABS, which had the lowest hardness. The figures are similar to the results previously reported for ABS/aluminium composites [37,41].

Also, the decrease in impact strength due to the incorporation of a filler was found to take place in conjunction with an increase in hardness [69]. It has been suggested that the hard particle filler absorbs the load and withstands plastic deformation, thus resisting indentation [6]. In this context, the metallic particles would change the polymer matrix's microstructure; thus, the introduction of more grain boundaries due to increasing the metal content would cause the composite to reach a greater hardness value [71].

However, the lower values of hardness recorded for the aluminium-reinforced composites relative to the copper-filled ones, in spite of the greater volume fraction of the aluminium particles, may suggest an inadequate dispersion of the filler and a poorer interface effectiveness between the ABS matrix and aluminium reinforcement. Statistical analysis of the data indicates that both the content and type of filler have significant effects on the surface hardness characteristics, and the interaction between the weight ratio and type of filler was also found to be a significant factor (ANOVA, $p < 0.01$). In addition, only ABS-60%Al and ABS-30%Cu displayed no significant differences ($p = 0.988$).

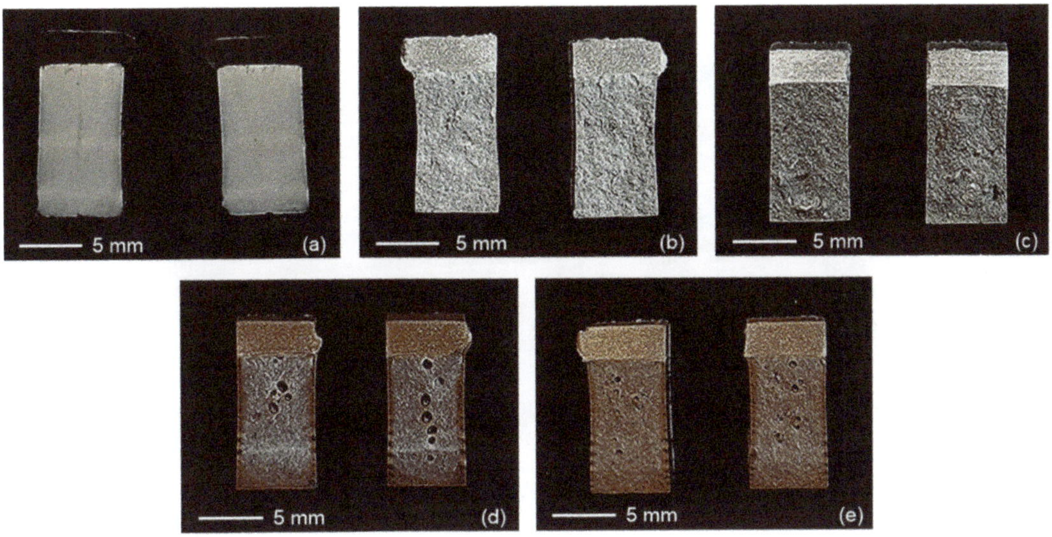

Figure 6. Representative fractographies from the Charpy impact test for (**a**) ABS, (**b**) ABS-30%Al, (**c**) ABS-60%Al, (**d**) ABS-30%Cu and (**e**) ABS-60%Cu; notches can be observed on the top of each cross-section.

3.3. Rheological Behaviour

The complex viscosity (η^*) mean values recorded at 195 °C for neat ABS and the metal-reinforced composites are presented in Figure 7 as a function of frequency (ω). There is a decrease of η^* with increasing ω, representing the shear thinning behaviour of ABS and the composites. The addition of metallic fillers did not change the overall η^* curve profile. Aluminium particles increased viscosity over the entire frequency range, while copper particles had a contrasting effect: ABS-30%Cu presented lower viscosity than ABS, indicating a possible plasticising effect due to the low volume fraction of Cu particles in the matrix, and ABS-60%Cu had a behaviour very close to that of neat ABS. Small amounts of a filler may reduce the viscosity as filler–filler interactions in the agglomerates would be weak and break under the shear force that orientates the flow direction [38]. At higher ω, the rate of decrease of η^* is greater, which is associated with the filler being forced to orient due to the shear stresses [51].

ABS presented a shear thinning parameter n of 0.4206. The changes in n for the composites are not large, with values ranging from 0.3962 for ABS-30%Cu to 0.4496 for ABS-60%Al, indicating low influence of the particles' dispersion on the rheological properties of ABS, which thus suggests low compatibility between the filler and polymer. In general, when there is a good interaction between the matrix and reinforcement, the slope of the η^* curve tends to increase, improving the shear thinning effect [52–56].

The increase in η^* of the ABS/aluminium composites accounts for the higher values of injection and holding pressure during the processing stage. Regarding the ABS/copper composites, despite a distinct trend of η^*, in order to achieve suitable moulding conditions those parameters were adjusted, preventing the polymer melt from freezing prematurely and to keep the mould pressurised.

The dependence of melt storage modulus (G') on ω is shown in Figure 8. The shoulder observed between 0.2 and 20 Hz can be attributed to the relaxation process of the butadiene domains within the styrene–acrylonitrile (SAN) phase. This typically represents the immiscibility of distinct phases of polymer blends and copolymers with discrete-continuous morphology [42]. The materials have nonterminal behaviour at both ends of the frequency range, which means that the long-range motion of backbone chains is hindered during flow [54].

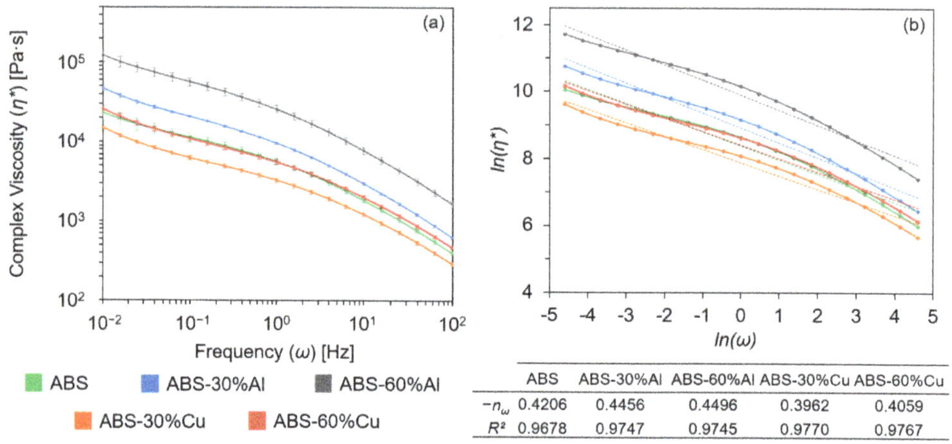

	ABS	ABS-30%Al	ABS-60%Al	ABS-30%Cu	ABS-60%Cu
$-n_\omega$	0.4206	0.4456	0.4496	0.3962	0.4059
R^2	0.9678	0.9747	0.9745	0.9770	0.9767

Figure 7. (**a**) The curves of η^* as a function of oscillation frequency; (**b**) the power law linear model for η^*, and the values of shear thinning parameter obtained from the linearisation of η^*.

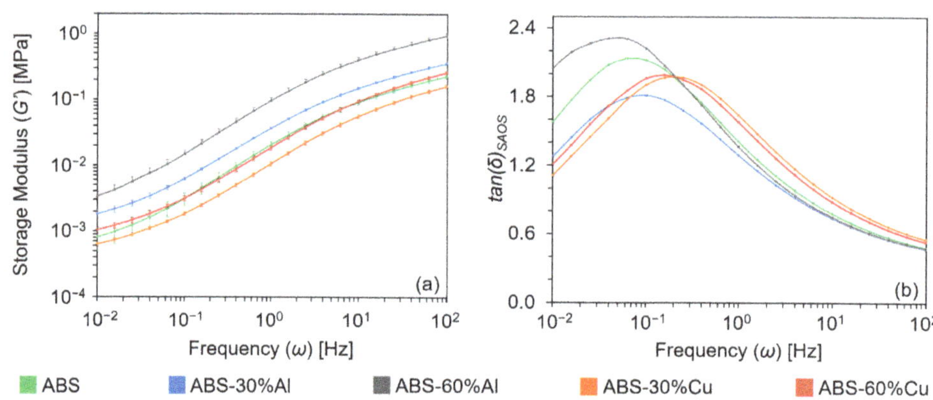

Figure 8. Curves of (**a**) melt storage modulus G' and (**b**) melt loss tangent $tan(\delta)_{SAOS}$ as functions of the oscillation frequency.

The change in G' at a given ω due to the type and content of reinforcement follows the same trend of change observed for η^*. Modulus shifts are observed on the ordinate axis, with minimal variations in the curve behaviour and no changes in the curve slope nor plateau developing at low ω. In composites with good matrix–filler interactions, G' rises with the filler content and develops a distinct plateau at low ω, which represents hindered chain relaxation due to the inability of filler particles to freely rotate [51,54,55]. ABS-60%Al displayed G' values ca. four times greater than those of ABS at all frequencies. The difference of G' between ABS and ABS-30%Al is greater at lower ω than at the higher end of the range, owing to a greater time to untangle chain entanglements at the lower end, but not enough time for chains to relax at higher ω [53]. Moreover, G' was found to overcome the loss modulus at higher ω, suggesting a viscous-to-elastic transition taking place [55].

Reinforcing a polymer matrix tends to reduce the melt $tan(\delta)_{SAOS}$ curve maxima, meaning a more elastic-like behaviour [52–54,56]. In this case, $tan(\delta)_{SAOS}$ decreased with the incorporation of a filler due to hindrance in chain motion and relaxation during flow, as noticed for ABS-30%Al, ABS-30%Cu and ABS-60%Cu relative to ABS, with their maxima shifting to higher ω. All samples displayed a slope transition within the low-frequency range, which may be associated with the hindered mobility of polymer chains, formation of three-dimensional network structures of fillers or a change in the filler–matrix interface interaction [53]. This greatly contrasts with the data regarding η^* and G', whose maximum values were achieved with ABS-60%Al. The lack of consistency of such results might further reveal that the affinity between the ABS matrix and the metallic reinforcement is poor.

Since MFB is a crucial parameter for 3D printing, SAOS has been employed to determine the appropriate processing conditions for high-performance thermoplastic AM feedstock, such as ABS. Ajinjeru et al. (2018) [56] assessed the effects of temperature, reinforcement and angular frequency on η^* of ABS and ABS/carbon fibre (ABS/CF) composites. It was reported that η^* reduced with the increase in the processing temperature from 230 to 270 °C and increase in ω, but was boosted with the incorporation of carbon fibres, enhancing the shear thinning effect. Finally, the authors claimed that the AM system's deposition temperature may have a significant impact on η^*, which is crucial when selecting the screw speed of large fabrication AM (LFAM) processes [56]. The increase in viscosity causes the torque on the extrusion screw to rise as well, so that the screw load excess may perform a counterbalancing action by adjusting the temperature, enabling a wide processing window by adjusting the flow rate [56]. Moreover, higher mixing screw speeds reportedly impart increased melting shear, thus significantly enhancing the filler dispersion, which in turn improves the composites' mechanical and thermal properties [72]. Regarding the processing temperature limits, ABS presents a main degradation step that starts from 300 °C, with an onset point at ca. 365 °C; therefore, no thermal degradation events are expected to take place within the temperature range utilised in the test [73,74].

3.4. Viscoelastic Properties

The storage modulus (E') mean values recorded at 30 and 90 °C, the glass transition temperature (T_g) of neat ABS and the metal-reinforced composites are presented in Table 5. Representative dynamic mechanical response curves of E' and the damping factor ($tan(\delta)_{DMA}$) as functions of temperature are displayed in Figure 9.

Table 5. Dynamic mechanical responses of ABS and its composites.

	ABS	ABS-30%Al	ABS-60%Al	ABS-30%Cu	ABS-60%Cu
E' (30 °C) [MPa]	2064.5 ±40.7	2878.0 ±190.7	4444.3 ±206.6	2636.3 ±164.7	3259.3 ±153.1
E' (90 °C) [MPa]	1097.8 ±41.2	1820.0 ±49.1	3201.3 ±178.1	1155.7 ±30.2	1666.0 ±42.9
T_g [°C]	110.0 ±0.61	114.1 ±1.50	117.8 ±1.82	107.0 ±0.21	109.7 ±0.52

For the neat ABS matrix, E' curve shows the glassy state plateau from 25 to 80 °C and the glass transition from 80 °C onwards. Copper-filled composites closely followed the same behaviour. On the other hand, aluminium-reinforced composites had the glass transition shifted by ca. 5 °C, with the glassy state plateau prevailing up to 85 °C and shifting the end of the glass transition to higher temperatures as well. At 30 °C, all the composites have E' values greater than that of the pure matrix, with the lowest being 28% higher at 2636.3 MPa for ABS-30%Cu, and the highest being 115% greater at 4444.3 MPa for ABS-60%Al.

In addition, E' was also found to have a linear dependence (R^2 = 0.9881) upon the volume fraction of the filler at 30 °C, mirroring the similar trend observed for Young's modulus. When analysing ABS and ABS/Al composites only, such linear dependence achieves R^2 = 0.9972, whereas ABS/Cu composites have a positive deviation respective to the linear regression model. Comparing ABS-30%Al and ABS-60%Cu, which have similar reinforcement volume fractions, it is noticed that the latter exhibits higher E'. These observations endorse the hypothesis of a better interface bonding between ABS and copper powder than between ABS and aluminium powder [75].

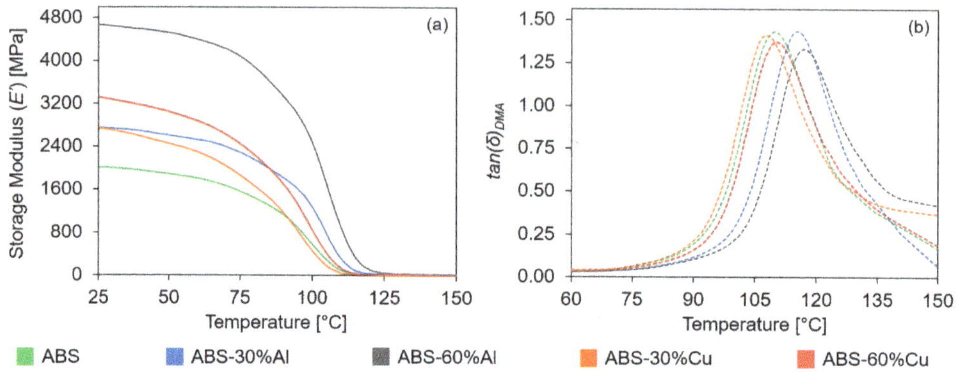

Figure 9. Effect of the metallic fillers on (**a**) storage modulus E' and (**b**) damping factor $tan(\delta)_{DMA}$ of ABS.

Overall, the composites' storage moduli are greater than that of ABS within the whole range of the glassy plateau, revealing that the metal reinforcement particles have a stiffening effect on the matrix, i.e., the metallic powders had a strong interlocking action in the ABS matrix, increasing the material's stiffness [7,55,74]. This effect is more pronounced for ABS/Al composites than for ABS/Cu, as the glassy state plateau lasts longer and the E' curves of ABS-30%Cu and ABS-60%Cu are overcome by ABS and ABS-30%Al, respectively, during the glass transition. At 90 °C, the temperature effects are evident as ABS-30%Cu displayed E' only 5% higher than ABS, and ABS-60%Cu had its E' reduced to 8.5% lower than that of ABS-30%Al. A possible reason for this would be a loss of effectiveness of the ABS/Cu interface due to increasing temperature [75].

The metal powders also affected the glass transition of the SAN phase of ABS, as shown in Table 5 and Figure 9. It is known that the T_g of ABS is at approximately 100 °C and relates to the glassy–rubbery transition of the SAN grafting block [25,33,39,57]. The maxima of the $tan(\delta)_{DMA}$ curves obtained with DMA displayed the same trend of change in ascending order as exhibited by G' and η^*.

Aluminium particles hindered the motion and rotation of ABS chains and retarded the onset of the viscous component contribution, as was observed with the increase in G', raising the T_g up to 117.8 °C for ABS-60%Al. The reinforcement introduces restrictions on the segmental mobility of interfacial SAN blocks, tending to increase the T_g due to mechanical locking [33,74]. Composites containing a 60 wt% filler have greater T_g than 30 wt% composites, owing to the confinement of the polymer within the filler particles and the consequential clamping effect [13]. An increase in the reinforcement particle size would also increase the T_g for the same loading levels [57].

ABS-30%Cu had the lowest T_g at 107.0 °C, which is a clear drop relative to 110.0 °C of ABS. This suggests that, at that filler volume fraction and with that particle size, the Cu powder had some plasticising effect into the SAN phase blocks by likely disrupting and weakening the intermolecular interactions between polymer chains, possibly due to a better dispersion of the smaller particles, likely combined with a probable loss of intensity in the polymer–metal interface bonding due to an increased free volume when the temperature was increased. This also relates to the reduction in η^* and G' observed for ABS-30%Al.

Increasing the copper content to 60 wt% T_g rose to 109.7 °C, representing a recovery of interlocking. As the T_g of ABS-60%Cu is close to that of ABS and lower than ABS-30%Al, the reinforcement loading is not the only factor affecting the glass transition, supporting the hypothesis that the ABS-copper particles' surface interactions display an adverse response brought about by temperature elevation. Therefore, despite the possible loss of the ABS-Cu interface's effectiveness with temperature, the size difference between aluminium and copper particles is likely related to the distinct T_g values recorded for ABS-30%Al and ABS-60%Cu.

3.5. Thermal Properties

Thermal conductivity results are shown in Figure 10. ABS has the lowest k at 0.460 W·K^{-1}·m^{-1}, while ABS-60%Al has the highest at 0.861 W·K^{-1}·m^{-1}, which is a significant increase of 87% ($p < 0.01$). For thermal conductivity, the filler volume fraction arises as a crucial factor since studies suggest that, beyond a minimum filling threshold, particle agglomerates would create a conductive path owing to the increased contact area, enhancing thermal conduction properties [4,6,7,71]. Increasing the filling content would reduce the space between reinforcement particles, hence increasing the probability of effective contact between neighbouring particles to form a conductive path [7].

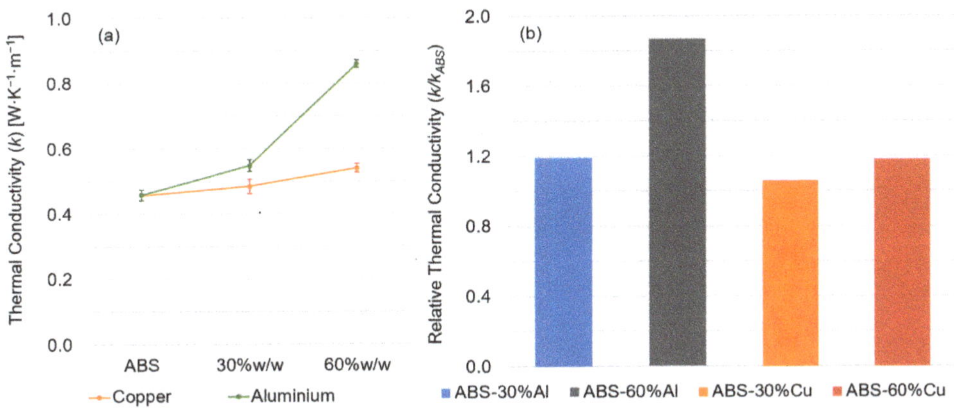

Figure 10. (**a**) Interaction plot of thermal conductivity of ABS and composites as a function of weight ratio; (**b**) relative thermal conductivity of the composites.

The results obtained by Masood and Song (2004, 2005) [9,10] suggested that the composites' thermal conductivity rose with the size and volume content of metal particles. This would indicate that copper powder at 30 wt% loading is insufficient to produce a conductive path within the polymeric matrix, and this also clarifies why ABS-30%Al and ABS-60%Cu have close values of k. In addition, the conductive path also depends upon the size of the reinforcement particles and the dispersion of voids, blisters and particle agglomerates [71]. Nikzad et al. (2011) [7] reported that volume fractions of copper particles lower than 0.10 cannot break the thermal resistance of the ABS matrix; thus, the threshold point of f_{met} would have to be between 0.1 and 0.2.

Contrastingly, an increase of 41% in the thermal conductivity of ABS loaded with 50 wt% Cu relative to the neat material has been described [1]. Values of k for ABS have been reported as 0.646 W·K^{-1}·m^{-1} [1], 0.17 W·K^{-1}·m^{-1} [4] and 0.145 W·K^{-1}·m^{-1} [6]. Such differences could be explained by the distinct specimen manufacturing techniques and measurement methods in each study. In injection-moulded particulate-reinforced composites, the moulding process is believed to align the filler in the melt flow direction, and both the contact area and orientation are promoted due to the high injection pressures; thus, heat and charge transport is expected to be superior to that seen in 3D-printed parts [4]. This also could help explain the higher thermal conductivity obtained for ABS/Al composites in spite of copper's well-known greater conductivity properties, as processing parameters such as the injection pressure had to be adjusted and accounted for the higher melt viscosity. Moreover, in general, the thermal conductivity of polymer composites depends upon various factors, e.g., filler content, filler shape and size, interfacial adhesion, resistance of the filler–polymer interface and filler structure quality within the matrix [76].

In addition, *k* demonstrated a linear dependence (R^2 = 0.9465) upon the volume fraction of the filler, following similar trends found for Young's and storage moduli, endorsing the results obtained by Masood and Song (2005) [10]. When analysing ABS and ABS/Cu composites only, the linear regression model yields R^2 = 0.9999. There are no significant statistical differences between ABS and ABS-30%Cu (0.027 $W \cdot K^{-1} \cdot m^{-1}$, $p = 0.263$), or between ABS-30%Al and ABS-60%Cu (0.006 $W \cdot K^{-1} \cdot m^{-1}$, $p = 0.985$). In addition, results indicate that both the content and type of filler have significant effects on *k*, and the interaction between the type of filler and weight ratio was also found to be a significant factor (ANOVA, $p < 0.01$).

3.6. Morphological Characterisation of the Composites

The SEM fractographies of the cross-section fracture surfaces resulting from the Charpy impact test are shown in Figure 11, allowing an assessment of the dispersion and agglomeration of the two metal powders within the ABS matrix. Aluminium particles were observed to be larger than copper particles. Moreover, the size of the aluminium particles in the composites seems bigger than when first introduced in the mixture at the processing stage. This suggests that this metal has a stronger filler–filler interaction, allowing a greater mobility in the molten matrix during the processing stages and a trend to aggregate, forming non-homogeneously dispersed agglomerations.

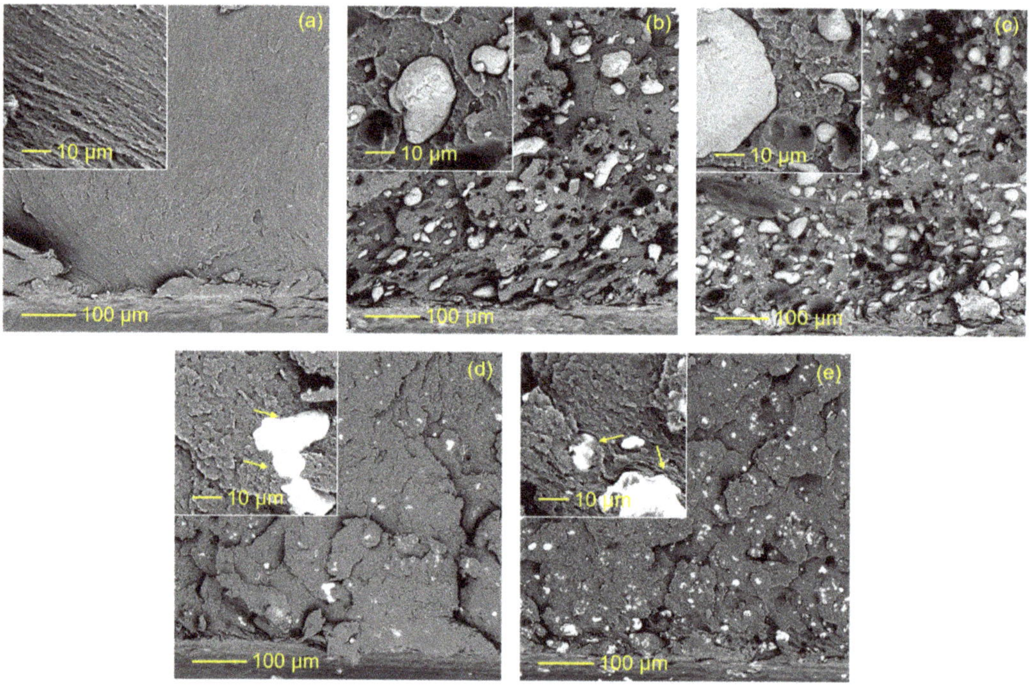

Figure 11. Representative SEM fractographies: (**a**) ABS, (**b**) ABS-30%Al, (**c**) ABS-60%Al, (**d**) ABS-30%Cu and (**e**) ABS-60%Cu. Notches can be observed on the bottom of each cross-section. Arrows indicate regions of matrix–particle adherence with bridging in ABS/Cu composites.

The quality of metal particle dispersion can also be verified through an elemental analysis that unveils the chemical composition of a surface through EDS mapping, as shown in Figure 12. Comparing ABS-30%Al and ABS-60%Cu and considering their comparable filler volume fraction f_{met}, the copper particles appear to manifest a better dispersion than aluminium particles. This supports the hypothesis of a better interaction of ABS with copper rather than with aluminium, resulting in a more effective polymer–filler interface. At lower filler contents, the composites contain a higher proportion of the ABS matrix compared to reinforcements. This hinders the contact between the metal particles in ABS, endorsing the dependency of thermal conductivity upon the content and volume ratio of the filler.

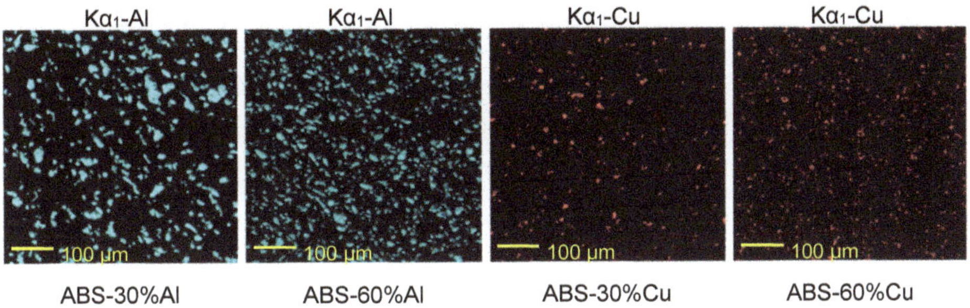

Figure 12. Dispersion of aluminium and copper particles within the composites' matrix obtained with EDS surface chemical composition mapping.

When undergoing a high strain rate test such as the impact one, polymer chains are subjected to extreme stresses until they undergo rupture, leading to the nucleation of microvoids in the matrix. SEM imaging enabled a further insight into the impact fracture surfaces (Figure 11), supporting the observations based on Figure 6 since indications of a plastic deformation during an impact rupture are once again clearly seen for neat ABS but not for the composites [69,74]. The occurrence of fibrils further endorses the extent of the plastic deformation in ABS, as the growth and propagation of the macroscopic crack (notch) is thought to be accompanied by the formation of additional plastic microcracks [74]. The SEM fractographies show clear differences between the aluminium- and copper-reinforced composites (Figure 11). The lack of bonding between the ABS matrix and aluminium particles becomes evident as there are considerable gaps (debonding) surrounding the Al filler. Moreover, hollow spaces (cupules), previously occupied by particles that were pulled out during the impact test, are visible. On the other hand, Cu particles display less debonding but are embedded in the ABS matrix, which is evinced by polymer segments attached to filler particles, instead of signs of pull-out as a result of the impact test. However, both composites present a layered structure comprising small aligned steps on the fracture surface, which indicates shear yielding [77].

Additionally, aluminium and copper composites presented distinct distributions of microvoids and crazes along the matrix. In the case of ABS/Al composites, cracks were initiated from microvoids on the weak matrix–filler interface and they coalesced towards the neighbouring particles, resulting in failure propagation through reinforcement debonding and pull-out and the formation of cupules [74,77,78]. On the other hand, in the case of ABS/Cu composites, metal particles plastically bridged the crazes due to their adherence to the matrix, preventing them from spreading and becoming cracks and favouring the nucleation of microvoids [77,79,80]. These effects are likely due to the different nature of each reinforcement as well as their particular characteristics and may have a significant impact on the mechanical properties of the composite, as its failure mode is affected by the wettability of the filler by the matrix [81,82]. In this sense, according to Ryder et al. (2018) [83], water-atomised metal particles, such as the copper ones in this study, are nonspherical, morphologically irregular and rough, which improves interfacial interactions between reinforcement particles and the polymer matrix.

4. Conclusions

Metal-reinforced thermoplastic composites are suitable materials for applications requiring special thermal, electrical or magnetic properties, enabling one to benefit from properties that usual polymer materials lack. Making these composites into filament wires for feedstock could potentially lead to high-performance, functional prototypes for a wide range of applications that could be manufactured with additive manufacturing (AM) processes, which would enable these materials to be quickly shaped into any design. Dynamic rheology is crucial to determine the appropriate processing conditions for AM feedstock as melt flow behaviour (MFB) is an important parameter for 3D printing. ABS composites reinforced with aluminium and copper particles were produced and injection-moulded and then characterised regarding their mechanical, rheological, viscoelastic and thermal properties.

Tensile testing revealed that the Young's modulus increased with the metal particle incorporation and followed the law of mixtures, displaying a linear dependence upon the filler volume fraction. However, there was an embrittling effect, evinced by the lower elongation at break and impact strength owing to a poor matrix–filler interface unable to properly transfer load.

Small angle oscillatory shear (SAOS) supported the low affinity between ABS and the metallic fillers as the complex viscosity (η^*) did not exhibit an enhancement of the shear thinning effect, with Al particles causing η^* to rise. DMA results unveiled that changes in viscoelastic properties are mainly due to the mechanical locking of backbone chains by filler particles.

Below T_g, ABS/Cu composites behave similarly to ABS/Al; however, as they undergo the glass transition, the copper filler acts as a plasticising agent, with composites featuring characteristics closer to those presented by neat ABS both in SAOS and DMA.

Thermal conductivity (k) was improved with the incorporation of the metals, especially aluminium, due to its lower density relative to copper, indicating that k also follows the law of mixtures and depends on the filler volume fraction. SEM imaging confirmed the poor interaction between the ABS matrix and aluminium and copper fillers (as debonding and pull-out were observed) and disclosed that the metals were associated with different fracture mechanisms.

Overall, the attained set of mechanical, rheological and viscoelastic properties offers a promising opportunity to optimise the processing–property relationship in spite of the poor matrix–filler interface, which can be improved using higher mixing speeds, considering the wide processing window available. Therefore, the obtained materials are potential candidates for metal-reinforced thermoplastic composites that could be processed through AM technologies.

Author Contributions: Conceptualisation, V.F.M., C.A.F. and D.M.D.; methodology, V.F.M.; validation, V.F.M. and H.P.; formal analysis, V.F.M. and H.P.; investigation, V.F.M. and H.P.; data curation, V.F.M.; writing—original draft preparation, V.F.M. and H.P.; writing—review and editing, V.F.M.; visualisation, V.F.M.; supervision, J.S.C., C.A.F. and D.M.D.; project administration, C.A.F. and D.M.D.; funding acquisition, D.M.D. All authors have read and agreed to the published version of the manuscript.

Funding: This publication contains results from research conducted with the financial support of Enterprise Ireland under the Technology Gateway Programme, grant number TG-2017-0114 (APT Ireland), and the Science Foundation Ireland (SFI), grant number SFI 16/RC/3918, co-funded by the European Regional Development Fund.

Data Availability Statement: All the data necessary to reproduce the results and support the conclusions are included within this paper. Further details may be provided upon request.

Acknowledgments: The authors would like to thank the APT Ireland staff, for assisting with the extrusion and injection moulding processes; Annelise Alves (UFRGS), for the BET analyses; Eduardo G. Barboza (UFRGS), for the granulometry tests; Lucas B. Gomes (UFRGS), for the XRD analyses; and Rodrigo A. Barbieri (UCS), for assisting with SEM/EDS analyses. The authors would also like to thank the funding agencies Enterprise Ireland, Science Foundation Ireland (SFI), Higher Education Authority (HEA), The Department of Further and Higher Education, Research, Innovation and Science (D/FHERIS), National Council for Scientific and Technological Development (CNPq) and CAPES Foundation for financial support.

Conflicts of Interest: The authors declare no conflict of interest.

References

1. Hwang, S.; Reyes, E.I.; Moon, K.; Rumpf, R.C.; Kim, N.S. Thermo-mechanical Characterization of Metal/Polymer Composite Filaments and Printing Parameter Study for Fused Deposition Modeling in the 3D Printing Process. *J. Electron. Mater.* **2015**, *44*, 771–777. [CrossRef]
2. Kalsoom, U.; Peristyy, A.; Nesterenko, P.N.; Paull, B. A 3D printable diamond polymer composite: A novel material for fabrication of low cost thermally conducting devices. *RSC Adv.* **2016**, *6*, 38140–38147. [CrossRef]
3. Carson, J.K.; Alsowailem, M. Thermal Diffusivity of Copper/Linear-low-density Polyethylene Composites. *Polym. Polym. Compos.* **2017**, *25*, 447–452. [CrossRef]
4. Quill, T.J.; Smith, M.K.; Zhou, T.; Baioumy, M.G.S.; Berenguer, J.P.; Cola, B.A.; Kalaitzidou, K.; Bougher, T.L. Thermal and mechanical properties of 3D printed boron nitride—ABS composites. *Appl. Compos. Mater.* **2018**, *25*, 1205–1217. [CrossRef]
5. Palmero, E.M.; Casaleiz, D.; de Vicente, J.; Hernández-Vicen, J.; López-Vidal, S.; Ramiro, E.; Bollero, A. Composites based on metallic particles and tuned filling factor for 3D-printing by Fused Deposition Modeling. *Compos. Part A Appl. Sci. Manuf.* **2019**, *124*, 105497. [CrossRef]
6. Hamzah, K.A.; Yeoh, C.K.; Noor, M.M.; Teh, P.L.; Aw, Y.Y.; Sazali, S.A.; Wan Ibrahim, W.M.A. Mechanical properties and thermal and electrical conductivity of 3D printed ABS-copper ferrite composites via 3D printing technique. *J. Thermoplast. Compos. Mater.* **2022**, *35*, 3–16. [CrossRef]
7. Nikzad, M.; Masood, S.H.; Sbarski, I. Thermo-mechanical properties of a highly filled polymeric composites for Fused Deposition Modeling. *Mater. Des.* **2011**, *32*, 3448–3456. [CrossRef]
8. Yu, J.H.; Cennini, G. Improving thermal conductivity of polymer composites in embedded LEDs systems. *Microelectron. J.* **2014**, *45*, 1829–1833. [CrossRef]
9. Masood, S.H.; Song, W.Q. Development of new metal/polymer materials for rapid tooling using Fused deposition modelling. *Mater. Des.* **2004**, *25*, 587–594. [CrossRef]
10. Masood, S.H.; Song, W.Q. Thermal characteristics of a new metal/polymer material for FDM rapid prototyping process. *Assem. Autom.* **2005**, *25*, 309–315. [CrossRef]
11. Moritz, V.F.; Bezerra, G.S.N.; Hopkins Jnr, M.; Fuenmayor, E.; Günbay, S.; Hayes, C.; Lyons, J.G.; Devine, D.M. Heat Dissipation Plays Critical Role for Longevity of Polymer-Based 3D-Printed Inserts for Plastics Injection Moulding. *J. Manuf. Mater. Process.* **2022**, *6*, 117. [CrossRef]
12. Coser, E.; Moritz, V.F.; Krenzinger, A.; Ferreira, C.A. Development of paints with infrared radiation reflective properties. *Polímeros* **2015**, *25*, 305–310. [CrossRef]
13. Merazzo, K.J.; Díez, A.G.; Tubio, C.R.; Manchado, J.C.; Malet, R.; Pérez, M.; Costa, P.; Lanceros-Mendez, S. Acrylonitrile Butadiene Styrene-Based Composites with Permalloy with Tailored Magnetic Response. *Polymers* **2023**, *15*, 626. [CrossRef] [PubMed]
14. Sofian, N.M.; Rusu, M.; Neagu, R.; Neagu, E. Metal Powder-Filled Polyethylene Composites. V. Thermal Properties. *J. Thermoplast. Compos. Mater.* **2001**, *14*, 20–33. [CrossRef]
15. Mamunya, Y.P.; Davydenko, V.V.; Pissis, P.; Lebedev, E.V. Electrical and thermal conductivity of polymers filled with metal powders. *Eur. Polym. J.* **2002**, *38*, 1887–1897. [CrossRef]
16. Luyt, A.S.; Molefi, J.A.; Krump, H. Thermal, mechanical and electrical properties of copper powder filled low-density and linear low-density polyethylene composites. *Polym. Degrad. Stab.* **2006**, *91*, 1629–1636. [CrossRef]
17. Tekce, H.S.; Kumlutas, D.; Tavman, I.H. Effect of Particle Shape on Thermal Conductivity of Copper Reinforced Polymer Composites. *J. Reinf. Plast. Compos.* **2007**, *26*, 113–121. [CrossRef]
18. Chifor, V.; Orban, R.L.; Tekiner, Z.; Turker, M. Thermal, Mechanical and Electrical Properties of High Density Polyethylene Composites Reinforced with Copper Powder. *Mater. Sci. Forum* **2011**, *672*, 191–194. [CrossRef]
19. Carson, J.K.; Noureldin, M. Measurements of the thermal diffusivity of linear-medium-density-polyethylene/aluminium composites using a transient comparative method. *Int. Commun. Heat Mass Transf.* **2009**, *36*, 458–461. [CrossRef]
20. Carson, J.K. Measurement and modelling of the thermal conductivity of dispersed aluminium composites. *Int. Commun. Heat Mass Transf.* **2011**, *38*, 1024–1028. [CrossRef]
21. Rahmati, S.; Dickens, P. Rapid tooling analysis of Stereolithography injection mould tooling. *Int. J. Mach. Tools Manuf.* **2007**, *47*, 740–747. [CrossRef]

22. Pontes, A.J.; Cousa, C.; Gomes, C.; Pouzada, A.S. Development of efficient composite systems for the production of resin cast moulding blocks. In *Proceedings of the PPS 24th Annual Meeting*; Polymer Processing Society: Salerno, Italy, 2008.
23. Ranjan, N.; Kumar, R.; Kumar, R.; Kaur, R.; Singh, S. Investigation of Fused Filament Fabrication-Based Manufacturing of ABS-Al Composite Structures: Prediction by Machine Learning and Optimization. *J. Mater. Eng. Perform.* **2023**, *32*, 4555–4574. [CrossRef]
24. Akrout, M.; Ben Difallah, B.; Kharrat, M.; Dammak, M.; Pereira, A.; Oliveira, F.J.; Duarte, I. On the Structural, Thermal, Micromechanical and Tribological Characterizations of Cu-Filled Acrylonitrile Butadiene Styrene Micro-Composites. *Materials* **2023**, *16*, 6428. [CrossRef] [PubMed]
25. Sa'ude, N.; Masood, S.H.; Nikzad, M.; Ibrahim, M.; Ibrahim, M.H.I. Dynamic mechanical properties of copper-ABS composites for FDM feedstock. *Int. J. Eng. Res. Appl.* **2013**, *3*, 1257–1263.
26. Dawoud, M.; Taha, I.; Ebeid, S.J. Mechanical behaviour of ABS: An experimental study using FDM and injection moulding techniques. *J. Manuf. Process.* **2016**, *21*, 39–45. [CrossRef]
27. Wang, X.; Jiang, M.; Zhou, Z.; Gou, J.; Hui, D. 3D printing of polymer matrix composites: A review and prospective. *Compos. Part B Eng.* **2017**, *110*, 442–458. [CrossRef]
28. Gong, K.; Liu, H.; Huang, C.; Cao, Z.; Fuenmayor, E.; Major, I. Hybrid Manufacturing of Acrylonitrile Butadiene Styrene (ABS) via the Combination of Material Extrusion Additive Manufacturing and Injection Molding. *Polymers* **2022**, *14*, 5093. [CrossRef]
29. Gong, K.; Xu, H.; Liu, H.; Cao, Z.; Fuenmayor, E.; Major, I. Hybrid manufacturing of mixed-material bilayer parts via injection molding and material extrusion three-dimensional printing. *J. Appl. Polym. Sci.* **2023**, *140*, e53972. [CrossRef]
30. Zhang, W.; Cotton, C.; Sun, J.; Heider, D.; Gu, B.; Sun, B.; Chou, T.-W. Interfacial bonding strength of short carbon fiber/acrylonitrile-butadiene-styrene composites fabricated by fused deposition modeling. *Compos. Part B Eng.* **2018**, *137*, 51–59. [CrossRef]
31. Sa'ude, N.; Ibrahim, M.; Ibrahim, M.H.I.; Wahab, M.S.; Haq, R.; Marwah, O.M.F.; Khirotdin, R.K. Additive manufacturing of copper-ABS filament by fused deposition modeling (FDM). *J. Mech. Eng.* **2018**, *5*, 23–32.
32. Abeykoon, C.; Sri-Amphorn, P.; Fernando, A. Optimization of fused deposition modeling parameters for improved PLA and ABS 3D printed structures. *Int. J. Light. Mater. Manuf.* **2020**, *3*, 284–297. [CrossRef]
33. Billah, K.M.M.; Lorenzana, F.A.R.; Martinez, N.L.; Wicker, R.B.; Espalin, D. Thermomechanical characterization of short carbon fiber and short glass fiber-reinforced ABS used in large format additive manufacturing. *Addit. Manuf.* **2020**, *35*, 101299. [CrossRef]
34. Akhoundi, B.; Modanloo, V. A multi-criteria decision-making analysis on the extrusion-based additive manufacturing of ABS/Cu composites. *Int. J. Interact. Des. Manuf.* **2023**, *17*, 1995–2003. [CrossRef]
35. Isa, N.M.A.; Sa'ude, N.; Ibrahim, M.; Hamid, S.M.; Kamarudin, K. A Study on Melt Flow Index on Copper-ABS for Fused Deposition Modeling (FDM) Feedstock. *Appl. Mech. Mater.* **2015**, *773–774*, 8–12. [CrossRef]
36. Sa'ude, N.; Ibrahim, M.; Ibrahim, M.H.I. Melt flow rate (MFR) of ABS-copper composite filament by fused deposition modeling (FDM). *ARPN J. Eng. Appl. Sci.* **2016**, *11*, 6562–6567.
37. Singh, R.; Kumar, R.; Ahuja, I. Mechanical, thermal and melt flow of aluminum-reinforced PA6/ABS blend feedstock filament for fused deposition modeling. *Rapid Prototyp. J.* **2018**, *24*, 1455–1468. [CrossRef]
38. Kumar, N.; Jain, P.K.; Tandon, P.; Pandey, P.M. Investigations on the melt flow behaviour of aluminium filled ABS polymer composite for the extrusion-based additive manufacturing process. *Int. J. Mater. Prod. Technol.* **2019**, *59*, 194. [CrossRef]
39. Alghadi, A.M.; Tirkes, S.; Tayfun, U. Mechanical, thermo-mechanical and morphological characterization of ABS based composites loaded with perlite mineral. *Mater. Res. Express* **2019**, *7*, 015301. [CrossRef]
40. Sezer, H.K.; Eren, O. FDM 3D printing of MWCNT re-inforced ABS nano-composite parts with enhanced mechanical and electrical properties. *J. Manuf. Process.* **2019**, *37*, 339–347. [CrossRef]
41. Moritz, V.F.; Ferreira, C.A.; Devine, D.M. Assessment of ABS/Aluminium composites for 3D printing filaments. In *Proceedings of the IMC37*; Irish Manufacturing Council: Dublin, Ireland, 2021.
42. Farias, N.C.; Major, I.; Devine, D.; Brennan Fournet, M.; Pezzoli, R.; Farshbaf Taghinezhad, S.; Hesabi, M. Multiple recycling of a PLA/PHB biopolymer blend for sustainable packaging applications: Rheology-morphology, thermal, and mechanical performance analysis. *Polym. Eng. Sci.* **2022**, *62*, 1764–1774. [CrossRef]
43. Wang, S.; Capoen, L.; D'hooge, D.R.; Cardon, L. Can the melt flow index be used to predict the success of fused deposition modelling of commercial poly(lactic acid) filaments into 3D printed materials? *Plast. Rubber Compos.* **2018**, *47*, 9–16. [CrossRef]
44. Ferg, E.E.; Bolo, L.L. A correlation between the variable melt flow index and the molecular mass distribution of virgin and recycled polypropylene used in the manufacturing of battery cases. *Polym. Test.* **2013**, *32*, 1452–1459. [CrossRef]
45. Mwania, F.M.; Maringa, M.; van der Walt, J.G. A review of the techniques used to characterize laser sintering of polymeric powders for use and re-use in additive manufacturing. *Manuf. Rev.* **2021**, *8*, 14. [CrossRef]
46. Ghanbari, A.; Mousavi, Z.; Heuzey, M.; Patience, G.S.; Carreau, P.J. Experimental methods in chemical engineering: Rheometry. *Can. J. Chem. Eng.* **2020**, *98*, 1456–1470. [CrossRef]
47. Nikzad, M.; Masood, S.H.; Sbarski, I.; Groth, A. A study of melt flow analysis of an ABS-Iron composite in fused deposition modelling process. *Tsinghua Sci. Technol.* **2009**, *14*, 29–37. [CrossRef]
48. Nikzad, M.; Masood, S.H.; Sbarski, I.; Groth, A.M. Rheological Properties of a Particulate-Filled Polymeric Composite through Fused Deposition Process. *Mater. Sci. Forum* **2010**, *654–656*, 2471–2474. [CrossRef]
49. Sun, Q.; Rizvi, G.M.; Bellehumeur, C.T.; Gu, P. Effect of processing conditions on the bonding quality of FDM polymer filaments. *Rapid Prototyp. J.* **2008**, *14*, 72–80. [CrossRef]

50. Ramanath, H.S.; Chua, C.K.; Leong, K.F.; Shah, K.D. Melt flow behaviour of poly-ε-caprolactone in fused deposition modelling. *J. Mater. Sci. Mater. Med.* **2008**, *19*, 2541–2550. [CrossRef] [PubMed]
51. Liu, B.; Zhang, Y.; Wan, C.; Zhang, Y.; Li, R.; Liu, G. Thermal stability, flame retardancy and rheological behavior of ABS filled with magnesium hydroxide sulfate hydrate whisker. *Polym. Bull.* **2007**, *58*, 747–755. [CrossRef]
52. Galvan, D.; Carneiro, F.; Mazzucco, M.; Bartoli, J.R.; D'Ávila, M.A.; Morales, A.R.; Fernandes, E.G. Effect of Organoclay Mixture on the Rheological Properties of ABS-Clay Nanocomposites. *Macromol. Symp.* **2012**, *319*, 167–172. [CrossRef]
53. Singh, P.; Ghosh, A.K. Torsional, tensile and structural properties of acrylonitrile–butadiene–styrene clay nanocomposites. *Mater. Des.* **2014**, *55*, 137–145. [CrossRef]
54. Gao, C.; Zhang, S.; Wang, F.; Wen, B.; Han, C.; Ding, Y.; Yang, M. Graphene Networks with Low Percolation Threshold in ABS Nanocomposites: Selective Localization and Electrical and Rheological Properties. *ACS Appl. Mater. Interfaces* **2014**, *6*, 12252–12260. [CrossRef]
55. Wei, W.; Hu, S.; Zhang, R.; Xu, C.; Zhang, F.; Liu, Q. Enhanced electrical properties of graphite/ABS composites prepared via supercritical CO2 processing. *Polym. Bull.* **2017**, *74*, 4279–4295. [CrossRef]
56. Ajinjeru, C.; Kishore, V.; Liu, P.; Lindahl, J.; Hassen, A.A.; Kunc, V.; Post, B.; Love, L.; Duty, C. Determination of melt processing conditions for high performance amorphous thermoplastics for large format additive manufacturing. *Addit. Manuf.* **2018**, *21*, 125–132. [CrossRef]
57. Jeong, S.; Song, Y.S.; Lim, E. Fabrication and Characterization of Aluminum Nanoparticle-Reinforced Composites. *Polymers* **2020**, *12*, 2772. [CrossRef] [PubMed]
58. Strano, M.; Rane, K.; Briatico Vangosa, F.; Di Landro, L. Extrusion of metal powder-polymer mixtures: Melt rheology and process stability. *J. Mater. Process. Technol.* **2019**, *273*, 116250. [CrossRef]
59. Hopkins, M.; Gunbay, S.; Hayes, C.; Moritz, V.F.; Fuenmayor, E.; Lyons, J.G.; Devine, D.M. Stereolithography (SLA) utilised to print injection mould tooling in order to evaluate thermal and mechanical properties of commercial polypropylene. *Procedia Manuf.* **2021**, *55*, 205–212. [CrossRef]
60. Chen, J.; Yan, W.; Townsend, E.J.; Feng, J.; Pan, L.; Del Angel Hernandez, V.; Faul, C.F.J. Tunable Surface Area, Porosity, and Function in Conjugated Microporous Polymers. *Angew. Chem. Int. Ed.* **2019**, *58*, 11715–11719. [CrossRef] [PubMed]
61. Kassym, K.; Perveen, A. Atomization processes of metal powders for 3D printing. *Mater. Today Proc.* **2020**, *26*, 1727–1733. [CrossRef]
62. Saheb, S.H.; Durgam, V.K.; Chandrashekhar, A. A review on metal powders in additive manufacturing. *AIP Conf. Proc.* **2020**, *2281*, 020018. [CrossRef]
63. Toby, B.H.; Von Dreele, R.B. GSAS-II: The genesis of a modern open-source all purpose crystallography software package. *J. Appl. Crystallogr.* **2013**, *46*, 544–549. [CrossRef]
64. Rahman, M.; Profili, J.; Moreau, C.; Stafford, L. Formation of self-organized patterns on aluminum substrates by atmospheric pressure plasma jet for surface engineering applications. *Surf. Coat. Technol.* **2022**, *448*, 128919. [CrossRef]
65. Dinaharan, I.; Albert, T. Effect of reinforcement type on microstructural evolution and wear performance of copper matrix composites via powder metallurgy. *Mater. Today Commun.* **2023**, *34*, 105250. [CrossRef]
66. Fu, S.-Y.; Feng, X.-Q.; Lauke, B.; Mai, Y.-W. Effects of particle size, particle/matrix interface adhesion and particle loading on mechanical properties of particulate–polymer composites. *Compos. Part B Eng.* **2008**, *39*, 933–961. [CrossRef]
67. Ozcelik, B.; Ozbay, A.; Demirbas, E. Influence of injection parameters and mold materials on mechanical properties of ABS in plastic injection molding. *Int. Commun. Heat Mass Transf.* **2010**, *37*, 1359–1365. [CrossRef]
68. Kinloch, A.J.; Young, R.J. *Fracture Behaviour of Polymers*; Springer: Dordrecht, The Netherlands, 1995; ISBN 978-94-017-1596-6. [CrossRef]
69. Braga, N.F.; Passador, F.R.; Saito, E.; Cristovan, F.H. Effect of Graphite Content on The Mechanical Properties of Acrylonitrile-Butadiene-Styrene (ABS). *Macromol. Symp.* **2019**, *383*, 1800018. [CrossRef]
70. Ou, Y.; Yang, F.; Yu, Z.-Z. A new conception on the toughness of nylon 6/silica nanocomposite prepared viain situ polymerization. *J. Polym. Sci. Part B Polym. Phys.* **1998**, *36*, 789–795. [CrossRef]
71. Hu, G.; Cao, Z.; Hopkins, M.; Lyons, J.G.; Brennan-Fournet, M.; Devine, D.M. Nanofillers can be used to enhance the thermal conductivity of commercially available SLA resins. *Procedia Manuf.* **2019**, *38*, 1236–1243. [CrossRef]
72. Venkatesh, C.; Chen, Y.; Cao, Z.; Brennan, S.; Major, I.; Lyons, J.G.; Devine, D.M. Influence of extrusion screw speed on the properties of halloysite nanotube impregnated polylactic acid nanocomposites. *J. Polym. Eng.* **2021**, *41*, 499–508. [CrossRef]
73. Roussi, A.T.; Vouvoudi, E.C.; Achilias, D.S. Pyrolytic degradation kinetics of HIPS, ABS, PC and their blends with PP and PVC. *Thermochim. Acta* **2020**, *690*, 178705. [CrossRef]
74. Kumar, P.; Singh, J.; Kumari, N.; Jurail, S.S.; Verma, D.; Maurya, A.K. Study of mechanical and thermal behavior of alkali modified groundnut shell powder reinforced ABS composites. *Polym. Compos.* **2022**, *43*, 4569–4587. [CrossRef]
75. Yu, T.; Ren, J.; Li, S.; Yuan, H.; Li, Y. Effect of fiber surface-treatments on the properties of poly(lactic acid)/ramie composites. *Compos. Part A Appl. Sci. Manuf.* **2010**, *41*, 499–505. [CrossRef]
76. Ervina, J.; Mariatti, M.; Hamdan, S. Mechanical, electrical and thermal properties of multi-walled carbon nanotubes/epoxy composites: Effect of post-processing techniques and filler loading. *Polym. Bull.* **2017**, *74*, 2513–2533. [CrossRef]
77. Greenhalgh, E.S. *Failure Analysis and Fractography of Polymer Composites*, 1st ed.; Woodhead Publishing Ltd.: Cambridge, UK, 2009; ISBN 9781845692179.

78. Torrado, A.R.; Shemelya, C.M.; English, J.D.; Lin, Y.; Wicker, R.B.; Roberson, D.A. Characterizing the effect of additives to ABS on the mechanical property anisotropy of specimens fabricated by material extrusion 3D printing. *Addit. Manuf.* **2015**, *6*, 16–29. [CrossRef]
79. Zheng, Y.; Shen, Z.; Cai, C.; Ma, S.; Xing, Y. Influence of nonmetals recycled from waste printed circuit boards on flexural properties and fracture behavior of polypropylene composites. *Mater. Des.* **2009**, *30*, 958–963. [CrossRef]
80. Carballeira, P.; Haupert, F. Toughening effects of titanium dioxide nanoparticles on TiO2/epoxy resin nanocomposites. *Polym. Compos.* **2009**, *31*, 1241–1246. [CrossRef]
81. Beura, S.; Chakraverty, A.P.; Thatoi, D.N.; Mohanty, U.K.; Mohapatra, M. Failure modes in GFRP composites assessed with the aid of SEM fractographs. *Mater. Today Proc.* **2021**, *41*, 172–179. [CrossRef]
82. Torrado Perez, A.R.; Roberson, D.A.; Wicker, R.B. Fracture Surface Analysis of 3D-Printed Tensile Specimens of Novel ABS-Based Materials. *J. Fail. Anal. Prev.* **2014**, *14*, 343–353. [CrossRef]
83. Ryder, M.A.; Lados, D.A.; Iannacchione, G.S.; Peterson, A.M. Fabrication and properties of novel polymer-metal composites using fused deposition modeling. *Compos. Sci. Technol.* **2018**, *158*, 43–50. [CrossRef]

designs

MDPI

Article

Development and Performance Evaluation of Fibrous Pseudoplastic Quaternary Cement Systems for Aerial Additive Manufacturing

Barrie Dams *, Paul Shepherd and Richard J. Ball

Department of Architecture & Civil Engineering, University of Bath, Bath BA2 7AY, UK; ps281@bath.ac.uk (P.S.); easrjb@bath.ac.uk (R.J.B.)
* Correspondence: bd272@bath.ac.uk

Abstract: Aerial additive manufacturing (AAM) represents a paradigm shift in using unmanned aerial vehicles (UAVs, often called 'drones') in the construction industry, using self-powered and untethered UAVs to extrude structural cementitious material. This requires miniaturisation of the deposition system. Rheological properties and known hydration times are important material parameters. Calcium aluminate cement (CAC) systems can be advantageous over purely ordinary Portland cement (OPC) binders as they promote hydration and increase early strength. A quaternary OPC/pulverised fuel ash (PFA)/CAC/calcium sulphate (CS) system was combined with polyvinyl alcohol (PVA) fibres and pseudoplastic hydrocolloids to develop a novel AAM material for miniaturised deposition. CAC hydration is affected by environmental temperature. Intending material to be extruded in situ, mixes were tested at multiple temperatures. OPC/PFA/CAC/CS mixes with PVA fibres were successfully extruded with densities of ≈1700 kg/m^3, yield stresses of 1.1–1.3 kPa and a compressive strength of 25 MPa. Pseudoplastic OPC/PFA/CAC/CS quaternary cementitious systems are demonstrated to be viable for AAM, provided mixes are modified with retarders as temperature increases. This study can significantly impact industry by demonstrating structural material which can be extruded using UAVs in challenging or elevated in situ construction, reducing safety risks.

Keywords: additive manufacturing; unmanned aerial vehicles; calcium aluminate cement; hydration; temperature; open time; rheology; PVA fibres

Citation: Dams, B.; Shepherd, P.; Ball, R.J. Development and Performance Evaluation of Fibrous Pseudoplastic Quaternary Cement Systems for Aerial Additive Manufacturing. *Designs* **2023**, 7, 137. https://doi.org/10.3390/designs7060137

Academic Editor: Obeidi Muhannad

Received: 28 October 2023
Revised: 17 November 2023
Accepted: 23 November 2023
Published: 27 November 2023

1. Introduction

The use of Additive Manufacturing (AM) methods in the construction industry is increasing [1], particularly with the use of 3D-printed concrete [2]. Construction-scale ground-based digital applications are realising structures via the layered extrusion of concrete [3,4], with controlled rheological properties [5] being of key importance while in the fresh state [6]. Layered extrusion can eliminate the need for traditional formwork [1], which can constitute up to 60% of the costs involved with building concrete structures [7]. Ground-based AM construction methods may consist of gantry frames [8–11], large robotic arms with multiple degrees of freedom [12,13] or coordinated robotic systems with multiple mobile agents [14,15]. There are numerous research groups and projects concerning AM in construction scenarios using cement-based material [16], where robotic agents can be utilised to realise cementitious-based structures both by layered extrusion but also by alternative methods such as spraying [17] or using temporary non-traditional formwork methods [18,19].

A key limitation of ground-based approaches, whether gantry frame or multiple-agent, is the building envelope being restricted by the dimensions and reach of the deposition equipment, particularly when considering height. Releasing AM from ground-based constraints allows the freedom to operate at a theoretically great height [20]. This freedom can be applied to construct a building taller than that capable of being constructed by a typical ground-based method, or the repair of tall structures—an inherently dangerous task in an industry with high numbers of falls-from-height-related serious accidents and fatalities [21].

The aerial additive manufacturing (AAM) project has demonstrated that multiple, coordinated, self-powered, non-tethered flying unmanned aerial vehicles (UAVs), each carrying and powering a small lightweight deposition device, can extrude cementitious pastes and mortars whilst following a pre-programmed, architecturally informed trajectory [22]. This represents a paradigm shift in the use of UAVs in the construction industry towards the potential for the construction of buildings in situ or elevated repair work to existing structures using multiple coordinated aerial agents. Previously, the use of UAVs in the construction sector was limited to surveillance [23] and data-gathering [24] applications.

Crucial material considerations for AM in construction are the choice of additives and rheology properties of materials while in the fresh state [25], particularly yield stress and recovery after shearing force has been applied [26]. The hydration or curing mechanism of a material is key [25], as, once deposition has occurred, it is desired for fresh material to harden as quickly as possible. Open time can be defined as the period of time following mixing in which fresh properties remain consistent [8] and the mix workable. Open time is a particularly important parameter for AAM since the mix has to remain fresh for a sufficient time period to allow for UAV loading, flight and deposition, with rapid hardening of material then being desired to mitigate post-extrusion deformation [27]. An in situ aerial deposition approach differs significantly from ground-based or ground-powered AM applications, requiring lower-density cementitious mixes and miniaturised, lighter deposition devices appropriate for carriage by self-powered UAVs [22,27].

Binders based upon ordinary Portland cement (OPC) alone possess an open time of $\approx$2 h [28] and hydration time of 12+ h, which can be considered too long for AM [29]. Ternary binder systems can achieve faster setting than a binder consisting solely of OPC [30].

Calcium aluminate cement (CAC) is an option for reducing open times and promoting hydration, but presents two challenges. Firstly, for AM applications, material must be able to flow through a miniature deposition system prior to accelerated exothermic CAC hydration reactions with a rapid rate of heat evolution [31,32] taking place following deposition. RMAs can be used to influence the rheology of fresh material mixes [33] and pseudoplastic RMAs such as hydroxyethyl methyl cellulose (HEMC) can possess retardation properties [34]. However, combining CAC with OPC promotes early ettringite formation and a flash-set [35]. Hydration times are directly affected by variations in quantities of OPC and CAC [36] and environmental temperature [37], with high temperatures promoting rapid hydration [38]. Therefore, acceleration (at low environmental temperatures) and retardation (at high temperatures) may need to be considered in OPC/PFA/CAC/CS binder-based mix formulations.

The second consideration is the status and reputation of CAC systems due to the historically misunderstood conversion reaction [39,40] leading to the potential reduction in strength to a stable long-term level when material experiences temperatures higher than those experienced during formation [32]. In CAC at low and intermediate temperatures, metastable hydrates CAH_{10} and C_2AH_8 form, whereas at higher temperatures stable hydrates C_3AH_6 and AH_3 (alumina gel) are formed. Material formed at low temperatures (<20 °C) possesses higher early strength, but the metastable hydrates convert to C_3AH_6 and AH_3 over time as temperature rises, leading to a relatively reduced longer term strength, which must be considered during design [38]. Conversion can take days to complete at higher temperatures of $\approx$60 °C [32], but may take years at ambient 20 °C temperatures [41,42]. Pure CAC systems were omitted from design standards for reinforced and pre-stressed structural concrete following roof and ceiling beam failures in the late 20th century, despite extensive inspections, revealing the vast minority of cases to be linked to loss of strength due to conversion. Instead, poorly detailed design or low-quality aggregates were determined as primary causes of collapse [32]. However, CAC has remained an important material for ground anchoring [43] and rapid repair work [31,36], particularly in marine and tidal environments [40], roads and airfield runways [44]. In OPC-rich blended cement systems such as OPC/CAC/CS, the smaller amount of CAC typically used does not lead to the formation of the metastable hydrates, and thus conversion would not occur [31]. It has also been shown that metastable hydrates do not form in a pulverised fuel ash (PFA)/CAC/CS ternary system [42].

This study investigates the feasibility of adding calcium aluminate cement (CAC) and calcium sulphate (CS) to OPC and PFA for AAM through the development of a novel fibrous quaternary OPC/PFA/CAC/CS binder system augmented with rheology-modifying admixtures (RMA) to promote ease of extrusion and rapid curing following extrusion. CAC acts as an accelerator [45], promoting early strength [40] and hardening of material, which would be an asset once material has passed through the deposition system to prevent deformation prior to hardening [46] due to either self-weight, trajectory imperfection or subsequently deposited layers. Fibres were required to add ductility to cured material and mitigate crack propagation, and pseudoplastic properties were desired to promote liquid-like properties in fresh material mixes while under stress within the miniature deposition system and solid-like properties with a significant increase in viscosity once deposited. PFA, consisting of spherical particles, is known to enhance workability (flow of fresh material) [47]. When added to OPC/CAC, CS anhydrite has been shown to further promote ettringite development [42] and stability [48], along with early strength [39]. In an OPC-rich quaternary OPC/PFA/CAC/CS binder system developed for AAM, the conversion reaction would not be expected to occur. Polyvinyl alcohol (PVA) fibres have been shown to be durable within the cementitious alkaline environment [49] and chopped PVA fibres have been included as constituents in formulated OPC/PFA/CAC/CS-based fresh mixes in this study. Individual 12 mm length fibres are appropriate additions in AAM mixes in low volumes [50], preventing brittle failure, resisting crack propagation and decreasing shrinkage in hydrated cementitious material. The feasibility of incorporating a fibre volume (FV) of 2%, typical for cementitious composites [51], is investigated for AAM. To suitably modify the rheology properties of fresh mixes, this study used a combination of HEMC and xanthan gum (XG), which have both been demonstrated to be effective RMA solutions for a cementitious material suitable for AAM [22].

Fresh mixes were subject to deformation, rheology and calorimetry tests, and hydrated mixes were tested for mechanical properties. Fresh mixes were extruded both autonomously using a miniature deposition device, and by hand, to analyse workability (defined in this study as the ability of a fresh material to be processed by a miniature deposition device) and buildability (defined in this study as the ability of the material to retain shape following extrusion and resist excessive deformation due to subsequently deposited layers). The potential applications of AAM with a quaternary binder system are discussed. This study presents the first time that CAC has been used in an AAM application, which is in itself a unique method of AM in construction, being the only study to use self-powered, untethered multiple UAV agents.

2. Materials and Methodology

AAM is intended for in situ applications, which, naturally, are subject to climatic and seasonal variation. With the use of CAC, it is to be expected that material mixes will perform differently in different environmental conditions and with differing water temperatures. The strategy undertaken in this study was that mixes should be adaptable to changeable climatic conditions, rather than to try and exert control over the environmental temperature or heat (or cool) the water used in order to satisfy the requirements of a singular mix.

Tests carried out involved the rheological properties of the fresh material mixes (yield stress, viscosity, complex modulus), extrudability through the AAM miniature deposition system and the power and force required to extrude, retard and deform fresh mixes, the calorimetry curves of fresh mixes and the mechanical strength of hydrated mixes. Tests were performed on five developed mixes, termed 1–5. Mix formulation was carried out in temperatures of 14.5 °C ± 2 °C with a water temperature of 15.5 °C ± 2 °C. It was decided to undertake mix formulation in moderate temperatures representing an external, in situ application scenario, keeping in mind that accelerating agents added in cooler environments will be substituted for retardation agents in warmer seasonal temperatures. Laboratory tests were conducted in internal shared facilities with the ambient temperature of the laboratory environment specified in each test subsection.

Mix 1, without CAC and CS, was based upon a material mix which had had been previously successfully extruded using a UAV in flight, carrying the material in a miniature deposition device [22]. Mixes 2–5 were OPC-rich OPC/PFA/CAC/CS quaternary binder-based mixes formulated to determine suitability for AAM; the formulations presented in this study feature a range of constituents and mixes which were presented and taken forward for further tests by trialling the fresh mixes with the miniature deposition device in order to confirm that they could be viably processed by the device while carried and powered by a flying UAV. It was intended for the formulations and constituents of mixes 2–5 to differ, in order to evaluate the effects and properties of cured mixes once they had all satisfied the base requirement of being viable for AAM deposition while fresh. Mix 1 essentially acted as a control mix, against which mixes 2–5 containing CAC and CS could be compared and evaluated.

This study used CEM I 42.5 R Portland cement (Dragon Alfa Cements Ltd., Gloucester, UK) with a particle size of 5–30 μm and bulk density of ≈1400–1500 kg/m^3, as the base binding constituent. The chemical composition of the CEM I, as determined by Rietveld quantitative phase analysis, is shown in Table 1.

The constituents and fresh densities of the five mixes are summarised in the process flowchart (Figure 1) and shown in detail by kg/m^3 in Table 2 along with specifications, suppliers, fibre volumes (FV) and sand/binder (S/B), water/binder (W/B), accelerator binder (A/B) and retarder/binder (R/B) ratios. The superplasticiser (SP, supplied by Imerys, UK) was sulphonated melamine-based, which works via electrostatic repulsion [32], which is the same mechanism as sulphonated napthalene-based superplasticisers which have been shown to promote shear thinning in slurries [52]. The 350 μm diameter (D) PVA fibres (supplied by Flint, London, UK) were 12 mm in length. The sand featured smoothed, sub-rounded particles to aid workability through a miniaturised deposition device.

Table 1. Rietveld quantitative phase analysis of the chemical composition of Dragon Alfa CEM I 42.5 R Portland cement shown as a percentage by weight.

CEM I Phase	% by wt.
Dicalcium silicate C_2S	14.6
Tricalcium silicate C_3S	71.5
Tricalcium aluminate C_3A	7.27
Tetra-calcium aluminoferrite C_4AF	4.46
Calcium sulphate phases	2.16

The EN 450 N grade type-F pulverised fuel ash (PFA) (supplied by Cemex, Bristol, UK), had a bulk density of 800–1000 kg/m^3 and particle size < 45 μm. The smooth-particle sand (supplied by British Playsand, UK, product number 365/0574), was kiln dried at a temperature of 105 °C prior to use for a period of twenty-four hours and possessed a dry bulk density of 1450 kg/m^3. The Ternal SE CAC (supplied by Imerys, UK Division) has a bulk density of 1100–1300 kg/m^3. The Ground Gypsum Superfine White CS (supplied by Industrial Plasters, Chippenham, UK) has a bulk density of 900–1100 kg/m^3.

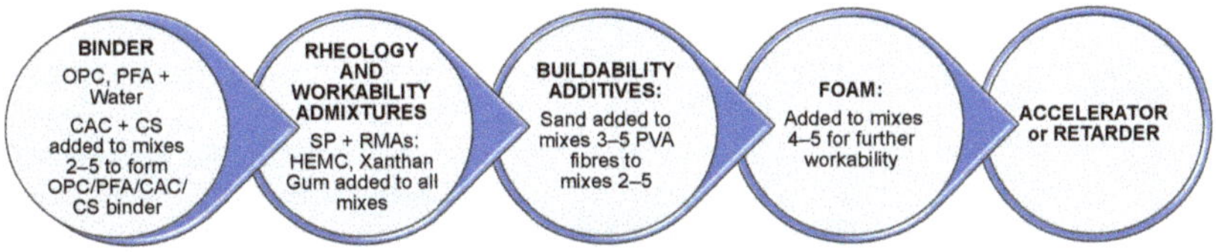

Figure 1. Mix process flowchart summarising the approach to mix design and constituents used. Please refer to the Abbreviations table for a full listing detailing acronyms and abbreviations used.

Table 2. Mixes 1–5 showing constituents in kg/m^3, and specification along with PVA fibre volumes (FVs) and mix ratios. Please refer to the Abbreviations table for a full listing detailing acronyms and abbreviations used. Mixes as shown were developed in a temperature of 14.5 °C ± 2 °C with a water temperature of 15.5 °C ± 2 °C.

kg/m^3 by wt.	Mix 1	Mix 2	Mix 3	Mix 4	Mix 5	Specification
Sand	-	-	287	326	298	<2 mm particles
OPC	770	553	460	391	358	CEM1 42.5R
PFA	415	298	248	211	193	Type 'F' EN-450
CAC	-	229	190	162	134	Ternal SE
CS	-	76.4	63.5	54.0	63.6	Anhydrite
SP	11.8	15.0	12.5	10.6	9.69	Peramin SMF10
Water	545	526	438	397	363	-
PVA fibres	-	13.9	9.70	13.2	24.7	350 μm diameter
Accelerator	-	-	10.5	6.11	5.59	Peramin AXL80
Retarder	-	1.38	-	-	-	citric, tartaric
HEMC	5.12	5.06	5.07	4.70	4.38	Walocel MKX6000
XG	10.9	8.99	9.01	10.1	8.76	-
Foam	-	-	-	49.5	76.3	Foaming agent
Total	1757	1726	1733	1634	1538	-
S/B	-	-	0.30	0.40	0.40	-
W/B	0.47	0.47	0.47	0.50	0.50	-
FV (%)	-	1.00	0.70	1.00	2.00	-
A/B (%)	-	-	1.10	0.75	0.75	-
R/B (%)	-	0.12	-	-	-	-

Mix *1* had an open time of ≈2 h, whilst mixes *2–5* had open times of 20–25 min in temperatures of 14.5 °C. It was anticipated that with increased rates of hydration in warmer temperatures above 15 °C, the accelerator quantities shown in Table 2 may need to be reduced, or even removed and substituted by suitable quantities of retarding agent. The accelerating agent used was Peramin AXL80 (supplied by Imerys, Dorset, UK) and the retarders were citric acid and tartaric acid (supplied by Sigma-Aldrich, Dorset, UK). HEMC MKX6000 was supplied by Dow Chemicals, UK Division, and xanthan gum by Minerals-Water, Purfleet, UK.

With the exception of water and foaming agent (supplied by EAB Associates, Manchester, UK), all mix constituents were of a dry, powdered consistency. Mixes were created by hand-mixing dry constituents and then adding water. Automated planetary motion mixing took place for three thirty-second periods, interspersed with manual gathering of constituents. If used, foam was then added to the slurry and integrated with two periods of thirty-second planetary motion mixing interspersed with manual gathering and mixing. Each fresh mix was subjected to a ten-second period of automated vibration to create a slurry ready for loading into the deposition device.

The AAM project miniature deposition device [22] with a 12 V DC motor suitable for UAV carriage is shown in Figure 2. Material is pushed through the cartridge by a plunger attached to a threaded rod. The device component, which determined mix formulation to the greatest extent, was the 560 mm length of flexible plastic tubing connecting the tip of the material cartridge to the 8 mm diameter nozzle. In Figure 2, the nozzle is depicted as being held and directed by a four degrees of freedom (DOF) Dobot magician robotic arm (Dobot, Shenzhen, China), representing the role performed by a stabilising delta arm robot, which attaches to the base of a UAV to guide, and provides lateral stability to the nozzle [22]. During the extrusion experiments carried out in this study, nozzle movement was also controlled by hand.

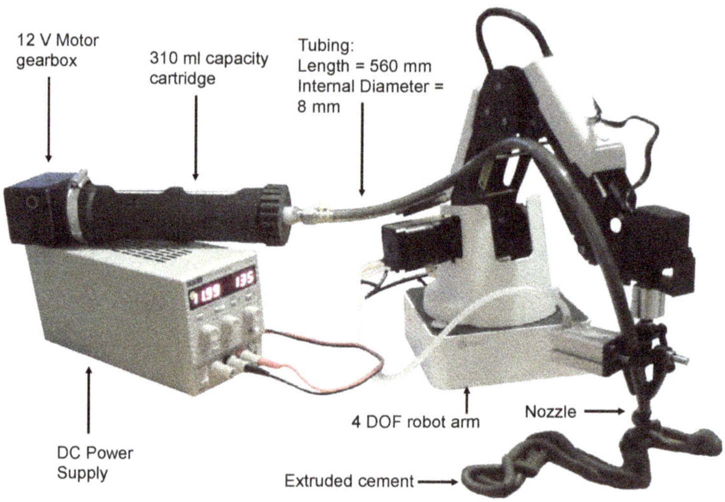

Figure 2. 310 mL capacity miniature deposition device [22] with length of flexible extrusion tubing held by a robotic arm.

2.1. Fresh Mix Extrusion

The five mixes were all tested for AAM suitability using a 310 mL cartridge capacity deposition device, shown in Figure 2, with mixes manufactured in environmental temperatures of 14.5 °C ± 2 °C with a potable water temperature of 15.5 °C ± 2 °C. It takes approximately 4 min for the deposition device to extrude a cartridge of material. Therefore, the aim of autonomous deposition is to commence extrusion 15–20 min following mixing (to allow for material loading into the cartridge and cartridge loading into the deposition device attached to the UAV) and complete extrusion at 20–25 min following mixing, at which point rapid hardening in the deposited material is desirable.

Deposition device tests extruded material in a trajectory consisting of alternate layers of three parallel lines and a peano curve layer, which has been demonstrated previously in Zhang et al., 2022 [22], in order to observe how extruded material may deform both in compression and sagging. The alternating layer trajectory approach was chosen for two reasons. Firstly, it had proved to have an effective compressive strength to amount of material used ratio when compared to alternative wall designs, including printing immediately adjacent layers to form a solid wall. Secondly, it was developed with the intention of extrusion by an untethered, self-powered autonomous flying UAV, which requires a stabilising delta robot to steady the extrusion nozzle and mitigate against enforced lateral deviation from programmed trajectories due to air movements and propeller thrust. Current UAV lateral deviations from programmed trajectories are within 1 mm, and this will improve with continuing development, but the alternating layer approach mitigates against the risk of an imprecise deposition of up to a millimetre of a layer on top of a previous layer of identical trajectory. Additional hand extrusions were also performed to examine other potential AAM applications of the material mixes, including printing directly on to a vertical surface and printing on support material, which was subsequently removed, to create a bespoke structure. A further area of AAM development is the ability for a UAV to land at elevation onto a vertical surface and extrude material, with the material in this study being demonstrated to be able to adhere to an inclined or vertical plane and be suitable for such potential applications of precision construction repair work in dangerous or elevated locations.

2.2. Power and Force Requirements

The force requirement tests for fresh mixes were conducted in the internal laboratory ambient temperature conditions of 20 °C ± 2 °C, with a potable water temperature of 17.5 °C ± 1 °C. Cartridges full of freshly mixed material were placed vertically upon an Instron Universal 2630-120/305632 device (Instron, High Wycombe, UK) and a rod was pushed down upon a plunger at a rate of 17 mm/min to simulate the movement of the deposition device plunger. The force required to process the material at this rate of displacement was analysed. While the deposition device processed the mixes, the current required to push the material through the tubing as indicated by the 12 V power supply was also recorded.

2.3. Fresh Mix Rheological Properties

To quantify the rheological properties of the fresh mixes (and how these are affected by temperature), flow and oscillation tests were conducted using upper and lower plate temperatures of 7 °C, 14.5 °C and 22 °C. A TA DHR-2 rheometer (TA Instruments, Hertfordshire, UK) was used to conduct the tests with a 25 mm diameter flat upper geometry plate and flat 40 mm diameter lower plate. Ambient internal laboratory temperatures were 20 °C ± 1 °C and laboratory potable water temperature was 17.5 °C ± 1 °C.

Flow tests were shear stress-controlled using linear stress ramps ranging from 300–6000 Pa and 900–50,000 Pa. Tests were designed to obtain yield stress and viscosity, which quantify flow and flow velocity, respectively [53]. Tests were repeated at regular time intervals during the open time of the fresh mixes, during which material was exposed to environmental temperatures prior to testing.

Oscillation tests were conducted over a period of 2000 s (covering the open time of all mixes) and were controlled using an angular displacement of 5×10^{-5} radians with frequency maintained at 1 Hz. Using the elastic modulus G', viscous modulus G'' and phase angle δ data, the complex modulus G^*, quantifying the stiffness of the fresh material, can be calculated as:

$$G^* = \frac{G'}{cos\delta} \tag{1}$$

where δ (radians) is:

$$\delta = \frac{G''}{G'} \tag{2}$$

2.4. Fresh Mix Calorimetry

All mixes were placed in a Calmetrix I-Cal 4000 isothermal calorimeter (Calmetrix, Boston, MA, USA) immediately following mixing and vibration. Each mix was tested three times, with the chamber temperature maintained at 7 °C, 14.5 °C and 22 °C, respectively. Tests were carried out over a period of forty-eight hours to analyse how the energy transferred and rate of heat evolution differed with constituent and temperature change during the hydration process. Ambient laboratory temperatures were 20 °C ± 1 °C and laboratory potable water temperature was 17.5 °C ± 1 °C.

2.5. Fresh Mix Deformation

Freshly extruded beads of material mixes were subjected to deformation on an Instron Universal 2630-120/305632 device. Ambient laboratory temperatures were 20 °C ± 2 °C and laboratory potable water temperature was 17.5 °C ± 1 °C. Freshly extruded beads of material 8 mm in diameter and 80 mm in length were placed between two steel plates and compressed at a rate of 2 mm/min to simulate an extruded layer of material being subjected to compressive loading from subsequent layers deposited on top. Extruded beads were tested at five minute intervals over the open time period of the material.

2.6. Hydrated Mix Mechanical Properties and Method of Flexural Failure

Fresh mixes were poured into 160 mm × 40 mm × 40 mm moulds in accordance with British Standard BS EN 1015-11:1999 [54] and mechanically vibrated. The density was calculated for each prism prior to testing. Flexural and compressive strength tests were carried out on the hydrated prisms using an Instron Universal 2630-120/305632 device to obtain an indication of 1-day and 28-day strength values. Ambient laboratory temperatures were 20 °C ± 2 °C and laboratory potable water temperature was 17.5 °C ± 1 °C. Four-point bending flexural tests were in accordance with British Standard BS EN 12390-5:2009 [55]. Compression tests were subsequently conducted in line with British Standard BS EN 1015-11:1999 [54].

2.7. Retardation and Further Rheological Tests

To further investigate the differing open times of the material arising through changes of ambient temperature, mix 2—retarder/binder (R/B) ratio 0.12%—was modified with the addition of retardation agents with R/B quantities of 0.52%, 0.78%, 1.04% and 1.56%. Both citric acid and tartaric acid were investigated as retarders independently. Citric acid is established as a commercial retarding agent for CAC systems, working by the mechanism of precipitation of gel-coatings around cement grains [43], thus inhibiting the formation of ettringite [30]. Tartaric acid possesses an affinity for aluminate surfaces and is also known to inhibit ettringite formation [56].

Further oscillation and flow rheology tests were undertaken. Ambient laboratory temperatures were 20 °C ± 2 °C and laboratory potable water temperatures were 17.5 °C ± 1 °C. Mix 2, as shown in Table 2, was modified with additional quantities of either citric or tartaric acid, and tested on the rheometer for the duration of the material open time. Flow tests were conducted at five-minute intervals following mixing, with the fresh mix subject to ambient environmental air. Oscillation tests were conducted with the retarded fresh material kept within the plates of the rheometer for the test duration.

3. Results

In the results section the following colour coding has been assigned to the mixes: blue (1), green (2), yellow (3), magenta (4) and grey-black (5).

3.1. Fresh Mix Extrusion

All five mixes were extruded by the 310 mL capacity deposition device when the mixes were manufactured in environmental temperatures of 14.5 °C ± 2 °C with a potable water temperature of 15.5 °C ± 2 °C. Figure 3 shows a range of extrusions both by hand and deposition device.

Printed by hand, Figure 3a shows a latticed dome structure consisting of mixes 2 and 3 printed upon supporting material, which was subsequently removed to prove the structure to be self-supporting (Figure 3b). A U-shaped 8-layered extrusion featuring mixes 1, 2 and 4 was deposited on a vertical surface (Figure 3c,d). Three layers of fresh material can be printed vertically in immediate succession during the open time of the material. The fourth and subsequent layers must take place after the material has hardened—Figure 3d shows three fresh layers extruded upon hardened previous layers.

Using the deposition device, Figure 3e shows alternating layers of three straight parallel lines and a peano curve shape using (top to bottom) mixes 2, 5 and 3. The straight line middle layer of the less dense, higher foam content mix 5 shows the most sagging, while mixes 2 and 3 exhibit minimal deformation.

Based on autonomous deposition success at lower temperatures, the open time of mix 1 is considered to be 2 h, while the open times of mixes 2–5 (as formulated in Table 2) are considered to be a working maximum of 25 min. Realistically, the quickest time that a deposition device can commence extrusion of material following mixing and vibration is 15 min and, at ≥14.5 °C, mixes 2–5 can become too stiff for extrusion within this time.

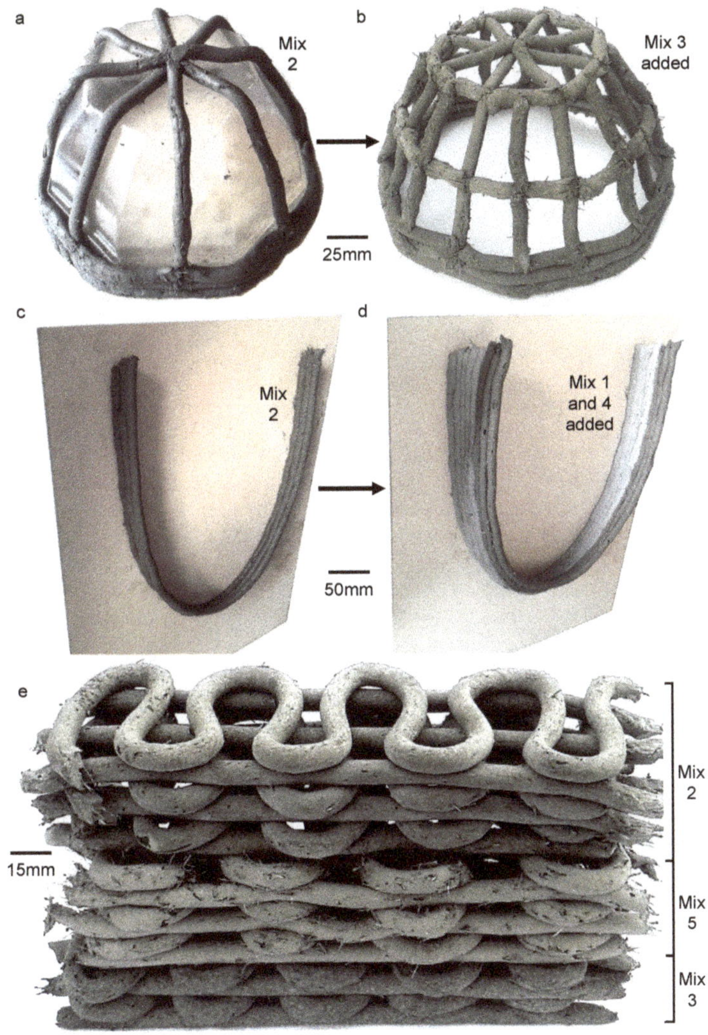

Figure 3. Hand and deposition device extrusions. (**a**,**b**) Hand-printed dome extrusion upon supporting material, subsequently removed, using mixes 2 and 3. (**c**,**d**) Extrusion by hand on to a vertical surface in a U-shaped design, using mixes 1, 2 and 4. (**e**) Alternate straight parallel lines and 'ruffle' layer design deposited by the deposition device using mixes 2 (top), 5 (middle) and 3 (bottom).

3.2. Power and Force Requirements

With a fixed 12 V supply, the deposition device was capable of pushing material through the plastic tubing drawing current within the range of 180 mA (where material enters the tubing) to 420 mA (material extrusion). More current is not viable for the deposition device, or for UAV carriage, and would indicate that the material lacks sufficient workability for AAM. Conversely, if less current were required, the material would not possess adequate buildability upon extrusion. Table 3 shows how power requirements increase as material is pushed through the tube and ultimately extruded on to a free surface for mixes 1–5, along with an empty cartridge for comparison. It can be seen that 5 W of power may be considered as an upper limit. Mixes were manufactured at temperatures below 14.5 °C and deposition device loading and full extrusion was completed within 25 min of material mixing and vibration. The OPC/PFA/CAC/CS mixes 2–5 required more power to process than the pure OPC-based mix 1.

Table 3. Power required for the device plunger to push mixes 1–5 (and an empty cartridge for comparison).

Fresh Mix	Tube-Start Power (W)	Tube-Middle Power (W)	Start of Extrusion Power (W)	End of Extrusion Power (W)
Empty	1.68	1.68	1.68	1.68
1	2.40	2.76	2.88	2.88
2	2.76	3.60	4.20	4.80
3	2.76	3.48	4.68	4.68
4	2.28	3.00	4.20	4.20
5	2.16	2.76	3.48	3.84

Force requirement tests revealed that an OPC/PFA/CAC/CS quaternary system can be problematic for a thermoplastic cartridge. During extrusion with the deposition device, warping of the threaded rod attached to the plunger had begun to take place. It was discovered that the exothermic hydration reactions create sufficient heat to cause the 310 mL capacity plastic cartridges containing the fresh, hydrating material to soften and become malleable to the extent that the threaded rod attached to the plunger, in addition to pushing down the cartridge, began to rotate about its longitudinal axis and push the cartridge radially outwards during rotation, thus gradually warping the threaded rod. The effects of the exothermic hydration of mixes 2–5 while in the cartridge can be seen in Figure 4, with the downward motion of the plunger forcing material outwards (a) and cartridge lateral expansion evident (b).

As a result, the only entirely successful test with this method was with mix 1 (load vs. extension profile with extrusion plateau shown in Figure 4c), which did not contain CAC or CS in the binder and required between 700 N–800 N of force to push the material through the tubing at 17 mm/min.

3.3. Fresh Mix Rheological Properties

Oscillation test results are presented in Figures 5–7. Figure 5 shows the complex modulus G^* and phase angle δ for mixes 1–5 at three different temperatures (7 °C, 14.5 °C and 22 °C) over the period of 15–25 min following the completion of mixing and vibration. Figures 6 and 7 show the complex modulus G^* and phase angle δ profiles, respectively, over a period of 2000 s following mixing to see how the mixes differ with temperature over the desired open time period.

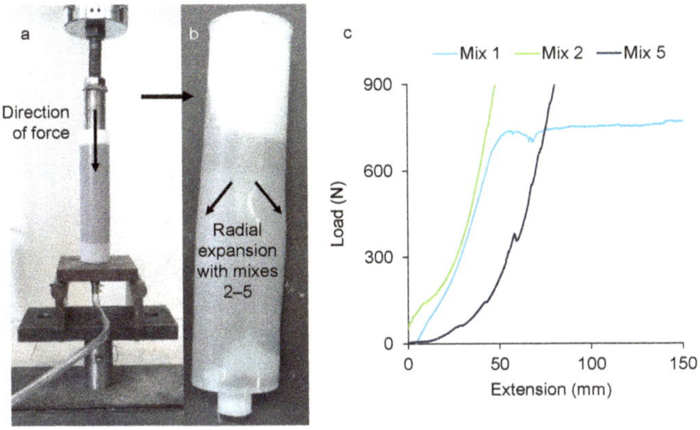

Figure 4. Force required to extrude material: (**a**) A restrained cartridge subjected to force upon the plunger. (**b**) Ensuing radial expansion of the cartridge due to exothermic hydration of mixes 2–5. (**c**) Force required curve showing the plateau of the successfully extruded mix 1 (without CAC and CS) and examples of the CAC mixes 2 and 5 which caused radial cartridge expansion and failed to extrude material.

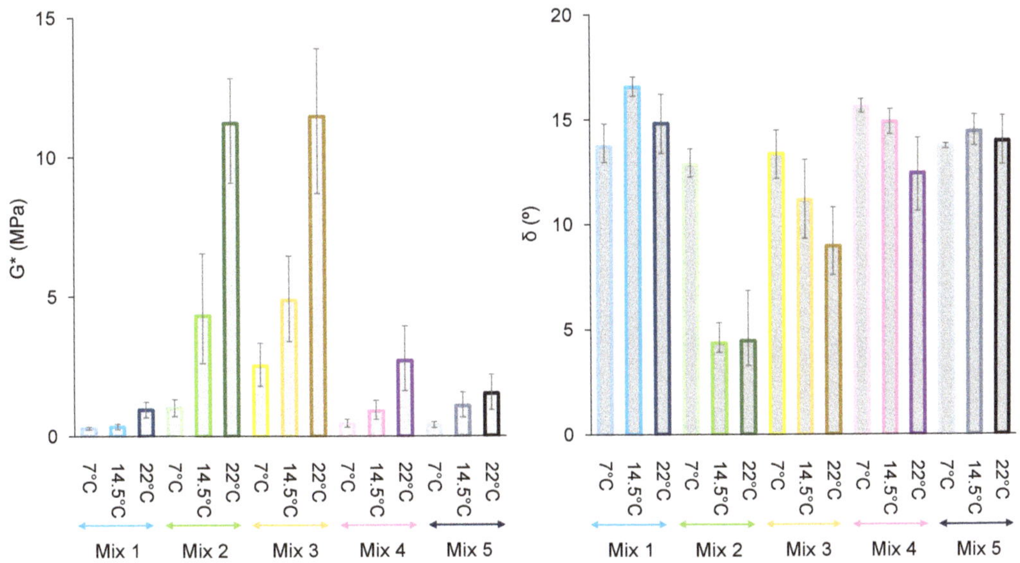

Figure 5. Oscillation complex modulus G^* and phase angle δ test results for mixes 1–5 over the period of 15–25 min following mixing. Tests were conducted at 7 °C, 14.5 °C and 22 °C. Error bars denote maximum and minimum values recorded within the time frame. For the three temperatures presented, mix colours are further defined from light-dark, for example mix 3 is lighter yellow (7 °C), medium yellow (14.5 °C) and darker yellow (22 °C).

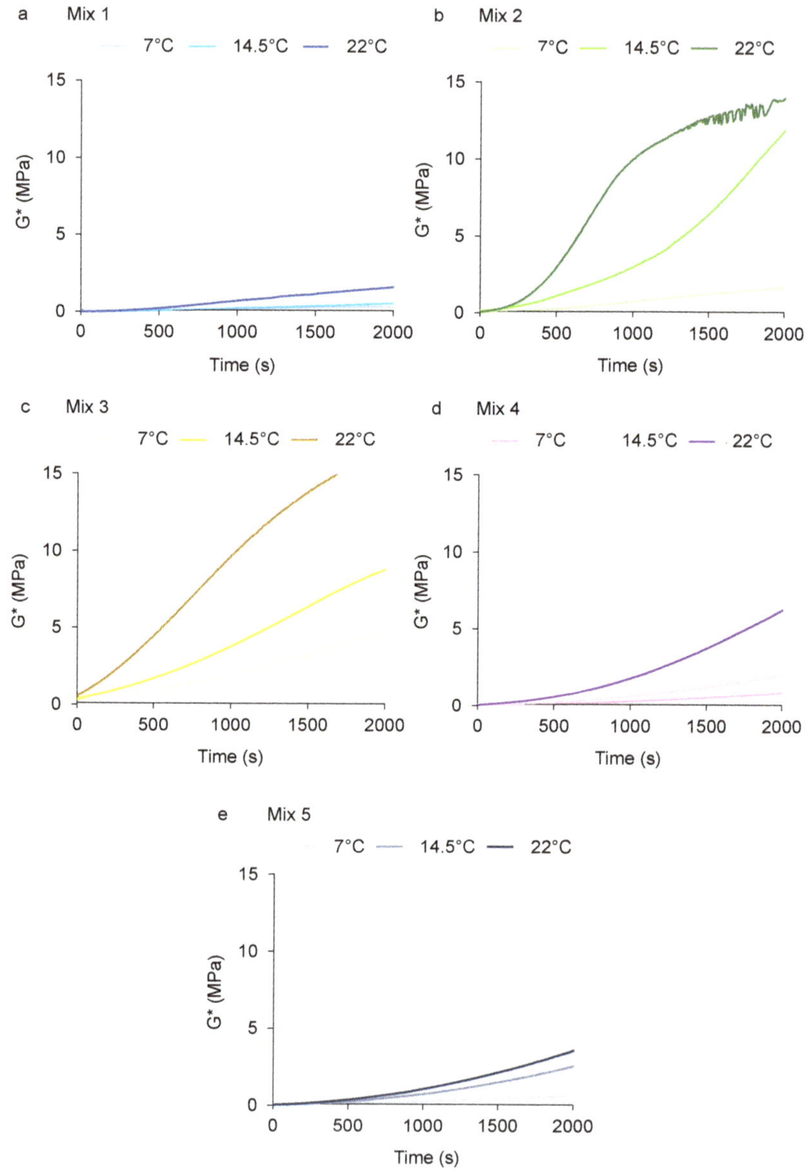

Figure 6. Oscillation complex modulus G^* profiles for mixes 1–5 over the period of 2000 s following mixing. Tests were conducted at 7 °C, 14.5 °C and 22 °C. For the three temperatures presented, mix colours are further defined from light-dark, for example mix 3 is lighter yellow (7 °C), medium yellow (14.5 °C) and darker yellow (22 °C).

As temperature rises, G^* increases and δ decreases over the open time period. Mixes 2 and 3 are the most stiff and solid-like, with the higher foam content mix 5 being rheologically closest to the purely OPC-based mix 1 and the only OPC/PFA/CAC/CS mix rheologically suitable at 22 °C. Results suggest that mixes 2 and 3 require additional retardation at 22 °C.

A G^* of ≈5 MPa is suitable for deposition device processing, with material becoming too stiff to pass through the tubing at 10 MPa or above.

Figure 8 shows how the complex modulus G^* and phase angle δ of mix 2 change with temperature over the open time period. G^* increased with time and temperature whilst δ decreased. Results show the times and temperatures during which mix 2, confined by the rheometer plates and not exposed to the environment, is viable for AAM. At 14.5 °C, mix 2 is viable for 1500 s (an ideal 25 min open time), but at 22 °C open time lasts for only 900 s, which is insufficient.

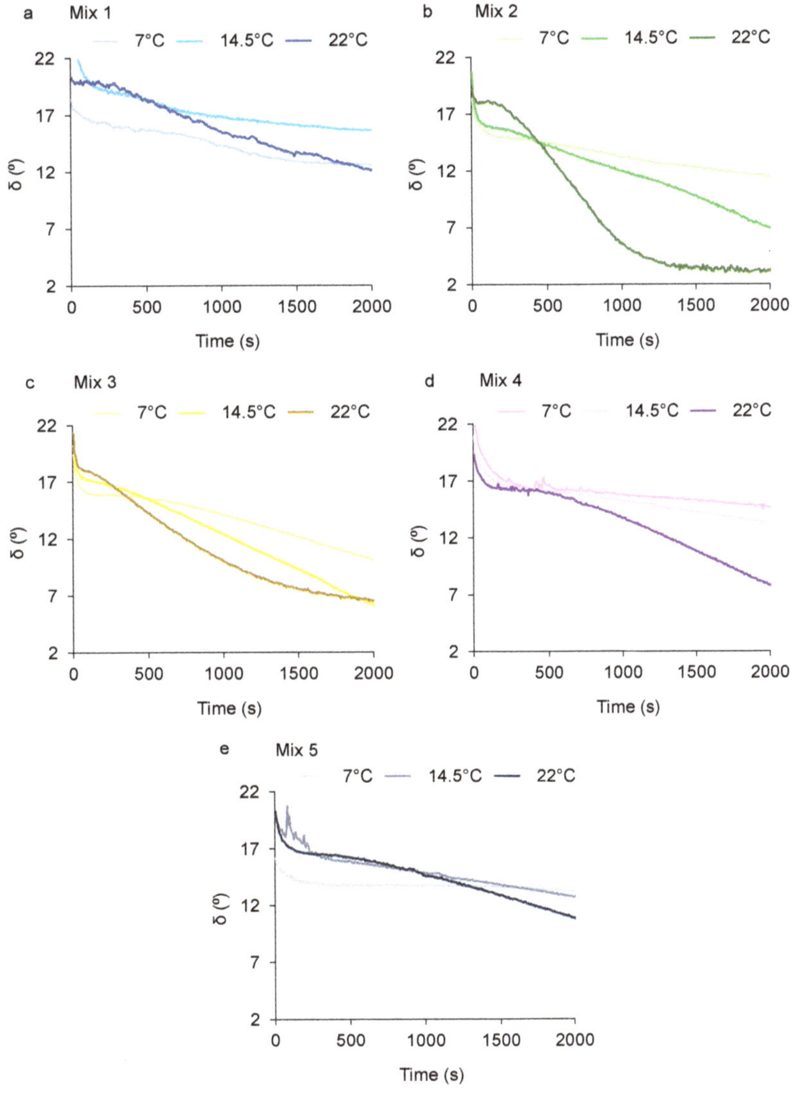

Figure 7. Phase angle δ profiles for mixes 1–5 over the period of 2000 s following mixing. Tests were conducted at 7 °C, 14.5 °C and 22 °C. For the three temperatures presented, mix colours are further defined from light-dark, for example mix 3 is lighter yellow (7 °C), medium yellow (14.5 °C) and darker yellow (22 °C).

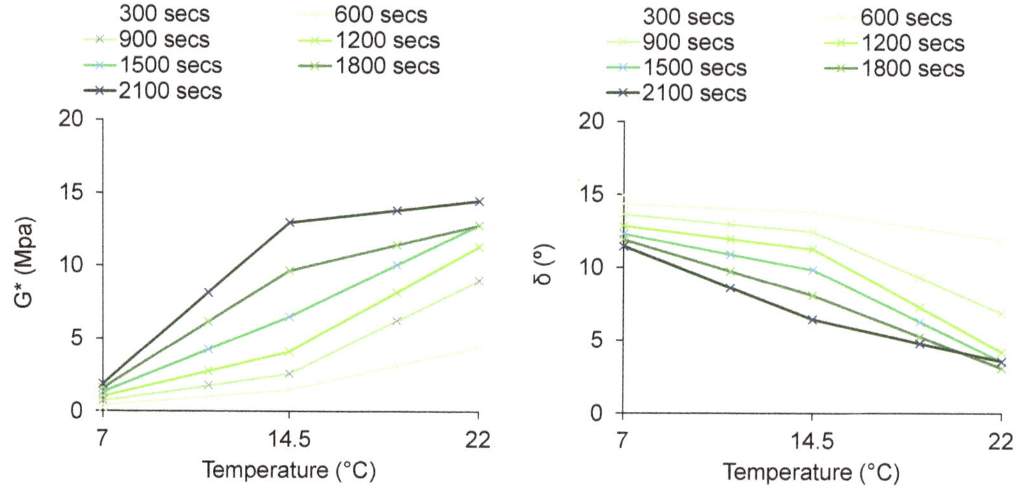

Figure 8. Mix 2 complex modulus G^* and phase angle δ changing with temperature over the open time period. Colour shading denotes time difference.

Flow test results are shown in Figure 9 (yield stress) and Figure 10 (viscosity profile) for mixes 1–5 at five-minute intervals over the open time period, with the rheometer plates kept at a constant 14.5 °C. Shear stress is plotted against viscosity to illustrate the yield stress of the fresh material, with the sudden reduction in viscosity demonstrating that the structure of the soft-solid material has been subjected to a level of stress sufficient to induce liquid-like flow and deformation. It can be seen in Figures 9f and 10f that the yield stress and viscosity profile for the same mix (in this case mix 2), tested immediately following mixing and vibration, does not vary with temperature as it does with time.

Yield stress and viscosity values increased with time for mixes 2–5 while mix 1, without CAC, did not increase within 25 min and can be autonomously deposited at all temperatures within this timescale. Rheometer plates were 14.5 °C but in between tests, material was exposed to the laboratory environment of 22 °C. It was not possible to test mixes 3, 4 and 5 for 25 min as the material, freely exposed to a 22 °C laboratory temperature, had become too stiff to test using plate rotation. Figure 9c shows that mix 3, with the highest concentrations of binder and accelerator, stiffens most rapidly. Viscosity profiles show the extent to which all mixes contain pseudoplastic properties suitable for AAM, with viscosity reducing by orders of magnitude as shear rate increases.

The viscosity profiles typically display a small 'hook' at the beginning of the tests at low shear rates. This is due to low angular velocity plate rotations resulting from initial low stresses struggling to displace material.

3.4. Fresh Mix Calorimetry

Calorimetry results are presented in Figure 11a–e (cumulative energy transferred—Ecemmat, Joules/gram) and Figure 12a–e (rate of heat evolution, Joules/gram/hour) for temperatures 7 °C, 14.5 °C and 22 °C. Mixes 2–5 all show rapid increases in energy transferred within the first couple of hours, whereas mix 1 (without CAC and CS) is very different, with increases between 16–48 h indicating the C-S-H hydration phase. All five mixes vary with temperature. Mixes 3–5 (with accelerator) show that less energy was transferred with higher temperature.

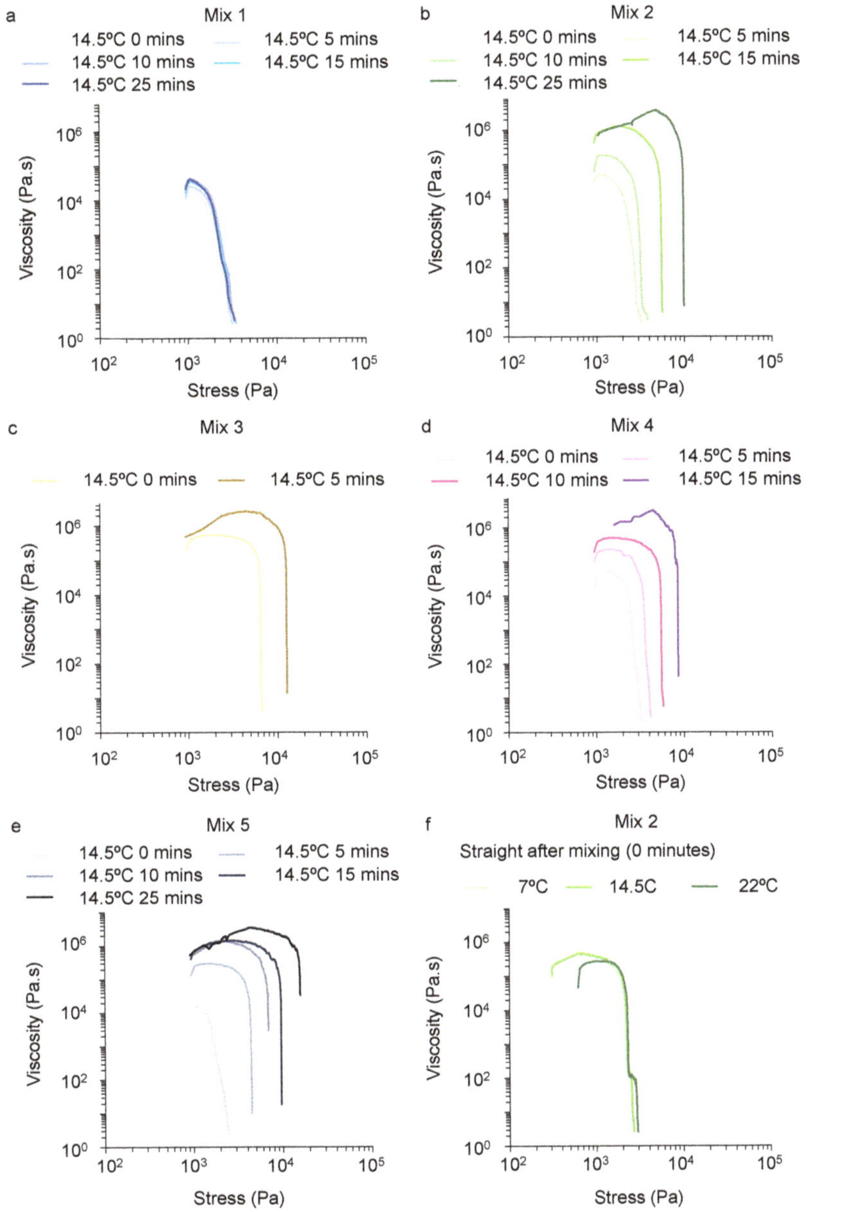

Figure 9. Open time yield stress results for mixes 1–5. (**a**–**e**) Rheometer plates kept constant at 14.5 °C. (**f**) Mix 2 tested at different temperatures following mixing. Colour shading denotes time differences.

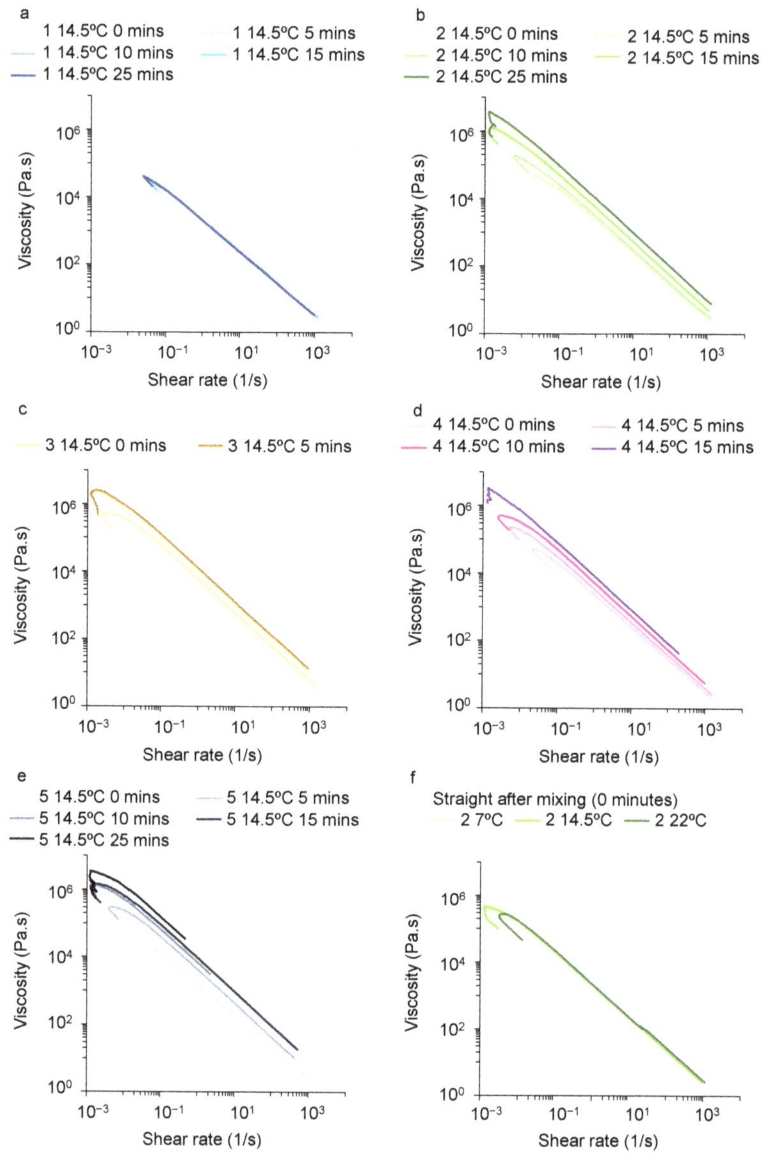

Figure 10. Open time viscosity flow profiles for mixes 1–5. (**a**–**e**) rheometer plates kept constant at 14.5 °C, colour shading denotes time differences. (**f**) Mix 2 tested at different temperatures following mixing, colour shading denotes temperature differences.

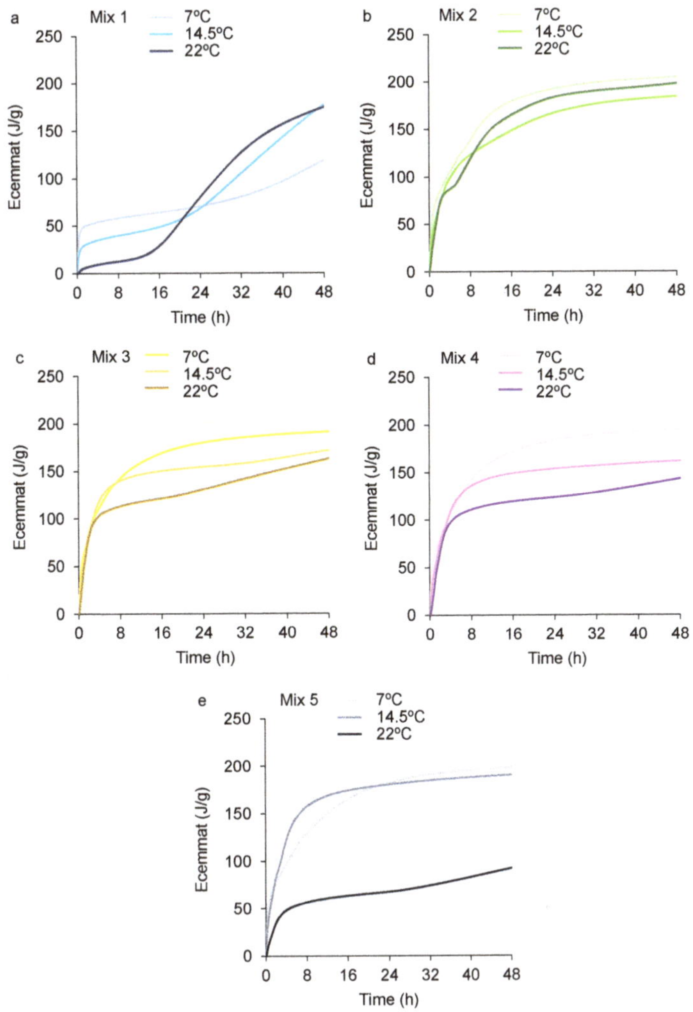

Figure 11. Calorimetry tests: cumulative energy transferred (Ecemmat) during the first 48 h for mixes 1–5 at 7 °C, 14.5 °C and 22 °C. For the three temperatures presented, mix colours are further defined from light-dark, for example mix 3 is lighter yellow (7 °C), medium yellow (14.5 °C) and darker yellow (22 °C).

Mix *1* (Figure 12a) is without CAC and CS, yet differs from a pure OPC rate of heat evolution profile, which should show a sharp peak at ≈12 h indicating the C_3S and, to a lesser extent, C_2S reactions forming C-S-H gel and $Ca(OH)_2$. For pure OPC, the sharp peak follows an initial C_3A reaction with gypsum immediately following mixing and ensuing an induction period of inhibited hydration [35]. With mix *1*, the C-S-H phase shows a broad peak between 16–32 h, and ettringite formation occurring between 32–40 h at 22 °C, as opposed to within 20 h with pure OPC. It is clear that mix *1* is retarded by the presence of HEMC and XG (relative to pure OPC paste) and also by decreasing temperatures, with the C-S-H phase much delayed at 7 °C in comparison with 22° C.

The rate of heat evolution profiles for pure CAC pastes consist of an initial peak with the formation of ettringite and CAH_{10} and AH_3 phases, followed by a dormant or 'induction' period of ≈6 h as AH_3 and ettringite particles cover the CAC particles forming a hydrate barrier, and an ensuing aluminate broad peak resulting from the decomposition of the unstable ettringite phase [35] and diffusion-controlled hydration [57]. The only indication of a pure CAC-characteristic broad monocalcium aluminate peak in Figure 12 is with mix 2, particularly at 22 °C. Mixes 3–5 (Figure 12c–e) contain sand, whereas mix 2 (Figure 12b) does not, thus possessing a relatively higher quantity of CAC and CS.

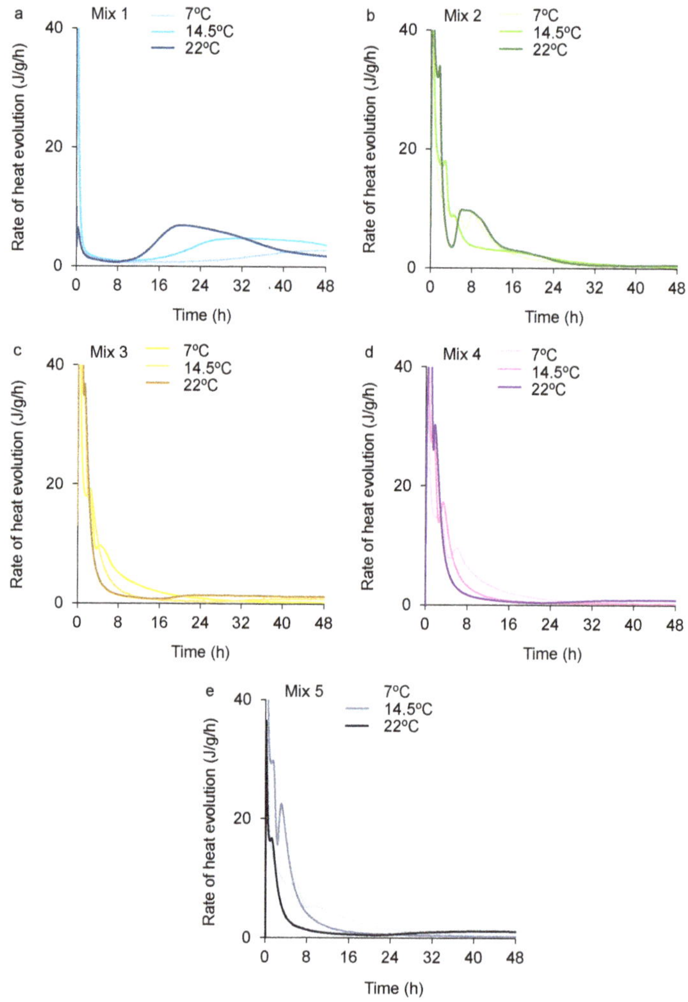

Figure 12. Calorimetry tests: rate of heat evolution during the first 48 h for mixes 1–5 at 7 °C, 14.5 °C and 22 °C. For the three temperatures presented, mix colours are further defined from light-dark, for example mix 3 is lighter yellow (7 °C), medium yellow (14.5 °C) and darker yellow (22 °C).

It is clear with mixes 3 and 4 (Figure 12c,d) that the initial reaction and formation of ettringite dominates hydration, with no further peaks after ≈8 h. The descending shoulder peaks following the initial reaction in mixes 2–5 are characteristic of the addition of CS [48]. Mix 5 (Figure 12e) has more foam and exhibits a sharp second peak at ≈4 h at 14.5 °C, which is more distinctive than those shown in mixes 2–4. Mixes 3–5 indicate the commencement of broad C-S-H peaks commencing after 16 h (mix 3, Figure 12c) and 24 h (mixes 4 and 5,

Figure 12d,e), with the peak itself occurring beyond 48 h. The addition of CAC to OPC clearly retards hydration of the C-S-H phases.

3.5. Fresh Mix Deformation

Deformation results illustrated in Figure 13 show how mixes 1–5, extruded into 8 mm diameter beads, settled when subjected to compressive loading representing subsequent layers deposited on top of the bead. The laboratory temperature was 22 °C during the tests, hence the swift hydration times, and tests were conducted for the duration of mix open times. Mix 1 is shown over a much longer time period, since its lack of CAC/CS means it has an open time of ≈2 h. It can be seen from Figure 13, particularly for mixes 2 and 3, that OPC/PFA/CAC/CS material, without retardation, hardens very quickly at 22 °C, which provides excellent buildability, but is challenging for deposition device extrusion.

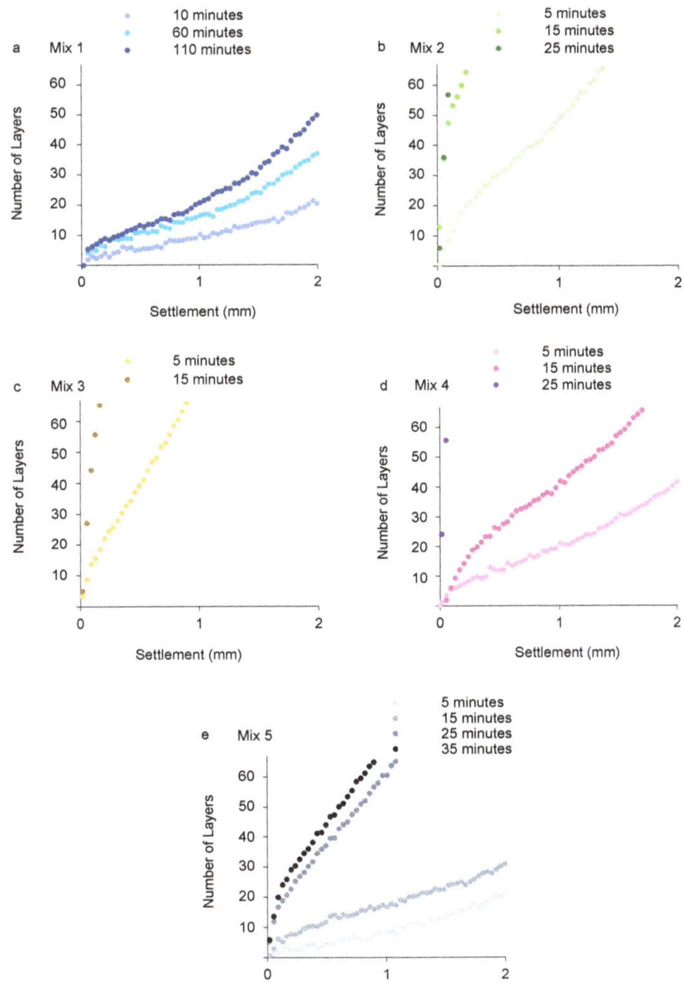

Figure 13. Deformation tests showing the extent to which mixes 1–5, in the form of extruded 8 mm diameter beads, settle when subjected to compressive loading. Tests were carried out at 22 °C. Colour shading denotes time differences.

3.6. Hydrated Mix Mechanical Properties and Method of Flexural Failure

Figure 14 shows compressive and flexural strength test results at both one day and twenty-eight days following mixing, along with the density of the hydrated specimens and the flexural stress–strain profiles at twenty-eight days. Coefficients of variation range between: Compressive strength 1-day 2–11%, Compressive strength 28-day 2–5%, Flexural strength 1-day 3–11%, Flexural strength 28-day 1–11%, Density 1-day and 28-day ≈1%, with 28-day flexural strength of the high-foam content mix 5 exhibiting the highest variation at 11.5%.

Mix 1, without CAC and CS, possesses the highest compressive strength at 28 days, but mixes 2 and 3 are also structurally viable at ≈25 MPa and possess higher 1-day strength than mix 1, though the difference is moderate. Mixes 4 and 5, containing foam, have lower density and failed to reach 20 MPa compressive strength. Mix 1, which has no fibres fails in a brittle manner whereas mixes 2–5 show the contribution PVA fibres make once the 28-day mortar matrix fails (Figure 14d). Mix 5, with higher FV, performed well in flexural tests.

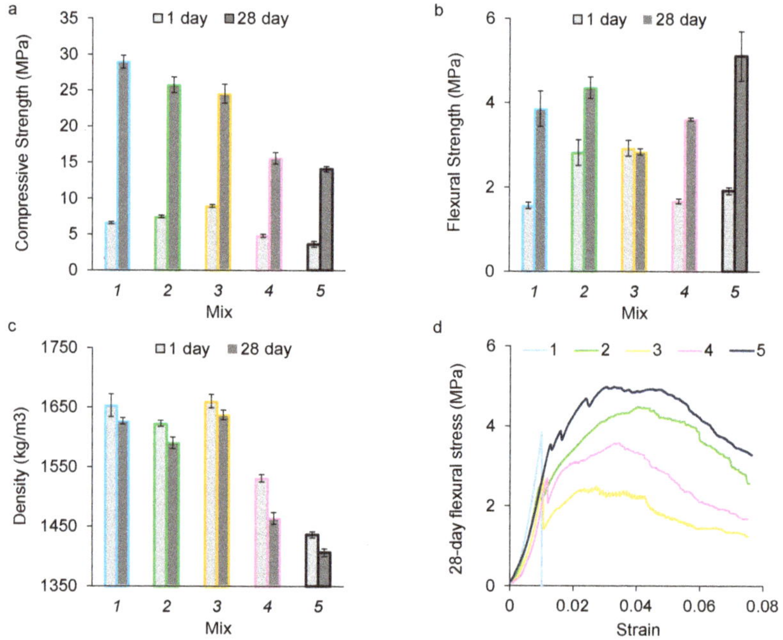

Figure 14. Mechanical tests indicating 1-day and 28-day strength for mixes *1–5*. (**a**) Compressive strength. (**b**) Flexural strength. (**c**) Density of the hydrated prisms. (**d**) Flexural failure of 28-day specimens with the legend denoting mix number. Error bars denote the standard deviation. Lighter colour shadings denote 1 day, darker shadings 28 days.

Figure 15 shows flexural failure of fibrous 28-day specimens of mixes 3 and 5. There is a significant difference made to crack propagation as FV increases, with mix 3 containing 0.75% FV deforming to a greater extent under loading than mix 5 with 2% FV. The failure mechanism of PVA fibres was pull-out, which can be seen in both mix 3 (Figure 15c) and mix 5 (Figure 15d). PVA fibres typically possess a tensile strength of 1400–1500 MPa [49,58], and therefore would not be expected to fracture in these tests, whereas a reduction in elastic modulus, typically 29,000 MPa [59], does occur.

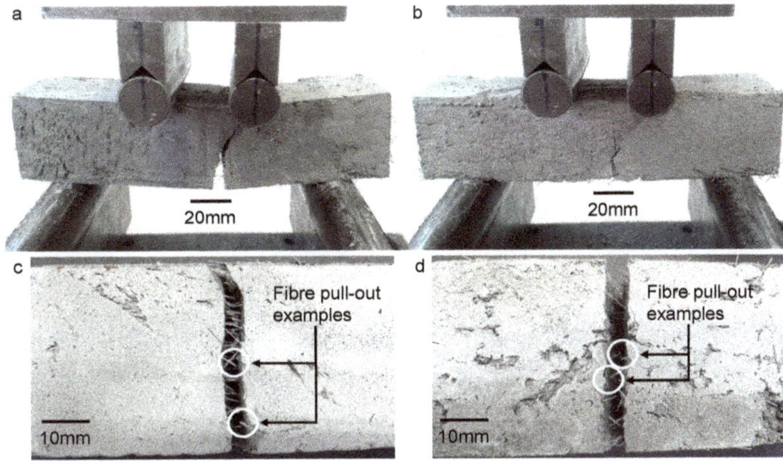

Figure 15. Flexural failure of fibrous 28-day specimens, with mixes 3 and 5 shown as fibrous example specimens. (**a**) Mix 3 with 0.75% FV. (**b**) Mix 5 with 2% FV. (**c**,**d**) PVA fibres showing pull-out for mix 3.

3.7. Retardation and Further Rheological Tests

Further rheology tests were conducted with mix 2 (0.12% R/B in Table 2) amended with additional amounts of retarder at 0.52%, 0.78%, 1.04% and 1.56% R/B. Flow and oscillation results featuring the retarder-modified mix 2 are shown in Figure 16. Oscillation tests show that tartaric acid was the more effective retarder, with mix 2, containing tartaric acid at 0.52% R/B possessing a lower complex modulus G^* and higher phase angle δ than with citric acid at 0.52% R/B.

Oscillation tests suggest that citric acid at 1.56% performs similarly to tartaric acid at 1.56% R/B. However, the flow results suggest that this is not the case. Using 1.1 kPa–1.3 kPa as a suitable range of yield stresses for deposition device extrusion, due to the performance of mix 1 being previously demonstrated as suitable for a UAV in flight [22], and 25 min as a desirable open time for AAM, the results in Figure 16f show that with 1.56% R/B citric acid, material at 10 min remains within this yield stress range but at 25 min is well beyond, therefore 1.56% R/B citric acid is an insufficient quantity and does not retard the fresh mix satisfactorily. Conversely, Figure 16e shows 1.04% R/B tartaric acid as an example of too much retarder being added, with the material still being effectively retarded at 40 min, which is beyond the desired open time for AAM.

Figure 16c,d suggest that tartaric acid at 0.78% R/B is a suitable quantity of retarding agent. Tartaric acid behaves unpredictably, with yield stresses showing variance and not necessarily increasing with time in an entirely uniform manner, but 0.78% R/B tartaric acid shows a material still being retarded within 25 min but increasing beyond 1.3 kPa at 40 min.

The results section concludes with Table 4, which summarises and gives an indication of the properties of an OPC/PFA/CAC/CS cementitious binder-based pseudoplastic material suitable for AAM extrusion.

Table 4. Indication of select material properties for an OPC/PFA/CAC/CS cementitious binder-based pseudoplastic material suitable for AAM extrusion. 28 d = 28 days. G^* = Complex Modulus. δ = Phase Angle.

Compressive Strength (28 d)	Flexural Strength (28 d)	Yield Stress (Fresh Mix)	G^* 2000 s at 14.5 °C (Fresh Mix)	δ 2000 s at 14.5 °C (Fresh Mix)
25 MPa	4 MPa	1.1–1.4 KPa	2–5 MPa	12–15°

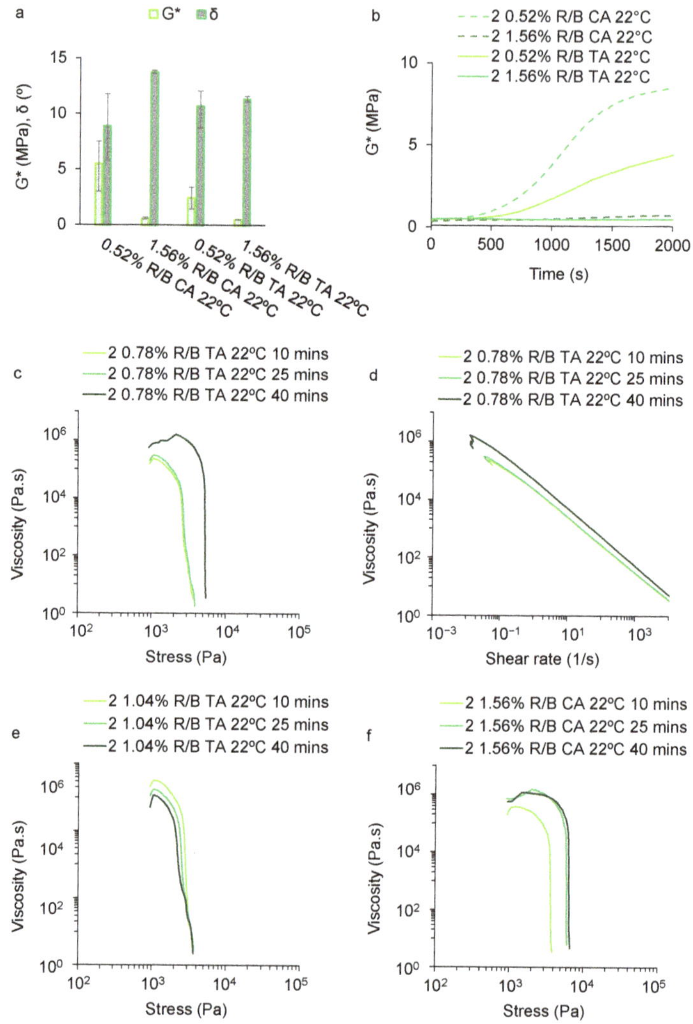

Figure 16. Rheology tests for the retarder-modified mix 2 tested at 22 °C. (**a**) Mix 2 0.52% and 1.56% R/B complex modulus G^* and phase angle δ for the period of 15–25 min following mixing (CA = citric acid, TA = tartaric acid). (**b**) Mix 2 0.52% and 1.56% R/B G^* profiles over a 2000 s period following mixing, CA and TA. (**c**) Flow test yield stresses for mix 2 0.78% TA (suitable retardation). (**d**) Flow test viscosity profiles for mix 2 0.78% R/B TA. (**e**) Flow test yield stresses for mix 2 1.04% R/B TA (too much retardation). (**f**) Flow test yield stresses for mix 2 1.56% R/B CA (insufficient or ineffective retardation). Colour shadings differentiate retarder type and quantities in (**a**,**b**), and time in (**c**–**f**).

4. Discussion

4.1. The Significance of Temperature and Rheology for AAM

A period of 20–25 min allows sufficient time following mixing and mechanical vibration for a 310 mL cartridge to be fully loaded and attached to a UAV, followed by the autonomous extrusion of the material. Developing an OPC/PFA/CAC/CS quaternary system to provide an open time of ≈25 min for AM proved a challenging task due to the extent that material hydration was affected by environmental temperature. In this study, below 14.5 °C, accelerator or only moderate use of retarder (up to 0.12% R/B) is required, as shown by mixes 2–5 in Table 2. At 14.5 °C and above, this would require modification with the complete removal of accelerating agent and addition of retarding agent. At 22 °C, open times without additional retardation are reduced from the desired ≈25 min at 7 °C, down to an insufficient 5–15 min. It is proposed that tartaric acid should be added in quantities of 0.78% R/B at temperatures of ≈14.5 °C and above while the mixes shown in Table 2 remain valid at lower temperatures.

The flow results in Figures 9, 10 and 16 can be considered the most suitable quantification of fresh material rheological properties for OPC/PFA/CAC/CS quaternary systems for AAM, due to the material being exposed to the environment prior to each five-minute interval test. Figure 9 highlights the difference on open time that temperature can make. Mix 3, with acceleration, could be processed with the deposition device at low temperatures, yet when exposed to higher temperatures, mix 3 cannot be used and accelerating agent must be substituted with retarding agent. Tartaric acid added at 0.78% R/B is shown by flow results to be capable of providing a suitable fresh mix open time at ≥14.5 °C.

CAC-free mix 1 is suitable for UAV extrusion and is not as significantly affected by environmental temperature. The yield stress and viscosity flow test results shown in Figures 9a and 10b further confirm a yield stress of 1.1–1.4 kPa as being suitable for AAM. Mixes 2–5 should be compared to mix 1; for example, the viscosity profile of mix 2 (at 22 °C) after 10 min is comparable to that of mix 1, whereas viscosity is significantly increased at 25 min, and is therefore unsuitable. Due to an open time of 120 min, the performance profile of mix 1 at 110 min in Figure 13 can also be considered a guide as to whether other developed material is appropriate for AAM. For example, mix 2, 15 min after mixing in a 22 °C environment, is significantly stiffer and therefore not suitable for a miniature deposition device.

The rheology oscillation results in Figures 6 and 7 do not entirely represent the behaviour of exposed material, as the material being tested is enclosed by the upper and lower plates of the rheometer. Material between oscillating plates did not harden as quickly as when open to the environment. Aiming for a 25 min window for the extrusion of a full cartridge and a G^* of ≈5 MPa, the results in Figure 8a suggest that at 14.5 °C and 7 °C extrusion is viable, whereas above 14.5 °C it is not. Practical extrusions using the deposition device suggest that at ≈14.5 °C, mix 2 can only be processed within 15 min maximum. Therefore, when using CAC, G^* cannot be considered to be an entirely reliable quantification of fresh material stiffness. However, oscillation tests do indicate how rheological properties differ with temperature within a confined environment such as an AAM deposition device tube or cartridge, and whether mixes shown in Table 2 require the substitution of accelerator with retarder at higher temperatures. For example, mixes 2 and 3 are very suitable at 7 °C and exhibit rheological behaviour close to mix 1, but at higher temperatures show very different properties to mix 1, possessing a much reduced open time, and require retardation with tartaric acid.

Therefore, for in situ AM construction printing, quaternary system mixes must be modified according to climatic variation as curing times and rheological properties of fresh material differ significantly with temperature, becoming stiffer sooner as temperature increases. In summer temperatures, retardation must be applied to prevent flash-setting occurring while fresh material is still within the deposition system carried by a flying UAV. In situ operatives will need to be aware of external conditions and administer retardation into mixes prior to material cartridge loading on to UAVs.

4.2. Heat of Hydration, Layer Settlement, Fibres and Strength

It became clear during force tests involving fresh material within the deposition device cartridge that the heat generated during the hydration of an OPC-rich quaternary system is still sufficient to compromise the thermoplastic material of the cartridge. Calorimetry results show that for mixes 2–5, CAC dominates the hydration reactions despite only being 8–13% of the mix. CAC phases consume the gypsum present in OPC and CS leading to rapid ettringite formation and retardation of OPC C-S-H hydration. With the deposition device used for this research, the cartridge is contained within a carbon fibre casing, but it is not firmly restrained. Therefore, it is feasible that some lateral movement of the cartridge occurred during autonomous deposition, causing gradual warping of the threaded rod.

Considering that the 1-day compressive strength of mixes 2 and 3 is not vastly higher than mix 1, in addition to the variable nature of CAC/CS systems according to environmental temperature, the primary motivation for choosing an OPC/PFA/CAC/CS binder system for an AAM material is to reduce deformation following extrusion with controlled, early hardening in a known time period. The extrusions shown in Figure 3 show minimal sagging between spans. With accelerator or retarder applied as appropriate for the environmental temperature, layer compression following subsequent extrusion is minimal for fewer than ten layers. When considering the immediate deposition of subsequent layers by multiple UAVs, it is reasoned that the hydration of the first layer in an OPC/PFA/CAC/CS system will be sufficiently advanced by the time of the deposition of the tenth layer following.

The approximate 72:28 ratio of OPC:CAC used in the mixes is not conducive to improving compressive strength, as greater compressive strength is typically found in 90:10 OPC:CAC blends, or alternatively pure OPC or CAC mixes [35]. However, as Figure 14 indicates, strength in low density mixes (relative to a traditional OPC mortar mix) with pseudoplastic hydrocolloids, is not excessively compromised, with mixes 2 and 3 remaining structurally viable compared to mix 1 in terms of compressive strength at 28 days. With mixes 2–5 containing no more than 8–13% CAC, subsequent further loss of compressive strength due to the conversion reaction of metastable hydrates is not expected to happen.

The tensile properties of PVA fibres played a major role in reducing crack propagation during mechanical tests and the difference between a FV of $\leq$1% and 2% was significant, with mix 5 (low density and high foaming agent content) indicating the greatest flexural strength.

It is reasoned that thermoplastic is not suitable for cartridges containing CAC/CS material systems for AAM unless coated with a suitable substance to protect the cartridge from the heat of hydration, or alternatively replaced with a thermoset plastic or heat-resistant material cartridge to address thermal-related issues affecting radial dimensions with continual use. The quaternary binder system has proved effective in mitigating in deformation following extrusion, which demonstrates the key choice for wanting to introduce CAC/CS into mixes—to quicken curing and mitigate post-extrusion deformation. Material must remain structurally viable in cured compressive strength however and results show that the addition of foam to mixes compromises compressive strength and structural viability. Fibres have proved effective in flexural strength, but if mixes contain a high quantity of fibres to the extent that foam needs to be introduced to induce sufficient workability for fresh material to be extruded, alternative approaches to material reinforcement may be investigated such as the placement of rods by UAVs on to deposited layers prior to the deposition of the next layer, or work in tandem with the extrusion of a reinforcing filament by another agent immediately following deposition while the material remains fresh and prior to CAC-induced flash-setting.

4.3. *Evaluation and Application of Results for Potential AAM Construction*

An OPC/PFA/CAC/CS binder-based composite mix solution for horizontal layered AAM extrusion could use different mixes in an alternating layer approach. Mix 2 has structurally viable 28-day compressive strength of $\approx$25 MPa with good workability within open time at 7 °C. It requires an increase in retarder from 0.12% R/B to 0.78% R/B at temperatures $\geq$ 14.5 °C. Mix 2 could be used in conjunction with mix 5, which has 2% FV and lower density to provide tensile capacity and ductile failure in alternate layers of AM extrusion. Accelerator or retarder should be modified to both mix 2 and mix 5 Table 2 specifications as climatically appropriate.

The quaternary binder-based material can be considered to be rapid-hardening and AAM deposition would be a suitable method for infrastructure repair work. The printing of the domed structure on supporting material shows the potential of AAM for printing or repairing lightweight grid-shell structures. Supporting material could consist of an inflatable object, with deflation occurring following cementitious hydration, or may be integrated directly as part of the AAM approach, depositing weaker temporary supporting material using UAVs. The ability to print on a vertical or inclined plane suggests particular suitability for elevated or difficult to access locations which pose a threat to human safety. UAVs could attach to, or hover close to, a vertical surface to administer material for precision repair work in elevated, marine, tidal or difficult-to-access applications. Prime examples of this potential are repairing cracks on concrete bridges, very tall buildings and marine structures with a rapid-hardening OPC/PFA/CAC/CS-based material.

Upscaling to the extent that a building may be constructed using AAM would necessitate the coordination of multiple flying UAV agents, each being capable of flying to a pre-programmed trajectory and aware of the location of other UAVs extruding material (and where material has been previously extruded). Lateral deviation from programmed trajectories by UAVs during flight due to external climatic conditions or resulting from UAV flying instability, would need to be reduced to an acceptable minimum. The material system developed in this study is designed to rapidly harden immediately in situ following extrusion to prevent deformation due to the weight of multiple subsequently printed layers or imperfections in the alignment of material extrusion due to UAV trajectory deviation. Post extrusion deformation would be further mitigated with the use of a rectangular nozzle; in this study, extrusion was demonstrated using a circular nozzle due to the robotic arm being unable to rotate about their own axis and UAV yaw precision. With continuing development concerning the yaw properties of UAVs and improvements in precision and rotation, a rectangular nozzle is envisaged to be a natural choice for an AAM wall system, with multiple UAVs printing complex wall designs intended to reduce the amount of material used in comparison to a solid traditional poured concrete wall. Only printing material specifically required for structural design is an ethos entirely in-keeping with the advantages of using AM in construction scenarios, with the aim of eliminating wastage arising from bulk volume construction and ultimately reducing the carbon footprint of the construction industry.

There is an absence of established international design codes specifically for AM material tests in the construction industry [45] and existing design codes are not directly applicable to cementitious material currently being investigated [1], although proposed frameworks are continually evolving. As AM technology and associated material development for the construction industry matures, it is suggested by this study that the future development of design codes purposefully for AM in construction should encompass a reappraisal of CAC as being suitable to be potentially part of a blended, structurally viable, conversion-free cementitious system capable of 3D-printing new structures in addition to precision, rapid-hardening repair work.

5. Conclusions

Mixes based upon a novel quaternary cement system with the addition of rheology-modifying admixtures and crack-mitigating fibres were formulated and evaluated for use in a miniaturised deposition system for Aerial Additive Manufacturing (AAM). Pseudoplastic OPC/PFA/CAC/CS mixes for AAM have hydrated densities of $\approx$1600 kg/m^3 and achieve a fibre volume (FV) up to 1% and structurally feasible 28-day compressive strengths of $\approx$25 MPa. To achieve a 2% FV mix suitable for a UAV-attachable miniature deposition device, mix density can be reduced from $\approx$1600 kg/m^3 to $\approx$1400 kg/m^3 by the addition of a foaming agent. This reduces compressive strength but increases flexural strength and tensile capacity. Alternate 1% FV compressive and 2% FV flexural layers may be extruded in an AAM application. Yield stress and viscosity results provide the most accurate quantification of rheological properties for OPC/PFA/CAC/CS mixes, due to the strong effect of environmental temperature on material hydration. Fresh material suitable for AAM should possess a yield stress of ideally 1.1 kPa, and certainly less than 1.3 kPa, for 25 min following mixing.

It was demonstrated that developed mixes required acceleration or retardation of 0.12% R/B at 7 °C, with greater retardation at 14.5 °C and 22.5 °C in order to promote a suitable material open time of ≈25 min. Tartaric acid was a suitable retarding agent. A fibrous pseudoplastic OPC-rich OPC/PFA/CAC/CS structurally viable composite material is suitable for AAM using flying, self-powered UAVs on the provision that mixes are modified with accelerating or retarding agents in accordance with different environmental and water temperatures, and deposition cartridges are protected from exothermic reactions while the material is in the fresh state.

Author Contributions: Conceptualization, B.D., P.S. and R.J.B.; methodology, B.D., P.S. and R.J.B.; software, B.D., P.S. and R.J.B.; validation, B.D., P.S. and R.J.B.; formal analysis, B.D.; investigation, B.D.; resources, B.D., P.S. and R.J.B.; data curation, B.D.; writing—original draft preparation, B.D.; writing—review and editing, P.S. and R.J.B.; visualization, B.D.; supervision, P.S. and R.J.B.; project administration, B.D., P.S. and R.J.B.; funding acquisition, R.J.B. All authors have read and agreed to the published version of the manuscript.

Funding: The Aerial Additive Manufacturing project is funded by the Engineering and Physical Sciences Research Council (EPSRC) [grant number EP/N018494/1]. Further support was provided by the EPSRC Centre for Decarbonisation of the Built Environment (dCarb) [grant number EP/L016869/1].

Institutional Review Board Statement: Not applicable.

Data Availability Statement: All data files supporting the results in this paper are available to access from the University of Bath data archive: Dams, B., 2023. Data relation to work concerning the "Development of a quaternary cementitious system for aerial additive manufacturing", 2023. Bath. Available from: https://doi.org/10.15125/BATH-00694 (accessed on 27 October 2023).

Acknowledgments: The authors express thanks to the technical support of the Department of Architecture and Civil Engineering laboratories, University of Bath, UK. Tomasz Omakowski (Technical developer, Imerys Aluminates Ltd., UK Division), Shan Yu (University of Bath, UK), Aerial Robotics Laboratory, Imperial College London, UK.

Conflicts of Interest: The authors declare that they have no known competing financial interests or personal relationships that could have appeared to influence the work reported in this paper. The funding bodies had no role in the design of the study; in the collection, analyses, or interpretation of data; in the writing of the manuscript; or in the decision to publish the results.

Abbreviations

The following abbreviations are used in this manuscript:

AAM	Aerial Additive Manufacturing
A/B	Accelerator/Binder ratio
AM	Additive Manufacturing
CA	Citric Acid
CAC	Calcium Aluminate Cement
CS	Calcium Sulphate
C-S-H	Calcium Silicate Hydrates
DOF	Degrees of Freedom
DC	Direct Current
Ecemmat	Energy transferred
FDM	Fused Deposition Modelling
PFA	Pulverised fuel ash (fly ash)
FV	Fibre Volume
HEMC	Hydroxyethyl Methyl Cellulose
J	Joules

mA	milliAmperes
MPa	MegaPascals
OPC	Ordinary Portland Cement
PVA	PolyVinyl Alcohol
R/B	Retarder/Binder ratio
s	Seconds
S/B	Sand/Binder Ratio
SEM	Scanning Electron Microscopy
SP	Superplasticiser
TA	Tartaric Acid
UAV	Unmanned Aerial Vehicles
V	Volts
W	Watts
W/B	Water/Binder Ratio
XG	Xanthan Gum

References

1. Wangler, T.; Roussel, N.; Bos, F.P.; Salet, T.A.; Flatt, R.J. Digital concrete: A review. *Cem. Concr. Res.* **2019**, *123*, 105780. [CrossRef]
2. Bi, M.; Tran, P.; Xia, L.; Ma, G.; Xie, Y.M. Topology optimization for 3D concrete printing with various manufacturing constraints. *Addit. Manuf.* **2022**, *57*, 102982. [CrossRef]
3. Reiter, L.; Wangler, T.; Roussel, N.; Flatt, R.J. The role of early age structural build-up in digital fabrication with concrete. *Cem. Concr. Res.* **2018**, *112*, 86–95. [CrossRef]
4. van den Heever, M.; Bester, F.; Kruger, J.; van Zijl, G. Numerical modelling strategies for reinforced 3D concrete printed elements. *Addit. Manuf.* **2022**, *50*, 102569. [CrossRef]
5. Bentz, D.P.; Jones, S.Z.; Bentz, I.R.; Peltz, M.A. Towards the formulation of robust and sustainable cementitious binders for 3-D additive construction by extrusion. *Constr. Build. Mater.* **2018**, *175*, 215–224. [CrossRef]
6. Marchon, D.; Kawashima, S.; Bessaies-Bey, H.; Mantellato, S.; Ng, S. Hydration and rheology control of concrete for digital fabrication: Potential admixtures and cement chemistry. *Cem. Concr. Res.* **2018**, *112*, 96–110. [CrossRef]
7. Rubio, M.; Sonebi, M.; Amziane, S. 3D Printing of Fibre Cement-Based Materials: Fresh and Rheological Performances. In Proceedings of the ICBBM 2017, 2nd International Conference On Bio-Based Building Materials, Clermont-Ferrand, France, 21–23 June 2017.
8. Lim, S.; Buswell, R.A.; Le, T.T.; Austin, S.A.; Gibb, A.G.F.; Thorpe, T. Developments in construction-scale additive manufacturing processes. *Autom. Constr.* **2012**, *21*, 262–268. [CrossRef]
9. Le, T.T.; Austin, S.A.; Lim, S.; Buswell, R.A.; Gibb, A.; Thorpe, T. Mix design and fresh properties for high-performance printing concrete. *Mater. Struct.* **2012**, *45*, 1221–1232. [CrossRef]
10. Bos, F.P.; Ahmed, Z.Y.; Wolfs, R.J.; Salet, T.A. 3D printing concrete with reinforcement. In *High Tech Concrete: Where Technology and Engineering Meet*; Springer: Maastricht, The Netherlands, 2018; pp. 2484–2493.
11. Wolfs, R.; Bos, F.; Salet, T. Early age mechanical behaviour of 3D printed concrete: Numerical modelling and experimental testing. *Cem. Concr. Res.* **2018**, *106*, 103–116. [CrossRef]
12. Labonnote, N.; Ronnquist, A.; Manum, B.; Ruther, P. Additive construction: State-of-the-art, challenges and opportunities. *Autom. Constr.* **2016**, *72*, 347–366. [CrossRef]
13. Buswell, R.A.; de Silva, W.L.; Jones, S.; Dirrenberger, J. 3D printing using concrete extrusion: A roadmap for research. *Cem. Concr. Res.* **2018**, *112*, 37–49. [CrossRef]
14. Keating, S.J.; Leland, J.C.; Cai, L.; Oxman, N. Toward site-specific and self-sufficient robotic fabrication on architectural scales. *Sci. Robot.* **2017**, *2*, eaam8986. [CrossRef]
15. Dörfler, K.; Dielemans, G.; Lachmayer, L.; Recker, T.; Raatz, A.; Lowke, D.; Gerke, M. Additive Manufacturing using mobile robots: Opportunities and challenges for building construction. *Cem. Concr. Res.* **2022**, *158*, 106772. [CrossRef]
16. Buswell, R.A.; da Silva, W.L.; Bos, F.P.; Schipper, H.; Lowke, D.; Hack, N.; Kloft, H.; Mechtcherine, V.; Wangler, T.; Roussel, N. A process classification framework for defining and describing Digital Fabrication with Concrete. *Cem. Concr. Res.* **2020**, *134*, 106068. [CrossRef]
17. Nuh, M.; Oval, R.; Orr, J.; Shepherd, P. Digital fabrication of ribbed concrete shells using automated robotic concrete spraying. *Addit. Manuf.* **2022**, *59*, 103159.
18. Lloret-Fritschi, E.; Wangler, T.; Gebhard, L.; Mata-Falcón, J.; Mantellato, S.; Scotto, F.; Burger, J.; Szabo, A.; Ruffray, N.; Reiter, L.; et al. From smart dynamic casting to a growing family of digital casting systems. *Cem. Concr. Res.* **2020**, *134*, 106071. [CrossRef]
19. Popescu, M.; Rippmann, M.; Liew, A.; Reiter, L.; Flatt, R.J.; Van Mele, T.; Block, P. Structural design, digital fabrication and construction of the cable-net and knitted formwork of the KnitCandela concrete shell. In *Proceedings of the Structures*; Elsevier: Amsterdam, The Netherlands, 2021; Volume 31, pp. 1287–1299.

20. Naicu, D.I.; Williams, C.J.K. The use of dynamic relaxation to solve the differential equation describing the shape of the tallest possible building. In Proceedings of the Textiles composites and inflatable structures VII: Proceedings of the VII International Conference on Textile Composites and Inflatable Structures, Barcelona, Spain, 19–21 October 2015; CIMNE: Barcelona, Spain, 2015; pp. 34–45.
21. Nadhim, E.A.; Hon, C.; Xia, B.; Stewart, I.; Fang, D. Falls from height in the construction industry: A critical review of the scientific literature. *Int. J. Environ. Res. Public Health* **2016**, *13*, 638. [CrossRef]
22. Zhang, K.; Chermprayong, P.; Xiao, F.; Tzoumanikas, D.; Dams, B.; Kay, S.; Kocer, B.B.; Burns, A.; Orr, L.; Choi, C.; et al. Aerial additive manufacturing with multiple autonomous robots. *Nature* **2022**, *609*, 709–717. [CrossRef]
23. Babel, J. Up in the Air: The emerging issue of drones in the construction industry. *XL Catlin Constr. Insid.* **2015**, *5*, 2015.
24. Dillow, C. Drones Take Off in the Construction Industry as a Cost-Saving Tool. 2016. Available online: https://fortune.com/ (accessed on 10 September 2022).
25. Ghaffar, S.H.; Corker, J.; Fan, M. Additive manufacturing technology and its implementation in construction as an eco-innovative solution. *Autom. Constr.* **2018**, *93*, 1–11. [CrossRef]
26. Jones, S.Z.; Hipp, J.B.; Allen, A.J.; Gagnon, C.V. Rheology and microstructure development of hydrating tricalcium silicate-implications for additive manufacturing in construction. *Cem. Concr. Res.* **2022**, *152*, 106651. [CrossRef]
27. Dams, B.; Chen, B.; Shepherd, P.; Ball, R.J. Development of Cementitious Mortars for Aerial Additive Manufacturing. *Appl. Sci.* **2023**, *13*, 641. [CrossRef]
28. Dams, B.; Wu, Y.; Shepherd, P.; Ball, R.J. Aerial Additive Building Manufacturing of 3D printed Cementitious Structures. In Proceedings of the 37th Cement and Concrete Science Conference UCL, London, UK, 11–12 September 2017; 55p.
29. Vaitkevičius, V.; Šerelis, E.; Kerševičius, V. Effect of ultra-sonic activation on early hydration process in 3D concrete printing technology. *Constr. Build. Mater.* **2018**, *169*, 354–363. [CrossRef]
30. Winnefeld, F.; Klemm, S. Influence of citric acid on the hydration kinetics of calcium sulfoaluminate cement. In Proceedings of the First International Conference on Sulphoaluminate Cement, Wuhan, China, 23–25 October 2013; pp. 288–308.
31. Scrivener, K. Calcium aluminate. *Adv. Concr. Technol.* **2003**, *1*, 2.
32. Domone, P.; Illston, J. *Construction Materials: Their Nature and Behaviour*; Domone, P., Illston, J., Eds.; CRC Press: Boca Raton, FL, USA, 2010.
33. Sonebi, M.; Perrot, A. Effect of mix proportions on rheology and permeability of cement grouts containing viscosity modifying admixture. *Constr. Build. Mater.* **2019**, *212*, 687–697. [CrossRef]
34. Cherop, P.T.; Kiambi, S.L.; Kosgey, E.K. Effect of Non-Ionic Cellulose Ethers on Properties of White Portland Cement. *Int. J. Appl. Eng. Res.* **2017**, *12*, 2502–2508.
35. Gawlicki, M.; Nocuń-Wczelik, W.; Bąk, Ł. Calorimetry in the studies of cement hydration: Setting and hardening of Portland cement–calcium aluminate cement mixtures. *J. Therm. Anal. Calorim.* **2009**, *100*, 571–576. [CrossRef]
36. Neville, A.M. *Properties of Concrete*; Longman: London, UK, 1995; Volume 4.
37. Antonovič, V.; Kerienė, J.; Boris, R.; Aleknevičius, M. The effect of temperature on the formation of the hydrated calcium aluminate cement structure. *Procedia Eng.* **2013**, *57*, 99–106. [CrossRef]
38. Scrivener, K.L.; Cabiron, J.L.; Letourneux, R. High-performance concretes from calcium aluminate cements. *Cem. Concr. Res.* **1999**, *29*, 1215–1223. [CrossRef]
39. Kirca, O. *Temperature Effect on Calcium Aluminate Cement Based Composite Binders*; Middle East Technical University: Ankara, Turkey, 2006.
40. Moffatt, E.G.; Thomas, M.D. Performance of rapid-repair concrete in an aggressive marine environment. *Constr. Build. Mater.* **2017**, *132*, 478–486. [CrossRef]
41. *BS EN 14647:2005*; Calcium Aluminate Cement. Composition, Specifications and Conformity Criteria. BSI: London, UK, 2005.
42. Fernández-Carrasco, L.; Vázquez, E. Reactions of fly ash with calcium aluminate cement and calcium sulphate. *Fuel* **2009**, *88*, 1533–1538. [CrossRef]
43. Rodger, S.; Double, D. The chemistry of hydration of high alumina cement in the presence of accelerating and retarding admixtures. *Cem. Concr. Res.* **1984**, *14*, 73–82. [CrossRef]
44. Gu, P.; Beaudoin, J. A conduction calorimetric study of early hydration of ordinary Portland cement/high alumina cement pastes. *J. Mater. Sci.* **1997**, *32*, 3875–3881. [CrossRef]
45. Khalil, N.; Aouad, G.; El Cheikh, K.; Rémond, S. Use of calcium sulfoaluminate cements for setting control of 3D-printing mortars. *Constr. Build. Mater.* **2017**, *157*, 382–391. [CrossRef]
46. Ashrafi, N.; Nazarian, S.; Meisel, N.; Duarte, J.P. A grammar-based algorithm for toolpath generation: Compensating for material deformation in the additive manufacturing of concrete. *Addit. Manuf.* **2022**, *55*, 102803. [CrossRef]
47. Khan, M.I.; Lynsdale, C. Strength, permeability, and carbonation of high-performance concrete. *Cem. Concr. Res.* **2002**, *32*, 123–131. [CrossRef]
48. Xu, L.; Wang, P.; Zhang, G. Calorimetric study on the influence of calcium sulfate on the hydration of Portland cement–calcium aluminate cement mixtures. *J. Therm. Anal. Calorim.* **2012**, *110*, 725–731. [CrossRef]
49. Noushini, A.; Vessalas, K.; Samali, B. Static mechanical properties of polyvinyl alcohol fibre reinforced concrete (PVA-FRC). *Mag. Concr. Res.* **2014**, *66*, 465–483. [CrossRef]

50. Dams, B.; Lumlerdwit, K.; Shepherd, P.; Ball, R.J. Fibrous cementitious material development for additive building manufacturing. In Proceedings of the 38th Cement and Concrete Science Conference, Coventry, UK, 10–11 September 2018; Tyrer, M., Ganjian, E., West, A., Eds.; Institute of Materials, Minerals and Mining: London, UK, 2018; p. 20.
51. Soltan, D.G.; Li, V.C. A self-reinforced cementitious composite for building-scale 3D printing. *Cem. Concr. Compos.* **2018**, *90*, 1–13. [CrossRef]
52. Lootens, D.; Hébraud, P.; Lécolier, E.; Van Damme, H. Gelation, shear-thinning and shear-thickening in cement slurries. *Oil Gas Sci. Technol.* **2004**, *59*, 31–40. [CrossRef]
53. Schmidt, W.; Sonebi, M.; Brouwers, H.J.; Kühne, H.C.; Meng, B. Rheology modifying admixtures: The key to innovation in concrete technology–a general overview and implications for Africa. *Change* **2013**, *4*, 5.
54. *BS EN 1015-11:1999*; Methods of Test for Mortar for Masonry—Part 11: Determination of Flexural and Compressive Strength of Hardened Mortar. BSI: London, UK, 1999.
55. *BS EN 12390-5:2009*; Testing Hardened Concrete Part 5: Flexural Strength of Test Specimens. BSI: London, UK, 2009.
56. Bishop, M.; Barron, A.R. Cement hydration inhibition with sucrose, tartaric acid, and lignosulfonate: Analytical and spectroscopic study. *Ind. Eng. Chem. Res.* **2006**, *45*, 7042–7049. [CrossRef]
57. Guan, B.; Lou, W.; Ye, Q.; Fu, H.; Wu, Z. Calorimetric study of calcium aluminate cement blended with flue gas desulfurization gypsum. *J. Therm. Anal. Calorim.* **2009**, *98*, 737. [CrossRef]
58. Hossain, K.; Lachemi, M.; Sammour, M.; Sonebi, M. Strength and fracture energy characteristics of self-consolidating concrete incorporating polyvinyl alcohol, steel and hybrid fibres. *Constr. Build. Mater.* **2013**, *45*, 20–29. [CrossRef]
59. Fehling, E.; Leutbecher, T.; Bunje, K. Design relevant properties of hardened ultra high performance concrete. In Proceedings of the International Symposium on Ultra High Performance Concrete, Kassel, Germany, 13–15 September 2004; Volume 1, pp. 327–338.

designs

Article

Enhancement of Fatigue Life of Polylactic Acid Components through Post-Printing Heat Treatment

Moises Jimenez-Martinez [1,*,†], Julio Varela-Soriano [1,†], Rafael Carrera-Espinoza [2], Sergio G. Torres-Cedillo [3] and Jacinto Cortés-Pérez [3]

1 Tecnologico de Monterrey, School of Engineering and Science, Via Atlixcayotl 5718, Mexico; j.varela@tec.mx
2 Departamento de Ingeniería Industrial y Mecánica, Universidad de las Américas Puebla, San Andrés Cholula 72810, Mexico; rafael.carrera@udlap.mx
3 Centro Tecnológico FES Aragón, Universidad Nacional Autónoma de México, Mexico City 57171, Mexico; sergiotorres98@aragon.unam.mx (S.G.T.-C.); jacop@unam.mx (J.C.-P.)
* Correspondence: moisesjimenezmartinez@gmail.com
† These authors contributed equally to this work.

Abstract: To reduce the carbon footprint of manufacturing processes, it is necessary to reduce the number of stages in the development process. To this end, integrating additive manufacturing processes with three-dimensional (3D) printing makes it possible to eliminate the need to use tooling for component manufacturing. Furthermore, using 3D printing allows the generation of complex models to optimize different components, reducing the development time and realizing lightweight structures that can be applied in different industries, such as the mobility industry. Printing process parameters have been studied to improve the mechanical properties of printed items. In this regard, although the failure of most structural components occurs under dynamic load, the majority of the evaluations are quasistatic. This work highlights an improvement in fatigue strength under dynamic loads in 3D-printed components through heat treatment. The fatigue resistance was improved regarding the number of cycles and the dispersion of results. This allows 3D-printed polylactic acid components to be structurally used, and increasing their reliability allows their evolution from a prototype to a functional component.

Keywords: fatigue; lightweight structures; additive manufacturing

Citation: Jimenez-Martinez, M.; Varela-Soriano, J.; Carrera-Espinoza, R.; Torres-Cedillo, S.G.; Cortés-Pérez, J. Enhancement of Fatigue Life of Polylactic Acid Components through Post-Printing Heat Treatment. *Designs* **2024**, *8*, 7. https://doi.org/10.3390/designs8010007

Academic Editors: Mahdi Bodaghi and Obeidi Muhannad

Received: 25 October 2023
Revised: 30 November 2023
Accepted: 19 December 2023
Published: 11 January 2024

1. Introduction

Three-dimensional (3D) printing is a computer-controlled additive manufacturing process that creates 3D objects by continuous material deposition. The mechanical properties of raw materials evolve into mechanical properties of a 3D-printed component, including the yield strength, ultimate tensile strength, fatigue, ductility, and brittle behavior. Overall, the manufacturing process depends on the component's geometry, the raw materials, the manufacturing parameters, and the post-treatment processes (e.g., thermal or thermomechanical processes). Traditional processes are usually either subtractive or additive manufacturing processes; however, traditional additive processes, such as foundry or molding, require tooling design, development, and validation. New additive processes in polymer printing, such as fused deposition modelling, eliminate the requirement for additional tooling stages because either only one material is added layer by layer or composite materials are developed in different ways, such as by adding reinforcements to the matrix. To improve the mechanical behavior of 3D-printed components, printing parameters, such as the velocity, bed temperature, extrusion temperature, and raster direction, have been examined because they can generate internal defects that can result in premature failure under quasistatic or dynamic loads [1–3]. Travieso-Rodriguez et al. [4] revealed the relationship between printing parameters and quasistatic response in terms of the stiffness and bending strength, but they did not define the parameters necessary to result in an improvement in fatigue strength.

There is still scope to improve 3D printing products, not only by evaluating the printing parameters, but also by adopting post-treatment processes to enhance the mechanical properties of the products [5]. Furthermore, the product's durability, in terms of fatigue life, should meet the requirements of dynamic loads and statistical parameters, and this should be validated in not just one sample; such validation should be included in lot production to reduce the scattered range found in all components made by considering the same parameters to improve the reliability. By combining printing parameters and optimized designs, the mechanical performance can be improved in a controlled manner [6,7]; however, the deposition process itself generates variability, and hence, there is scope to improve the mechanical performance [8]. The additive metals and plastics are subjected to post-treatment processes, such as thermal treatments, to improve their properties [9–11]; one parameter that changes is the hardness of the sample because of microstructural changes. However, the best improvement is observed in homogenization throughout the whole transverse section or as a function of the layer position similar to that observed in dual-phase materials. Homogenization in the microstructure enhances the fatigue strength of the material.

Heat treatments of reinforced plastics or polymer-based compounds show improvements in static and fatigue strengths [9,12]. Different heat treatments may influence the fracture behavior but have minimal effects on tensile strength [13]. The change observed in the mechanical properties depends on the type of heat treatment, as seen in the case of metals. In polymers, thermal treatments can be normalized and annealed [14].

To evaluate this proposal, the effect of heat treatments on hardness was analyzed by performing fatigue tests. The flexibility of additive manufacturing introduces variation among designs, which can generate diverse sets of printing variables. The inherent uncertainty in 3D printing technology, arising from the complicated interaction of heat transformation and dissipation, leads to more pronounced dispersion compared to conventional additive manufacturing methods, such as injection molding, casting, or subtractive processes. Since the material undergoes a phase change during the printing process, a temperature gradient emerges, giving rise to distinct structures across the cross-section. This gradient of hardness and properties contributes to a dispersion in the strength of the 3D-printed component. To achieve greater reliability, we heat-treated 3D samples at different temperatures, and the effects of these treatments were measured via hardness measurements and fatigue tests to evaluate the dynamic behavior of the printed product [?]. Polymeric materials are highly likely to be used in 3D-printed components due to their potential for different applications in various industries, such as mobility (automotive and aircraft), medical, and manufacturing industries. Enhancing their behavior requires normalizing the material in the transverse direction across its thickness. This can be achieved through the application of a thermal treatment. Considering the component as a closed system, the process can be modeled by incorporating the conservation laws of mass, linear momentum, and energy.

$$\frac{d\rho}{dt} + \rho(\nabla \cdot \vec{v}) = 0 \tag{1}$$

$$\rho\frac{d\vec{v}}{dt} = -\nabla P + \nabla \cdot \vec{\tau} + \rho\tilde{g} \tag{2}$$

$$\rho C_p\frac{dT}{dt} = \beta T\frac{dP}{dt} + \eta\dot{\gamma}^2 + \nabla \cdot \tilde{q} \tag{3}$$

where ρ is the density; P is the pressure; $\vec{v}$ is the velocity vector; $\vec{\tau}$ is the viscous stress tensor; and the specific heat, thermal expansion coefficient, and the heat flux vector are expressed by C_p, β, and $\tilde{q}$, respectively.

Although the printing process has similarities to injection processes, one of the biggest differences between the two is that in the printing process, material is deposited directly from the nozzle onto the component. Such deposition affects the temperature gradient. This effect can be evaluated by a shift function (a_t) using the polymer parameters and the reference temperature (T_r), expressed by the Williams–Landel–Ferry relation [16] as follows:

$$loga_t = \frac{-C_1(T - T_r)}{C_2 + T - T_r} \tag{4}$$

In this case, the proposed heat treatment is as follows: perform normalization of each one of the pieces, considering the result reported in previous works that the hardness on the bedside is greater than that on the printing side [17]. The effect of heat treatment has been evaluated by hardness measurements and fatigue life [18]. It is believed that the component's position in the furnace influences the treatment process, and placing the printing side in contact with the furnace does not cause deformation by heat treatment. Thus, we can define the position in the process so that the side of the printing bed is not in direct contact with the furnace. This study aims to contribute to this growing area of research in the use of polylactic acid (PLA) printed components as structural components. This investigation contributes significantly through a comprehensive approach to printed PLA using FDM. The treatment involves increasing the printing bed temperature by 5 °C. The experimental results showed that the temperature influences the mechanical behaviors of the 3D-printed PLA materials. Hence, it is important to consider multiple printing times because this is the time required for the manufacturing process. The same printing time must be used for the heat treatment to achieve standardization in the component structure. When the oven time equals the printing time, the experimental results' durability and dispersion are improved.

2. Materials and Methods

We adopted a case study approach to evaluate the mechanical behavior of 3D-printed PLA products. Specimens with a dog-bone geometry and a thickness of 5 mm were printed according to ASTM D7791 [19].

2.1. Additive Manufacturing

Test components were fabricated on a commercial 3D printing machine provided by Ender using a red PLA filament obtained from the brand Color Plus with a diameter of 1.75 mm. The process was carried out at 200 °C in the liquefier chamber and 50 °C in the build platform, with a raster angle of 45°. The manufacturing process of each component took 8 h. A schematic printing process is shown in Figure 1.

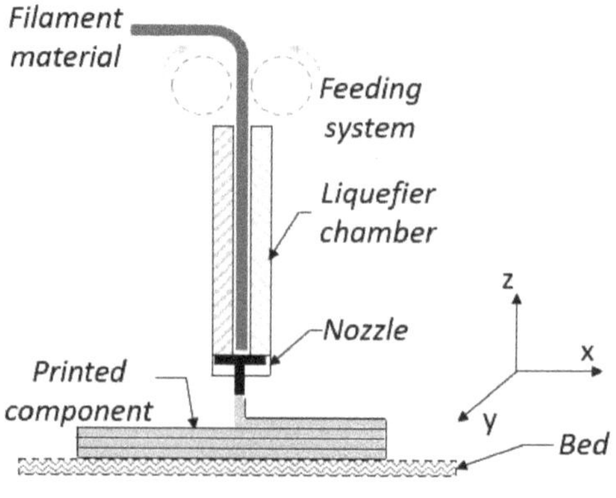

Figure 1. Schematic 3D printing process.

Printing was performed at a temperature of 205 °C and 50 °C on the printing side and build platform, respectively. Important parameters for components under cyclic loads are the raster angle, set at 45°, and the infill percentage, set to a 100%, as reported in [20]. The specimens were printed at room temperature. Ultimaker Cura was used for slicing the 3D printing model (in STL format). Then, the same software was used to control the settings of the 3D printing process and finally export the G code compatible with the manufacturing machine. Heat treatments were conducted using a *Terlab* precision furnace, preheated for 2 h to stabilize the treatment temperature (Figure 2). The component side that had contact with the printing bed was oriented toward the furnace chamber. An annealing treatment was applied to achieve uniform hardness, improve ductility, and alleviate residual forces generated during printing.

Additionally, treatments based on the glass transition temperature were implemented to align the polymer structure. It is important to mention that the heat treatment mechanism can be quenching, normalizing, and annealing. In the quenching process, hardness is heightened; however, the material loses toughness and promptly fractures, leaving minimal time for the material to undergo plastic deformation. In the case of annealing, the specimens tend to develop buckling or warpage, resulting in a complete deformation of the part and the acquisition of both hardness and brittleness, often leading to the loss of the original geometry. The normalizing mechanism effectively controls brittleness and warpage while enhancing the resistance to cyclic loading.

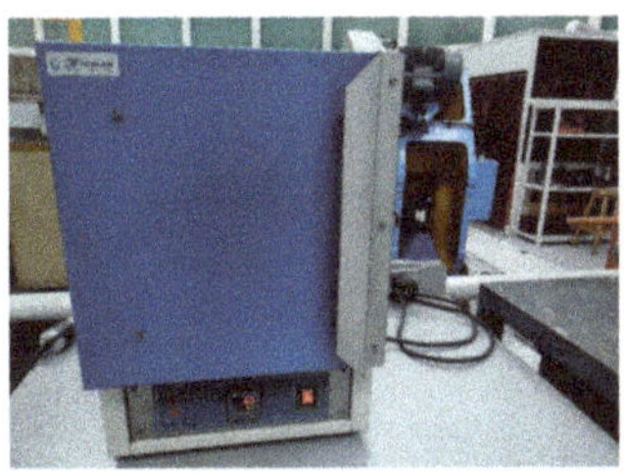

Figure 2. Heat treat furnace.

2.2. Mechanical Fatigue

Durability tests were performed on an Instron uniaxial test machine (Figure 3) under the following load conditions at room temperature according to the standard ASTM D7791 [19]: amplitude, 2000 N; frequency, 2 Hz; and R = −1.

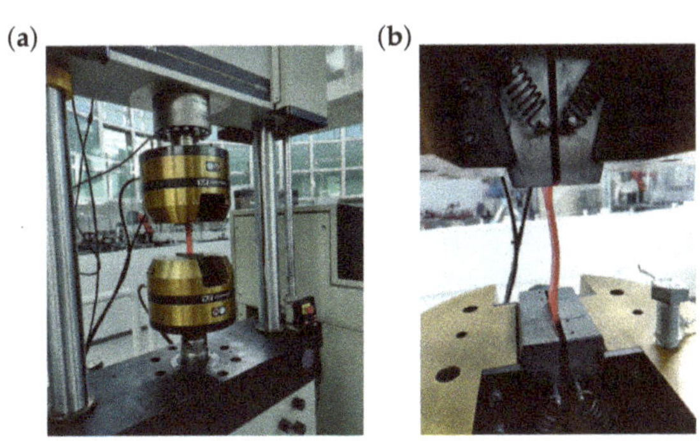

Figure 3. Uniaxial fatigue test (**a**) test bench and (**b**) component at compression load.

The reliability was monitored by obtaining the average number of samples per treatment using the following expression.

$$\mu = \frac{\sum_{i=1}^{n}}{x_i} \tag{5}$$

where x_i is the result of hardness measurement and cycles, and n is the sample size per variable.

Since each component yields precisely the same results, even under identical manufacturing conditions, assessing the scatter through an analysis of the standard deviation is imperative.

$$s_{log} = \sqrt{\frac{\sum(x-\mu)^2}{n}} \tag{6}$$

Hardness measurements were taken using an Instron microhardness instrument under a load of 1000 g in 11 a according to ASTM E384 [21]. The first step in this process was to evaluate the printed component without treatment and any support—at least five measurements were taken for each component to evaluate the hardness. The result was defined as the raw component, and the hardness was 18HV. Fatigue assessments were performed until crack propagation. The experimental results were 22,518, 52,307, 35,887, 43,494, and 26,652 cycles. Equations (5) and (6) were employed to analyze the obtained experimental results, which are summarized in Table 1.

Table 1. Standard deviation in fatigue tests of components without thermal treatment.

Cycles	log
22,518	4.3525
52,307	4.7185
35,887	4.5549
43,494	4.6384
26,652	4.4257
μ	36,172
s_{log}	0.1499

This study evaluated the changes in hardness and durability after being subjected to axial fatigue on 46 printed components. Heat treatments play a very important role in the polymer and metal industry. The support effect has a direct relationship with the hardness and fatigue life, and the evaluation of this effect on printed components with thicknesses of 1, 5, 10, 15, and 20 mm is shown schematically in Figure 4. The supports were removed by applying a turn between the component and the bracket section; when the support with a height of 20 mm was separated, a disassembler had to be used to start separating the components.

Table 2 summarizes the different heights of the supports and their effects on fatigue strength. The supports have a square geometry of 2.5 mm and they are 0.5 mm thick along the entire component; the difference is the height, which ranges from 1 to 20 mm. A case study approach was used to obtain further in-depth information on the effect of the use of supports on the thermal history of the component. To understand the relationship between the height of the support and the hardness generated in the component, different supports were evaluated to analyze whether they generate a standardized hardness and, at the same time, to understand the mechanical effect of the use of supports on fatigue resistance. In observational studies, there is a potential for bias resulting from different support configurations. Hence, all the supports used in this study had identical parameters to prevent any bias.

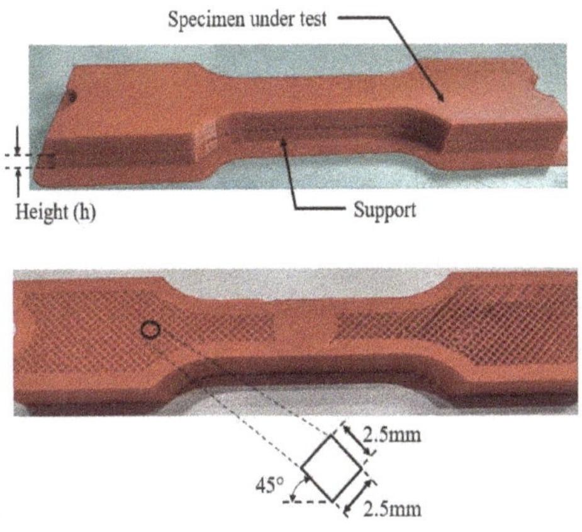

Figure 4. Component printed with support: (**a**) as printed and (**b**) support parameters.

Table 2. Durability strength and hardness measurements of specimens with supports.

Support Height (mm)	Cycles	Hardness HV
1	20,791	14
5	18,141	14
10	16,950	12
15	17,080	13
20	16,908	12
μ	17,974	
s_{log}	0.03824	

As shown in Table 2, the support had an unwanted effect on durability by reducing the mean durability by almost 50%. Notably, the standard deviations are reduced. Although the durability due to the use of supports did not increase, the dispersion of the experimental loads decreased, suggesting that the effect of temperature was reduced. Additionally, there is no difference in hardness for brackets with a size of 10 mm and 20 mm, and the duration of cycles differs only slightly by 0.2%. These results, therefore, need to be interpreted with caution. The obtained hardness was reduced by up to one-third of the component without support and without treatment. A comparison of the supports, hardnesses, and cycle durations does not reveal any direct relationships. The highest value of cycles was obtained with 1 mm-thick supports, yielding a hardness of 14 HV. Although the lowest hardness value coincides with the lowest durability values for the product with brackets, this result was observed only for the 10 and 15 mm brackets.

The effect of the heat treatment is a result of a combination of the effects of the oven temperature, time in the oven, cooling medium, and its velocity [22]. Based on previous research presented by Mayen et al. [23], components with a raster angle of 45° and post-heat treatment of 100 °C show a fatigue life of 18,505 cycles. Bakar et al. [13] evaluated the heat treatment near the glass transition temperature (70–80 °C) of PLA to enhance the durability of the specimen. Another studied temperature range for heat treatment was 80–160 °C for very short periods of time, namely, 60–180 s, and the best improvement was achieved at 120 °C [14].

3. Results and Discussion

For designing a heat treatment method to improve fatigue strength, we considered a combination of temperature, time in the oven, and cooling, as previously described. Cooling was performed at room temperature (23 °C ± 2°). Heat treatment was carried out at 60 °C, 80 °C, and 120 °C for 1 h. Table 3 summarizes the changes in fatigue life after heat treatment at different temperatures.

Table 3. Experimental results from specimens with heat treatment

Temperature °C	Fatigue Life Cycles
60	6211
60	14,956
80	3208
80	14,142
120	7742
120	13,840

In the 60–120 °C range, the best approach of using 60 °C was evaluated for different oven times. The hardness at this temperature was 20 HV for oven times of 4, 8, 12, and 24 h. Note that there may be possible biases in these responses. Although the hardness was improved at temperatures above 60 °C, this resulted in a brittle component, as seen from the fatigue life evaluation shown in Figure 5.

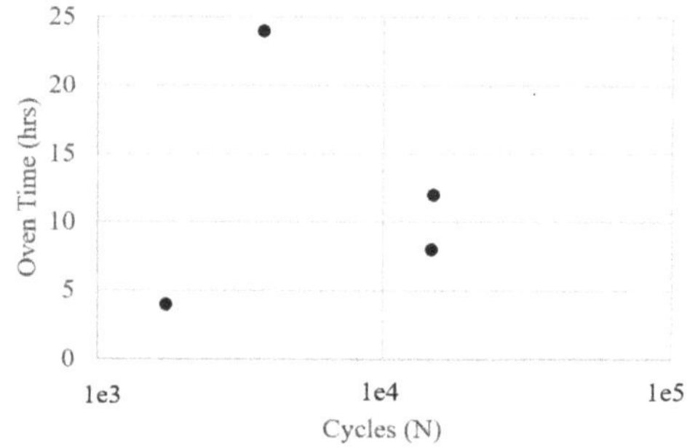

Figure 5. Semi log test results with a heat treatment of 60 °C.

A comparison of the durability of components with and without treatments showed no enhancement in the behavior above 60 °C. Based on these evaluations, the oven temperature was updated to 55 °C, and the minimum and maximum oven times were 8 h and 168 h, respectively. It is believed that the internal homogenization of the component is not possible within a short time.

Figure 6 shows the duration of each treatment regarding fatigue under the conditions above. The graph shows that the longer the oven time, the higher the durability in some cases. The improvement in resistance to fatigue is noteworthy as it correlates with the duration of oven exposure, attributed to the alignment of polymer chains. This is a consequence of the printed components' volumetric nature; it is necessary to extend the oven time. Therefore, short treatment durations only alter the edges of the component, leading to brittle failure, even more so than without any treatment.

Heat treatment enhances the fatigue strength not only in terms of durability but also in terms of scatter reduction. In cases where the results exhibit a positive trend in durability, a minimum of three components with identical characteristics undergo testing. Conversely, two components were tested when the trend indicated a decline in durability, leading to failure. By its inherent nature, the process of accumulated mechanical fatigue damage is a statistical phenomenon dependent on factors such as the design, the load, the material, and the manufacturing process. Design variations can originate from tolerances that allow for a deviation from the nominal value. Loads were then determined by a load spectrum involving all load cases and variables such as the frequency, load sequence, and environmental conditions. Materials can also exhibit variations in mechanical properties, even from the same supplier or within the same production batch.

Moreover, manufacturing processes can improve components' resistance dispersion for subtractive and additive methods. These parameters collectively contribute to an increased strength variation, leading to failure when the most critical load is applied to the component with the lowest strength, as shown in Figure 6.

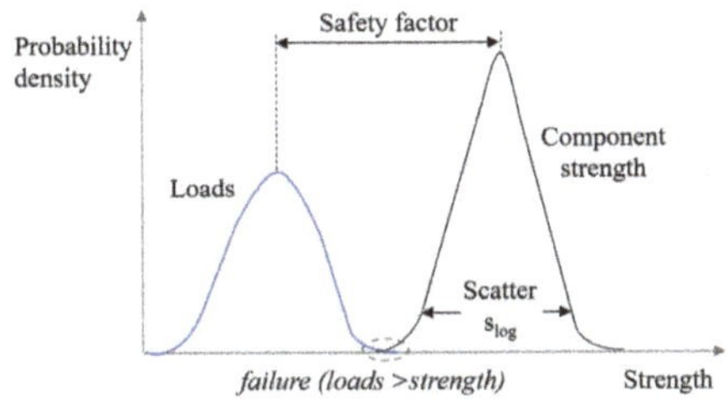

Figure 6. Failure probability.

Hence, decreasing scatter in components exposed to cyclic loads is necessary to mitigate the risk of failure.

The first objective of this study was to improve fatigue strength. For this purpose, the mean value of cycles was evaluated at different oven times with the same printing time (8 h). The mean value of cycles improved 1.25-fold, but importantly, the scatter was reduced by 24.2%, increasing the hardness to 19 HV. The results are summarized in Table 4; the treatment time is described as the oven time.

A temperature of 55 °C (Figure 7) is optimal for the treatments at a time when the material has more life or resistance to fatigue. It is believed that this behavior provides greater durability to the material as it can withstand more load cycles of tension and compression. The material possesses ductility until it ruptures; on the other hand, test components with a shorter duration tend to be brittle.

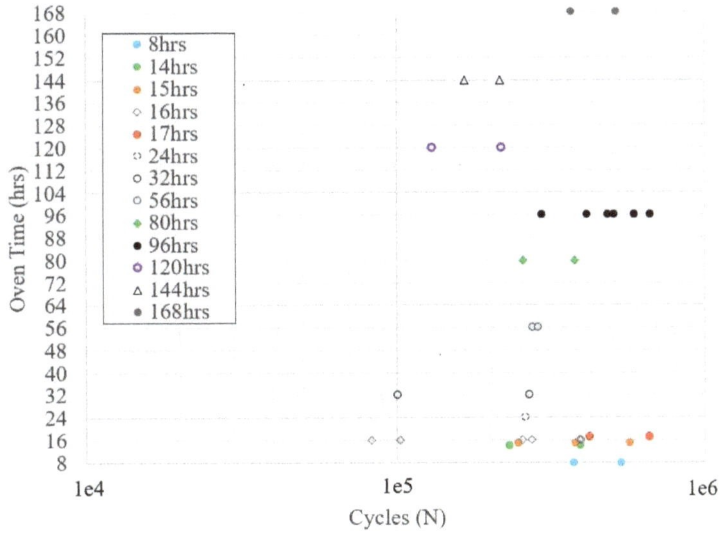

Figure 7. Semi-log test results with a heat treatment of 55 °C.

Table 4. Experimental results obtained for different oven times at 55 °C.

Oven Time (Hours)	μ	s_{log}	Hardness (HV)
8	45,272	0.10983	19
14	31,151	0.16085	18
15	47,290	0.12726	17
16	24,868	0.29187	18
17	54,023	0.13881	19
18	19,430	0.25816	18
24	26,085	NA	20
32	18,467	0.30229	19
56	28,071	0.01189	17
80	31,543	0.11767	18
96	46,351	0.1619	18
120	17,449	0.15981	19
144	19,025	0.08140	19
168	43,972	0.10434	18

During additive manufacturing processes, changes are generated in the properties of the used input material. These changes arise from the state transformation and the layer-by-layer material addition process. This process is based on thermo-mechanical processes and the temperature differences between the printing bed, the printing layer's temperature, and the molten material being added. Internally, the material develops resistance based on the obtained structural characteristics, while externally, hardness is achieved. For materials produced with these differences, therefore, in this work, the aim is to define an approach that allows the implementation of a thermal treatment standardizing the structure of the component in its cross-section. Initially, the printing time is thought to correlate directly with the overall processing time. The printing time was initially considered to be directly related to the time required for processing. This assumption was based on the results obtained for an oven time of 16 h.

Interestingly, the durability decreased and scatter increased, but importantly, the hardness did not vary. To understand the effect of dispersion, two ranges were defined around this value of 16 h, differing from this value by 1 and 2 h. The same hardness was obtained for an oven time of 14 h, but the component's durability did not improve without treatment, and a greater dispersion was observed. In 15 h, the fatigue resistance increased 1.30-fold, dispersion decreased by 15.1%, and hardness was reduced to 17 HV. Although at 17 h, the durability was increased 1.49-fold, the dispersion was only reduced by 7.4%; the hardness increased to 19 HV. While analyzing the duration, cycles alone may appear optimal; this approach yields the peak value. However, the integration of dispersion to enhance reliability does not consider of the global dispersion in load, material, design, and manufacturing processes. This could lead to premature failure with increased variability. Ultimately, the hardness was sustained at an oven time of 18 h, but the fatigue resistance decreased and dispersion increased.

These results show similarities with those reported by Shbanah et al. [5], who evaluated the improvement in mechanical properties at 55 °C and 65 °C for 5 h. They reported the best mechanical behavior at 65 °C; however, they adopted a quasistatic evaluation and considered a bed platform temperature of 60 °C. Although the oven time was 5 h, the printing time was 4.5 h. More surprisingly, the oven time was correlated with the printing time. It is possible to define a relationship between the oven time and the same printing time, as shown in this work. Comparing the durability of untreated components with treated components, there is a reduction of 41.5% at 60 °C, 52% at 80 °C, and 78. 6% at 120 °C. The optimum parameters for heat treatment are a temperature of 55° and the same printing time (8 h). This increases the mean duration value at cyclic loads 1.25 times, reducing dispersion by 26.7%

To analyze the effect of the heat treatment on the standardization of the behavior of the component, an optical microscope was used. Three different materials were analyzed:

a component without heat treatment, the part with the best durability, and a part with low durability.

Figure 8a shows the specimen without heat treatment. Ductile and brittle behavior can be observed in the same component. On the boundary, it has a lighter color due to ductile behavior. The reddish color in the middle of the cross-section of the component is a result of brittle behavior. Figure 8b shows a specimen after heat treatment for 96 h at 55° with a zoom of 0.8x. Homogenization is shown as whitish bands, resulting in an improved fatigue strength. This component has a fatigue life two times longer than that without treatment. With a treatment of 60 °C for one hour, the structure generates brittleness, as is shown in Figure 8c. The durability diminishes by 92.1%. Figure 8d shows both failure mechanisms; the lower part of the image shows the brittleness and the upper part shows the ductile failure mechanism. In fragile zones, it is observed that failure did not cause any type of deformation, only detachment, i.e., the component separates. In the case of failure due to ductile behavior, plastic deformations and slower crack propagation occur until catastrophic failure occurs.

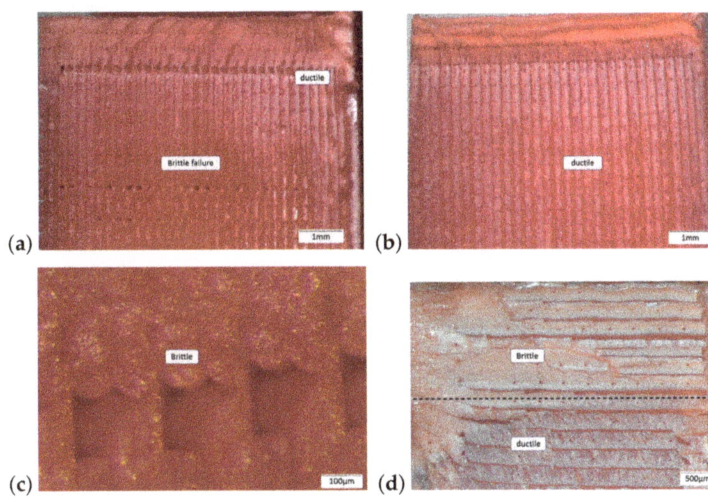

Figure 8. Optical microscopy of components (**a**) without treatment, (**b**) with ductile behavior, and (**c**) with brittle behavior and (**d**) a comparison between ductile and brittle failure.

To analyze the behaviour, the component with the worst fatigue strength (Figure 8c) was analyzed via SEM. Figure 9a,b show the ductile behavior on the boundary generated by homogenization. Figure 9c presents the failure in the internal void. This observation may support the hypothesis that the best durability is achieved by homogenizing the component structure along the cross-section.

The results of this study show that the fatigue strength of PLA 3D-printed materials can be improved, and the temperature of treatment increases with the temperature of the printing bed at 5 °C. The same printing time must be used to achieve standardization in the component structure. With very short durations, only the edges of the printed component are treated, generating a ductile structure, but within the cross-section, a fragile structure remains. Hence, it is important to state that heat treatments are necessary for additive manufacturing to achieve a homogenized microstructure, reduce scatter, and improve the reliability of the printed components.

The process of accumulated fatigue damage was analyzed by comparing the applied loads with the component strength. This resistance considers the dispersion parameters in design, the material, and the manufacturing processes. The probability of failure is also reduced by reducing the scatter of the component resistance.

Overall, these results indicate that it is possible to evaluate fatigue life strength enhancements as a quantitative value and by using the number of the cycles. Qualitatively, the best improvement was achieved in terms of the scatter. When printing polymers, predicting the dynamic response of fatigue life with a direct parameter such as hardness is impossible.

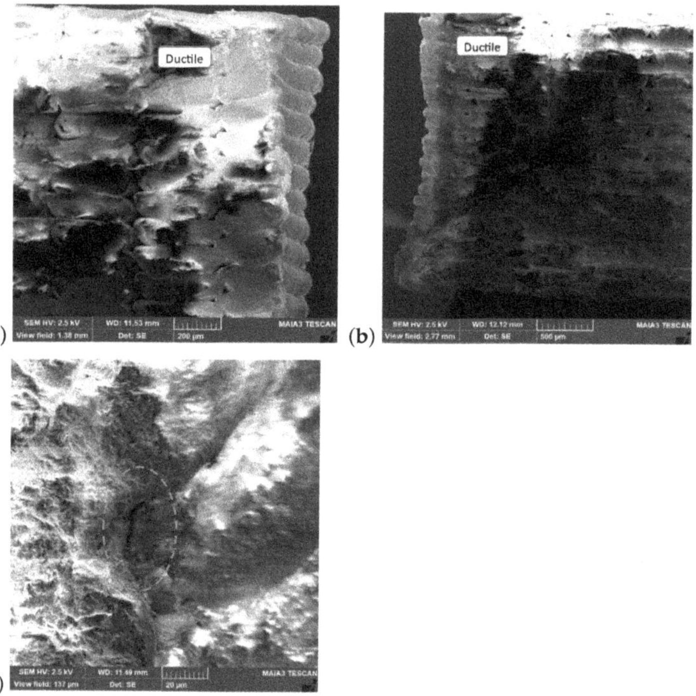

Figure 9. SEM analysis. (**a**) Top side of the component, (**b**) bottom side, and (**c**) internal failure.

4. Conclusions

This study aimed to assess the feasibility of post-treatment to improve the dynamic behavior of printed products in the additive manufacturing of polymers. The results support the idea that heat treatment can be applied to 3D-printed PLA components to improve their mechanical performance under dynamic loads. The following conclusions were drawn from this research study:

- The hardness of a component can be improved by thermal treatment. However, obtaining a direct relationship to estimate the fatigue life only from the hardness itself was impossible. Defining the oven time as a function of the printing time is necessary. The best performance was achieved using the same printing time; this generated an improvement not only on the surface but across all the components. A shorter time does not allow for the generation of a homogeneous structure.
- The position of the printed part in the furnace for treatment also affects the mechanical properties of the printed part. This effect originates from the difference in hardness achieved by printing the hardest side (the one that is in contact with the bed) at the exact moment that the printing side is placed in the furnace.
- Normalization is the only treatment that positively affects PLA. The best bed temperature for heat treatment is +5 °C/0 °C.
- The minimum heat treatment time is the printing time. Therefore, short treatment times generate non-homogeneous structures, producing brittle behavior.

Hence, post-processing treatments are necessary in additive manufacturing to achieve a homogenized microstructure, reduce the scatter, and improve the reliability of the printed components.

Author Contributions: Conceptualization, M.J.-M.; methodology, M.J.-M.; validation, J.V.-S. and M.J.-M.; formal analysis, R.C.-E., J.V.-S. and M.J.-M.; investigation, R.C.-E. and M.J.-M.; resources, M.J.-M.; data curation, R.C.-E., J.V.-S., S.G.T.-C., J.C.-P. and M.J.-M. ; writing—original draft preparation, J.V.-S. and M.J.-M.; writing—review and editing, M.J.-M.; visualization, R.C.-E., J.V.-S., S.G.T.-C. J.C.-P. and M.J.-M.; supervision, M.J.-M. All authors have read and agreed to the published version of the manuscript.

Funding: This research received no external funding.

Acknowledgments: The authors are grateful for the support of the students of the lightweight structures lab of Tecnologico de Monterrey, and to the department of Mechanics and Advanced Materials for all the support provided for the development of the applied research projects.

Conflicts of Interest: The authors declare no conflicts of interest.

References

1. Cox, J.R.; Kipling, I.; Gibbons, G.J. Ensuring supply chain integrity for material extrusion 3D printed polymer parts. *Addit. Manuf.* **2023**, *62*, 103403. [CrossRef]
2. Lampron, O.; Lingua, A.; Therriault, D.; Lévesque, M. Characterization of the non-isotropic tensile and fracture behavior of unidirectional polylactic acid parts manufactured by material extrusion. *Addit. Manuf.* **2023**, *61*, 103369. [CrossRef]
3. Palmieri, M.; Zucca, G.; Morettini, G.; Landi, L.; Cianetti, F. Vibration Fatigue of FDM 3D Printed Structures: The Use of Frequency Domain Approach. *Materials* **2022**, *15*, 854. [CrossRef] [PubMed]
4. Travieso-Rodriguez, J.A.; Jerez-Mesa, R.; Llumà, J.; Traver-Ramos, O.; Gomez-Gras, G.; Roa Rovira, J.J. Mechanical Properties of 3D-Printing Polylactic Acid Parts subjected to Bending Stress and Fatigue Testing. *Materials* **2019**, *12*, 3859. [CrossRef]
5. Shbanah, M.; Jordanov, M.; Nyikes, Z.; Tóth, L.; Kovács, T.A. The Effect of Heat Treatment on a 3D-Printed PLA Polymer'r Mechanical Properties. *Polymers* **2023**, *15*, 1587. [CrossRef]
6. Dizon, J.R.C.; Espera, A.H.; Chen, Q.; Advincula, R.C. Mechanical characterization of 3D-printed polymers. *Addit. Manuf.* **2018**, *20*, 44–67. [CrossRef]
7. Amza, C.G.; Zapciu, A.; Constantin, G.; Baciu, F.; Vasile, M.I. Enhancing Mechanical Properties of Polymer 3D Printed Parts. *Polymers* **2021**, *13*, 562. [CrossRef]
8. Santosh, S.; Nithyanandh, G.; Ashwath, J.; Kishore, K.L. Comparison of internal friction measurements on Ni-Ti reinforced smart composites prepared by additive manufacturing. *J. Alloys Compd.* **2022**, *924*, 166027. [CrossRef]
9. Diniță, A.; Neacșa, A.; Portoacă, A.I.; Tănase, M.; Ilinca, C.N.; Ramadan, I.N. Additive Manufacturing Post-Processing Treatments, a Review with Emphasis on Mechanical Characteristics. *Materials* **2023**, *16*, 4610. [CrossRef] [PubMed]
10. Somireddy, M.; Czekanski, A.; Singh, C.V. Development of constitutive material model of 3D printed structure via FDM. *Mater. Today Commun.* **2018**, *15*, 143–152. [CrossRef]
11. Lim, L.T.; Auras, R.; Rubino, M. Processing technologies for poly(lactic acid). *Prog. Polym. Sci.* **2008**, *33*, 820–852. [CrossRef]
12. Jayanth, N.; Jaswanthraj, K.; Sandeep, S.; Mallaya, N.H.; Siddharth, S.R. Effect of heat treatment on mechanical properties of 3D printed PLA. *J. Mech. Behav. Biomed. Mater.* **2021**, *123*, 104764. [CrossRef]
13. Bakar, A.A.B.A.; Zainuddin, M.Z.B.; Adam, A.N.B.; Noor, I.S.B.M.; Tamchek, N.B.; Alauddin, M.S.B.; Ghazali, M.I.B.M. The study of mechanical properties of poly(lactic) acid PLA-based 3D printed filament under temperature and environmental conditions. *Mater. Today Proc.* **2022**, *67*, 652–658. [CrossRef]
14. Guduru, K.; Srinivasu, G. Effect of post treatment on tensile properties of carbon reinforced PLA composite by 3D printing. *Mater. Today Proc.* **2020**, *33*, 5403–5407. [CrossRef]
15. Vorkapić, M.; Mladenović, I.; Ivanov, T.; Kovačević, A.; Hasan, M.S.; Simonović, A.; Trajković, I. Enhancing mechanical properties of 3D printed thermoplastic polymers by annealing in moulds. *Adv. Mech. Eng.* **2022**, *14*, 16878132221120737. [CrossRef]
16. Yi, S.; Oh, N.; Min, K.E.; Shin, J.S.; Kim, C. Thermo-Viscoelastic Characterization of 3D Printing Polymers. *Appl. Sci.* **2023**, *13*, 2876. [CrossRef]
17. Jimenez-Martinez, M.; Varela-Soriano, J.; Carreón, J.J.R.; Torres-Cedillo, S.G. Mechanical fatigue of PLA in additive manufacturing. *Eng. Fail. Anal.* **2023**, *149*, 107273. [CrossRef]
18. Nazir, A.; Gokcekaya, O.; Md Masum Billah, K.; Ertugrul, O.; Jiang, J.; Sun, J.; Hussain, S. Multi-material additive manufacturing: A systematic review of design, properties, applications, challenges, and 3D printing of materials and cellular metamaterials. *Mater. Des.* **2023**, *226*, 111661. [CrossRef]
19. ASTM. D7791. Uniaxial Fatigue Testing of Plastics. 2012; pp. 1–5. Available online: https://step-lab.com/astm/astm-d7791/ (accessed on 1 October 2023)

20. Domingo-Espin, M.; Travieso-Rodriguez, J.A.; Jerez-Mesa, R.; Lluma-Fuentes, J. Fatigue Performance of ABS Specimens Obtained by Fused Filament Fabrication. *Materials* **2018**, *11*, 2521. [CrossRef]
21. *E384-17-03-01*; Standard Test Method for Microindentation Hardness of Materials. ASTM: West Conshohocken, PA, USA, 2017; pp. 1–40.
22. Arash Afshar, D.M. Enhancing durability of 3D printed polymer structures by metallization. *J. Mater. Sci. Technol.* **2020**, *53*, 185–191. [CrossRef]
23. Mayén, J.; Del Carmen Gallegos-Melgar, A.; Pereyra, I.; Poblano-Salas, C.A.; Hernández-Hernández, M.; Betancourt-Cantera, J.; Mercado-Lemus, V.; Del Angel Monroy, M. Descriptive and inferential study of hardness, fatigue life, and crack propagation on PLA 3D-printed parts. *Mater. Today Commun.* **2022**, *32*, 103948. [CrossRef]

MDPI
St. Alban-Anlage 66
4052 Basel
Switzerland
www.mdpi.com

Designs Editorial Office
E-mail: designs@mdpi.com
www.mdpi.com/journal/designs

www.ingramcontent.com/pod-product-compliance
Lightning Source LLC
Chambersburg PA
CBHW040311270926
51850CB00007BC/1032
* 9 7 8 3 7 2 5 8 1 0 3 1 4 *